Springer Theses

Recognizing Outstanding Ph.D. Research

Aims and Scope

The series "Springer Theses" brings together a selection of the very best Ph.D. theses from around the world and across the physical sciences. Nominated and endorsed by two recognized specialists, each published volume has been selected for its scientific excellence and the high impact of its contents for the pertinent field of research. For greater accessibility to non-specialists, the published versions include an extended introduction, as well as a foreword by the student's supervisor explaining the special relevance of the work for the field. As a whole, the series will provide a valuable resource both for newcomers to the research fields described, and for other scientists seeking detailed background information on special questions. Finally, it provides an accredited documentation of the valuable contributions made by today's younger generation of scientists.

Theses are accepted into the series by invited nomination only and must fulfill all of the following criteria

- They must be written in good English.
- The topic should fall within the confines of Chemistry, Physics, Earth Sciences, Engineering and related interdisciplinary fields such as Materials, Nanoscience, Chemical Engineering, Complex Systems and Biophysics.
- The work reported in the thesis must represent a significant scientific advance.
- If the thesis includes previously published material, permission to reproduce this must be gained from the respective copyright holder.
- They must have been examined and passed during the 12 months prior to nomination.
- Each thesis should include a foreword by the supervisor outlining the significance of its content.
- The theses should have a clearly defined structure including an introduction accessible to scientists not expert in that particular field.

More information about this series at http://www.springer.com/series/8790

Enrico Morgante

Aspects of WIMP Dark Matter Searches at Colliders and Other Probes

Doctoral Thesis accepted by
the University of Geneva, Switzerland

 Springer

Author
Dr. Enrico Morgante
Deutsches Elektronen-Synchrotron
Hamburg
Germany

Supervisor
Prof. Antonio Riotto
Département de Physique Théorique
Université de Genève
Geneva
Switzerland

ISSN 2190-5053   ISSN 2190-5061   (electronic)
Springer Theses
ISBN 978-3-319-67605-0   ISBN 978-3-319-67606-7   (eBook)
https://doi.org/10.1007/978-3-319-67606-7

Library of Congress Control Number: 2017952502

Printed on acid-free paper

This Springer imprint is published by Springer Nature
The registered company is Springer International Publishing AG
The registered company address is: Gewerbestrasse 11, 6330 Cham, Switzerland

Supervisor's Foreword

This thesis is a report of the work carried on by Enrico Morgante on the theoretical aspects of WIMP Dark Matter (DM) searches, with a particular emphasis on LHC physics. This constitutes the main topic on which his research has focussed so far, even if his interests and activity have spanned other topics in between particle physics and early universe cosmology.

As far as LHC searches are concerned, the present work starts with a critical discussion of the EFT approach, whose limitations were highlighted by our group in a series of papers which have had a major impact in the field. The techniques that we proposed to overcome these limitations and get sensible results in terms of effective operators—that are thoroughly described in the thesis—had become a standard tool for the LHC collaborations. The discussion then turns to simplified models, which are introduced and carefully analysed, paying attention to their consistent construction and use, which is essential in order to obtain reliable results, in particular when comparing results coming from different searches. An example of a simplified model is introduced and analysed, drawing conclusions whose relevance goes well beyond the particular model under scrutiny.

A chapter of the thesis is devoted to DM searches through cosmic rays, and in particular on the possibility of discovering a DM signal with a measure of the antiproton flux. It is pointed out that a possible antiproton signal may not be regarded as a smoking gun for DM annihilation in the halo because of its degeneracy with a purely astrophysical process in which secondary antiparticles are produced at the acceleration region of supernova remnants, resulting in a spectral shape very similar to that predicted by DM models. If such a signal is observed, further studies of this phenomenon would be necessary to disentangle it from the desired signal, taking into account the information from other complementary search channels.

An additional value of this thesis is given by its long and detailed introduction to the DM puzzle, its possible solutions and the experimental searches for WIMP DM. The introduction is complemented by precise references to the original works and to other review articles, which could serve as a useful reference for researchers active in the field and particularly for young researchers.

It gave me immense pleasure that the work of Enrico Morgante was selected by the University of Geneva and Springer to be published in the Springer theses series.

Geneva, Switzerland   Antonio Riotto
June 2017

Abstract

The problem of Dark Matter (DM) is one of the most longstanding problems in cosmology and particle physics, and surely one of the most interesting. A number of astrophysical and cosmological evidence point to the existence of a large amount of non-luminous matter, whose nature is still a mystery. The simplest possible solution to this puzzle is in terms of new neutral and stable particles with weak-scale mass and couplings (named WIMPs for weakly interacting massive particles), which would not only reproduce naturally the observed cosmological DM abundance, but also fit very well in many well-motivated theories for physics beyond the standard model. In this thesis, I discuss the present status of WIMP searches and of our understanding of the theoretical issues involved in them, with a particular emphasis to searches at the Large Hadron Collider.

Preface

Among the natural sciences, the oldest is probably astronomy. Its origins date back to the rise of the first civilizations, when astronomical observations allowed for producing the first calendars. For thousands of years, cosmogonies were directly related to religious beliefs, showing a connection with the deepest questions of human heart that keep fascinating generations of modern scientists. It was not until the seventeenth century that astronomy started assuming its present aspect, thanks to the pioneers Nicolaus Copernicus, Galileo Galilei, Tycho Brahe, Johannes Kepler and Isaac Newton. It was due to their work that we understood that the orbits of celestial bodies are governed by mathematical relations; if any deviation is seen, either new, unseen objects are present to modify the equation of motion, thanks to their gravitational attraction, or a modification of the laws of gravity is necessary for some regime.

An example of the first approach is the discovery of Neptune, the eighth planet of the Solar System, whose existence was hypothesized by Urbain Le Verrier and, independently, by John Couch Adams in 1846, in order to account for the anomalous motion of Uranus. Neptune was first observed by Johann Galle and Heinrich d'Arrest of the Berlin Observatory on 23 September of the same year, the same night when Galle received a letter from Le Verrier asking to confirm his predictions. Neptune was found within $1°$ of the predicted location, providing a strong confirmation of the laws of celestial mechanics. The second approach proved to be useful in the case of another planet, Mercury. The precession of its orbit was reported by Le Verrier himself, who pointed out in 1859 that its characteristics were at odds with the Newtonian laws of gravitation. His faith in Newtonian mechanics, and the success with Neptune, suggested him that the solution could be a new unseen planed (which he named Vulcan) or a series of smaller "corpuscules". These attempts proved to be unsuccessful, and the explanation was finally provided by Albert Einstein in 1916, with the introduction of the theory of General Relativity.

The problem of DM is conceptually not different from that of the motion of planets in the Solar System. A number of anomalies were observed, starting back in 1933, from the galactic scales to that of the largest structures, that can be explained only by the existence of some new non-luminous component which constitutes more than 80% of the total mass of galaxies and galaxy clusters, or else by some further modifications of General Relativity. Which of the two solutions is the correct one is still matter of debate, even if a modification of the matter content is commonly regarded as more likely, due to the difficulty of modified gravity theories of deal with the formation of large-scale structures. Furthermore, it is a well-established result that exotic astrophysical objects can only constitute a small fraction of the total DM content of the Universe, with only some small windows left open by astrophysical observations. Because of this, the most plausible solution to the puzzle seems to be in the form of subatomic particles, and DM can be regarded as a particle physics problem as well as an astrophysical one.

Our present understanding of particle physics is based on the so-called Standard Model (SM), which describes the interactions of all the known elementary particles in terms of a $SU(3) \times SU(2) \times U(1)$ gauge symmetry, spontaneously broken to the $SU(3) \times U(1)$ symmetry of the strong and electromagnetic interactions. Given the fact that DM is stable and that it does not seem to interact electromagnetically, the only DM candidate in the SM is the neutrino. As we will see in Sect. 1.3, neutrinos cannot account for all the DM, and the introduction of new stable particles is necessary.

Two very exciting facts occur at this point. On the one hand, a simple calculation of the relic abundance of DM in the thermal hot Big Bang scenario indicates that a particle with a mass around the weak-scale and weak coupling strength would have the right present-day abundance. This coincidence is often referred to as the "WIMP miracle", for *weakly interacting massive particle*. On the other hand, the SM suffers from the so-called "Higgs naturalness problem", which is the fact that the electro-weak scale is unstable against quantum corrections and, assuming the point of view that the SM is just the low energy limit of some more fundamental physics, maybe at the Plank scale, a very fine-tuned choice of the parameters is necessary in order for the EW scale to remain light. Solutions to this problem typically involve new physics at the TeV scale, as in the case of phenomenological supersymmetric models as the MSSM, in extra dimension theories as the Arkani-Hamed, Dimopoulos & Dvali model or the Randall-Sundrum one, *etc.* Interestingly, most of these theories include quite naturally stable neutral particles, with the right quantum numbers and properties to be a perfectly suitable WIMP candidate.

These two observations together catalysed a huge attention in the last decades around the WIMP DM paradigm and the experimental techniques for a possible detection. A number of probes exist to test the WIMP hypothesis. They are most commonly classified as direct, indirect and collider searches. Direct WIMP searches look for tiny energy deposits when DM particles of the galactic halo scatter off atomic nuclei in ultra-sensitive, low-background underground detectors. Indirect searches instead aim at observing annihilation products of DM particles such as neutrinos, antiprotons, positrons and gamma rays from galactic regions of increased density. Finally, collider searches look for the production of DM particles among the final states of proton–proton collisions (in the case of the LHC). In the collider setup, DM particles leave no trace in the detector and show themselves as unpaired momentum in the centre of mass frame, with the recoil of some SM particles which are necessary to tag the event.

In order to experimentally constrain the properties of DM, it is necessary to fix a reference model to compare with data. As we will discuss in detail in Sect. 5.3, this is not a trivial task in general. In direct searches, the very low momentum transfer involved in the scattering process allows one to write down a set of non-relativistic effective operators which capture all the possible feature of the DM–nucleons interaction in any DM model. Limits are therefore model independent, provided that the correct non-relativistic operator is considered. Quite the opposite happens at colliders: the very high-energy reach (larger than the naïve WIMP energy scale) makes it of primary importance to carefully consider which is the best model to consider: on the one hand, searches can be built to constrain the parameter space of beyond the Standard Model (BSM) theories as the MSSM or other realizations of supersymmetry. On the opposite side of the spectrum, one can rely on a set of effective operators, taking into account the strong limitations posed by the fact that, for an Effective Field Theory (EFT) approach to be valid, the energy involved in the process must be smaller than the Wilson coefficient of the operators. Finally, somehow in between of these two approaches, one can construct a set of simplified models, in which the SM is complemented with a DM candidate and some particle to mediate the DM interactions, in such a way to avoid the energy limitations of the EFT approach and to grasp the most relevant features of a given BSM model.

This thesis is organized as follows: in Chap. 1, we will introduce the DM problem by reviewing; in Sect. 1.1 the observational evidences of the existence of DM; in Sect. 1.2 the possible mechanisms for obtaining the present relic DM abundance in the early universe; in Sect. 1.3 the most studied DM candidates from BSM theories.

In Chaps. 2 and 3, we will introduce direct and indirect DM searches in general, and describe the present state of the art. In Chap. 4, we will discuss the implication of a signal in the antiprotons channel by the AMS-02, showing how this could not have been unambiguously attributed to DM annihilation events.

We will discuss DM searches at the LHC in Chap. 5. We will start by describing the characteristics of DM searches, with the prototypical example of the mono-jet one, and then discuss in general the three classes of DM models to which we referred above. The limitations of the EFT approach are described in Chap. 6. Chapter 7 contains a broad discussion on simplified DM models, complemented in Chap. 8 with a discussion of the role of the relic abundance calculation in the context of LHC searches.

In Chap. 9, we will consider a simplified model with a Z' additional gauge boson, and we will show how different experiments provide complementary bounds, highlighting in particular the role of the IceCube experiment in the context of spin-dependent DM–nucleon interactions. Our conclusions are then presented in the last chapter.

Hamburg, Germany   Dr. Enrico Morgante

List of Publications

Publications Related to this Thesis

1. G. Busoni, A. De Simone, E. Morgante and A. Riotto, *On the Validity of the Effective Field Theory for Dark Matter Searches at the LHC*, Phys. Lett. B **728** (2014) 412, http://arxiv.org/abs/1307.2253.
2. G. Busoni, A. De Simone, J. Gramling, E. Morgante and A. Riotto, *On the Validity of the Effective Field Theory for Dark Matter Searches at the LHC, Part II: Complete Analysis for the s-channel*, JCAP **1406** (2014) 060, http://arxiv.org/abs/1402.1275.
3. G. Busoni, A. De Simone, T. Jacques, E. Morgante and A. Riotto, *On the Validity of the Effective Field Theory for Dark Matter Searches at the LHC Part III: Analysis for the t-channel*, JCAP **1409** (2014) 022, http://arxiv.org/abs/1405.3101.
4. V. Pettorino, G. Busoni, A. De Simone, E. Morgante, A. Riotto and W. Xue, *Can AMS-02 discriminate the origin of an anti-proton signal?*, JCAP **1410** (2014) no.10, 078, http://arxiv.org/abs/1406.5377.
5. G. Busoni, A. De Simone, T. Jacques, E. Morgante and A. Riotto, *Making the Most of the Relic Density for Dark Matter Searches at the LHC 14 TeV Run*, JCAP **1503** (2015) no.03, 022, http://arxiv.org/abs/1410.7409.
6. E. Morgante, D. Racco, M. Rameez and A. Riotto, *The 750 GeV Diphoton excess, Dark Matter and Constraints from the IceCube experiment*, JHEP **1607** (2016) 141, http://arxiv.org/abs/1603.05592.
7. T. Jacques, A. Katz, E. Morgante, D. Racco, M. Rameez and A. Riotto, *Complementarity of DM Searches in a Consistent Simplified Model: the Case of Z'*, JHEP **1610** (2016) 071, http://arxiv.org/abs/1605.06513.

Other Journal Papers

1. M. Biagetti, A. Kehagias, E. Morgante, H. Perrier and A. Riotto, *Symmetries of Vector Perturbations during the de Sitter Epoch*, JCAP **1307** (2013) 030, http://arxiv.org/abs/1304.7785.
2. P. Ciafaloni, D. Comelli, A. De Simone, E. Morgante, A. Riotto and A. Urbano, *The Role of Electroweak Corrections for the Dark Matter Relic Abundance*, JCAP **1310** (2013) 031, http://arxiv.org/abs/1305.6391.
3. J. R. Espinosa, G. F. Giudice, E. Morgante, A. Riotto, L. Senatore, A. Strumia and N. Tetradis, *The cosmological Higgstory of the vacuum instability*, JHEP **1509** (2015) 174, http://arxiv.org/abs/1505.04825.

White Papers

1. J. Abdallah et al., *Simplified Models for Dark Matter and Missing Energy Searches at the LHC*, http://arxiv.org/abs/1409.2893.
2. J. Abdallah et al., *Simplified Models for Dark Matter Searches at the LHC*, Phys. Dark Univ. **9-10** (2015) 8, http://arxiv.org/abs/1506.03116.
3. D. Abercrombie et al., *Dark Matter Benchmark Models for Early LHC Run-2 Searches: Report of the ATLAS/CMS Dark Matter Forum*, http://arxiv.org/abs/1507.00966.

Conference Proceedings

1. E. Morgante, *On the validity of the effective field theory for dark matter searches at the LHC*, Nuovo Cim. C **38** (2015) no.1, 32, http://arxiv.org/abs/1409.6668.
2. E. Morgante, *Cosmological History of the Higgs Vacuum Instability*, Frascati Phys. Ser. **61** (2016) 115.

Acknowledgements

Non nobis.

I am grateful to my supervisor, Toni Riotto, for the 4 years I spent working with him in Geneva. He has been a great teacher, an exceptional scientist and a close friend for me and my family. Without his patience, his help and the attention he had with me, this work would not have been possible.

A special thank goes to Malcolm Fairbairn, Nicolao Fornengo and Andrey Katz for the many valuable comments about the contents of this thesis.

I want to thank my fellow students, together with whom I had the privilege of beginning my journey in the fascinating world of physics. In particular, I cannot help mentioning Matteo Biagetti and Davide Racco, who have been good friends before being collaborators, as well as Stefano Bolzonella, Mattia Sormani, Emanuele Sobacchi, Federico Fabiano, Stefano Recanatesi, Giacomo De Palma, Laura De Stefanis, Alessandro Moia, Lorenzo Barattini, Bruno Bertini, Nicolò Grilli, Simone Surdi, Giulio Pelaggi, Vito Gagliardi, Manuel Marchiò, Antonello Scarcella, Marco Morelli and Alessandro Principi.

I also have to thank all my collaborators for all I could learn from them: Giorgio Busoni, Paolo Ciafaloni, Denis Comelli, Andrea De Simone, José Ramon Espinosa, Nayara Fonseca, Gian Francesco Giudice, Johanna Gramling, Thomas Jacques, Andrey Katz, Alex Kehagias, Hideki Perrier, Valeria Pettorino, Mohamed Rameez, Leonardo Senatore, Géraldine Servant, Alessandro Strumia, Nikolaos Tetradis, Augustin Vanrietvelde and Wei Xue.

I am deeply indebted to my teacher Riccardo Sangoi, who first introduced me to the realm of science. I still remember the day when, aged 15, he made me compute the trajectory of a small iron sphere falling from his desk, asking me to measure the impact point on the ground and to compare it with my prediction: perfect agreement! Oh wonder! On that day I learned a lesson I will never forget: that physics is not just a long list of complicated formulas, but it is the journey towards those laws that exist behind our reality, a noble adventure that we have the privilege of living.

Finally, I want to thank my wife and my daughters, and also my parents for all they had done for me and for always supporting me. There are no words to express my gratitude to them.

Hamburg, Germany   Enrico Morgante
June 2017

Contents

Chapter 1
Introduction to Dark Matter

1.1 Evidences of the Existence of Dark Matter

The history of our understanding of dark matter is long and fascinating. It is not easy to indicate a precise starting point for these studies, since the question about the amount of low-luminosity material in the Milky Way was asked numerous times already in the 19th century. An important step, which is often mentioned as the first observation of the presence of dark matter, was achieved by Fritz Zwicky in 1933 [1], when he observed a large velocity dispersion within the galaxies of the Coma cluster, inferring from this observation the presence of a large amount of dark matter (he used this very word, which was already in use within the astronomical community at that time to indicate any form of matter which was not visible with telescopes). A nice and thorough review of the history of DM, from the first observations to present day DM searches, is presented in [2].

What is probably the most striking evidence of DM is the fact that its existence may be inferred from a number of different observations which span a large range of scales and redshifts. The observation of the motion of galaxies, the behaviour of galaxy clusters, our knowledge of structure formation and measurements of the temperature anisotropy of the Cosmic Microwave Background, all put towards the existence of a non luminous matter component which, with common notation, goes under the name of dark matter. In this section we are going to briefly review these evidences from our modern perspective, and what lessons we can learn from them in terms of understanding the properties of DM.

1.1.1 The Galactic Scale

Starting from the '70s, astronomers noticed that the rotation curves of galaxies, i.e. the circular velocity of star and gas as a function of the distance from the galactic centre, have a characteristic flat behaviour at large distances, beyond the edge of the

© Springer International Publishing AG 2017
E. Morgante, *Aspects of WIMP Dark Matter Searches at Colliders and Other Probes*, Springer Theses, https://doi.org/10.1007/978-3-319-67606-7_1

visible disks (see e.g. [3]). The rotation curve is determined by means of the Doppler shift of the Hydrogen 21 cm line, and can be compared with the one predicted from the amount of observed material in the galaxy. In Newtonian dynamics the circular velocity is expected to be

$$v(r) = \sqrt{\frac{GM(r)}{r}} \, , \tag{1.1}$$

where $M(r)$ is the integral of the mass density $\rho(r)$ inside the sphere with radius r. The velocity $v(r)$ should be falling with $1/\sqrt{r}$ beyond the optical disc. The fact that it is approximately constant on even larger distances implies the existence of a non luminous halo with $\rho \propto 1/r^2$, and $M \propto r$.

The first paper where the discrepancy between the predicted and observed rotation curves is explicitly discussed was by Freeman in 1970 [4], followed by many others in subsequent years (see in particular [2, 5, 6] for a detailed historical account). By the end of the '70s the existence of missing mass in the outer regions of spiral galaxies was well established.

Other less standard indications of the existence of DM and of its distribution in our galaxy may be found in the literature: for example, there exist indications that a DM disk exists [7] in addition to the standard halo, possibly with an inner thin disk originated by self interacting DM [8, 9], which could explain the periodicity of comet impacts on Earth [10] or even paleoclimatic data through an oscillatory component in the motion of the solar system with respect to the galactic plane [11].

1.1.2 The Scale of Galaxy Clusters

As we already saw discussing the original observations by Zwicky, the dynamics of galaxy clusters enforce the evidence for the existence of DM. The total mass of a cluster can be inferred in many ways: applying the virial theorem (as originally done by Zwicky), by weak gravitational lensing and by studying the profile of the X-ray emission of the hot gas in the cluster. In each case, the total amount of matter which is necessary to explain the cluster's dynamics is roughly a factor of 10 larger than those of baryons, which can be inferred by means of the so-called Sunyaev-Zel'dovich effect, by which the spectrum of CMB photons is distorted by their Compton scattering on the hot electrons of the intra-cluster gas.

Another striking evidence at the cluster scale was provided by the observation of the so-called "bullet cluster" 1E0657-56 [12–14]. This was the first example in which a spatial separation within the three components that make up structures in our universe was detected. It is composed by two clusters that, about 150 million years ago, collided in a direction close to perpendicular to our line of sight. The trajectories followed by its components were then dictated by their individual properties: galaxies acted as point-like non interacting particles and continued along their ways, and are now separated; intra-cluster gas, which has a large self interaction cross section, was spread throughout the two incident clusters; finally, the DM fluid passed through the interaction region, and its distribution, inferred by weak lensing measurements, now coincide with that of the galaxies of the two clusters. This, incidentally, gives us a bound on the self interaction cross section of DM particles of $\sigma/m_\chi \lesssim 1.25\,\mathrm{cm}^2\,\mathrm{g}^{-1} \sim 2 \times 10^{-24}\,\mathrm{cm}^2\,\mathrm{GeV}^{-1}$ [15], where m_χ is the mass of the DM. Unfortunately, events like those seen in the bullet clusters are rare, and very few examples are known up to now in which gravitational lensing measurements can help in separating the DM and the gas component of cluster mergers [16–18] (see also [19] for a review of lensing techniques in DM astrophysics).

1.1.3 Dark Matter in Large-Scale Structures

The universe on very large scales shows a hierarchical organization: galaxies are gathered in clusters, clusters are part of superclusters, which in turns are arranged into large filaments surrounded by voids. Our current understanding of the formation of such scale has collisionless dark matter as one of its key ingredients, which acts gravitationally to start the collapse of structures as we see them today. Indeed, up to the time of matter-radiation decoupling, baryons and photons behaved as a single fluid, with a large pressure that opposed gravitational collapse. Given the extreme smallness of CMB fluctuations, ordinary matter can have formed the structures we see today only if large inhomogeneities in the gravitational potential were already present, and those could only have been generated by a component decoupled from the thermal plasma. The observations from large-scale structure (LSS) surveys as the 2-degree Field Galaxy Redshift Survey (2dFGRS, [20]) and the Sloan Digital Sky Survey (SDSS, [21]) are matched well by N-body numerical simulations such as the Millennium simulation [22].

In addition to providing an evidence for its existence, LSS teach us another important lesson about the velocity dispersion of DM. The observation of very high redshift galaxies suggests that structure formation has proceeded bottom-up: small structures were formed first and then merged by gravitational attraction into larger and heavier ones. This would not have been possible if the DM consisted of particles relativistic at the time of structure formation (which therefore were, if thermally distributed, hot). Indeed, relativistic DM would have had a free-stream length larger than the size of a galaxy, smearing all the small scale overdensities which would therefore have formed after the large ones. This observation leads to the exclusion of the most obvious particle DM candidate: the neutrino. On the contrary, we know that DM has to be cold (or, more precisely, non relativistic).

In recent years, this last point has been questioned in view of the fact that, while numerical simulations with collisionless cold DM typically yield a cuspy shape in the centre of DM halos, galaxy observations indicate that the profile is smooth. The tension can be solved if the DM particles are assumed to have a certain velocity dispersion, large enough to smooth the distribution at the centre of galactic halos but not too large in order not to spoil the global picture of structure formation. Lower limits on the mass of this "warm dark matter" are of the order 1–$10\,\mathrm{keV}$

(see e.g. [23]), and many concrete realizations are excluded unless non standard scenarios are invoked for determining their relic abundance. In addition, self-interacting DM can help in release the tension. In must be noted, anyway, that the subject is still debated, and that the problem can be simply due to the insufficiency of current N-body simulations in describing the correct physics. In particular, numerical simulation typically do not include the effect of hadrons and of luminous feedback, and there are indication that, when this is taken into account, the tension is relaxed.

1.1.4 Cosmological Scales

The total amount of dark matter in the universe can not by measured by the observation of gravitationally bound objects only. Nevertheless, this can be inferred by means of the cosmic microwave background (CMB). The CMB is the redshifted spectrum of photons that permeate the universe, almost freely streaming since the time they decoupled from matter due to the expansion of the universe. The CMB spectrum follows with amazing precision that of a black body with a temperature 2.7255 K, and small fluctuations at the level of 10^{-5}.

The temperature anisotropies of the CMB can be expanded in spherical harmonics as

$$\frac{\delta T}{T}(\theta, \phi) = \sum_{\ell=0}^{+\infty} \sum_{m=-\ell}^{l} a_{\ell m} Y_{\ell m}(\theta, \phi). \tag{1.2}$$

The monopole $\ell = 0$ represents the average temperature and the dipole $\ell = 1$ (which represents the larger anisotropy of the order 10^{-3}) is interpreted as a doppler shift due to our motion with respect to the CMB rest frame. Therefore they are not included in the analysis of the anisotropies. A fundamental quantity is the variance of the $a_{\ell m}$ coefficients:

$$C_\ell \equiv \langle |a_{\ell m}|^2 \rangle = \frac{1}{2\ell + 1} \sum_{m=-\ell}^{l} |a_{\ell m}|^2 . \tag{1.3}$$

If the temperature fluctuations are distributed with gaussian probability (as they appear to be), all the information contained in the CMB map is encoded in the power spectrum of C_ℓ, whose most recent measure may be found among the 2015 Planck results [24].

The typical oscillating shape of the spectrum is produced by the phenomenon of acoustic oscillations in the matter-radiation plasma before decoupling. In this phase, the opposite action of gravitational attraction and radiation pressure generated density fluctuations in the plasma. Oscillation modes that are caught at minima or maxima of their oscillation correspond to peaks in the power spectrum. Since modes start collapsing as soon as they enter the horizon, the first peak (which corresponds to the largest scale, which entered the horizon the latest) corresponds to a mode that underwent only the first half cycle: it entered the horizon, compressed the plasma, and decoupling took place. The second peak corresponds to the first complete cycle (compression and successive rarefaction) and so on. This simple picture is affected by the precise value of all cosmological parameters, which both modify the spectrum of photons at decoupling and its propagation to Earth (an excellent description is given in [25, 26]). In particular, the energy density of ordinary non-relativistic matter (which in cosmology is referred to generically as baryons) affects the ratios between odd and even peaks in the spectrum, and it can be fixed by a measurement of the ratios between the second and the first and third peaks, while the total matter density (baryons + DM) is determined by the overall peak amplitude.

Through a fit of these data a value $\Omega_{DM}h^2 = 0.1197 \pm 0.0022$ [24], where h is the reduced Hubble radius which was measured to be $H_0 \equiv 100h \, \text{km s}^{-1} \, \text{Mpc}^{-1} \simeq 67 \, \text{km s}^{-1} \, \text{Mpc}^{-1}$ and $\Omega_{DM} = \rho_{DM}/\rho_{crit}$ is the DM density defined in terms of the critical density $\rho_{crit} \simeq 1.88 \times 10^{-29} h^2$, i.e. the total energy density necessary to have a flat universe. The DM density is then proved to be a factor of 5–6 larger than the density of baryons: $\Omega_b h^2 = 0.02222 \pm 0.00023$ [24].

A measure of the total amount of baryons independent of the CMB one is provided by Big Bang Nucleosynthesis (BBN), which is the process of formation of stable nuclei from protons and neutrons during the very first minutes of the Universe's expansion. After the temperature of the Universe dropped below $T \sim 0.3 \, \text{GeV}$, quarks and gluons confined and hadrons were formed. In particular, protons and neutrons were kept in equilibrium by weak interaction processes

$$
\begin{aligned}
p + e^- &\leftrightarrow n + \nu , \\
p + \bar{\nu} &\leftrightarrow n + e^+ , \\
n &\to p + e^- + \nu .
\end{aligned}
\tag{1.4}
$$

Neutrino decoupling made the first two interactions extremely rare, and only neutron decays remained active. At this moment, the mass difference between n and p made protons three times more abundant than neutrons. Neutrons started decaying, and only a portion of them formed a stable deuterium bound state through the reaction

$$
p + n \leftrightarrow D + \gamma .
\tag{1.5}
$$

This reaction is the real bottleneck for the start of nucleosynthesis. Indeed, for the first minutes deuterium nuclei (which have a binding energy of $E_b^D = 2.225 \, \text{MeV}$) are immediately disintegrated by energetic photons. It takes roughly three minutes (and a temperature $T_D \approx 7 \times 10^{-5} \, \text{GeV}$) for the temperature to drop enough for having only very few photons with energy $\lesssim E_b^D$. At this point deuterium formation becomes efficient. The number density of protons and neutrons are $n_p/n_n \approx 1/7$. Deuterium nuclei immediately combine with other protons and neutrons to form tritium, ^3He, ^4He, ^6Li, ^7Li and ^7Be. Because of its high binding energy, roughly all neutrons present at this stage form ^4He, while only traces of the other elements are present. Heavier elements can not be formed, because no stable nucleus exists with mass number 5 or 8, and therefore collisions of helium with protons or helium is ineffective.

A more detailed calculation shows that the precise abundance of D and, to a lesser extent, other rare primordial nuclei, is strongly sensitive to the number density of baryons with respect to photons n_b/n_γ. The best determination of this fraction comes therefore from the measurement of the density of primordial deuterium abundance. This is performed from the observation of Ly-α absorption lines of deuterium in regions with low metallicity (i.e. with a low concentration of heavy elements, in the astrophysical language) which are therefore believed not to be influenced by stellar nucleosynthesis and to reproduce the primordial concentrations. When this is compared with the measure of n_γ by CMB observations, an independent determination of Ω_b is obtained, in agreement with the fact that an additional component is needed.

1.2   Dark Matter Production in the Early Universe

We are now going to discuss the possible production mechanisms of DM in the early universe, assuming that it is made of some unknown particle not included in the Standard Model. This is clearly a big assumption, and possible alternatives (primordial black holes, faint astrophysical objects, or modification of gravity) will be briefly discussed in the next section.

1.2.1   Freeze-Out

The simplest and most studied mechanism for the realization of the DM abundance is the so-called "thermal freeze-out". In synthesis, in the thermal freeze-out mechanism DM is a heavy stable particle that interacts feebly with the SM. At early times, DM particles (that we are going to indicate with χ from now on) are kept in thermal equilibrium with the SM by the process

$$\chi\chi \leftrightarrow ff \tag{1.6}$$

where f stands for any SM particles which may interact with DM. The DM will then follow a Boltzmann distribution, and its density will gradually decrease as temperature drops off. In the very same way as other SM particles like neutrinos, cosmic expansion will gradually reach the point in which the number density of DM particles is so low that creation/annihilation interactions do not happen any more, and the particle goes out of equilibrium. Its "relic" density is now fixed (or, since the temperature cools down with cosmic expansion, "frozen") and it will just evolve according to the expansion of the universe. As a rough estimate, the freeze-out will happen when the annihilation rate equals the expansion rate.

A detailed description of the freeze-out mechanism was given in [27]. Let us summarize the key points here, leaving more details in Appendix D. The evolution of the number density of all species present at a certain stage of the universe's evolution is described by means of a set of coupled Boltzmann equations. In the case of the DM this can be written as

$$\frac{\mathrm{d}n}{\mathrm{d}t} = -3Hn - \langle \sigma v \rangle \left(n^2 - n_{\mathrm{eq}}^2 \right) , \qquad (1.7)$$

where n is the DM number density (in physical coordinates), H is the Hubble rate $\dot{a}/a$ and $a(t)$ is the scale factor of the universe, σ is the annihilation cross section of two DM particles and v is their Møller velocity, defined in terms of the velocities of the two particles as $v = [|\vec{v}_1 - \vec{v}_2|^2 - |\vec{v}_1 \times \vec{v}_2|^2]^{1/2}$, and a factor of $1/2$ should be added in front of the cross section if the DM is not a self-conjugate particle (a Dirac fermion, for example). The first term on the right hand side describes the effect of the cosmic expansion, while the second takes into account annihilation and production of DM particles, where n_{eq} is the equilibrium number density. A semi-analytical solution of this equation is possible. The freeze-out temperature is defined as the temperature at which thermal equilibrium is lost (more precisely, the temperature at which the number density normalized to the total entropy density of the universe $Y = n/s$ equals some given number (typically $c = 1/2$) times the equilibrium value. This is obtained by the non algebraic equation

$$e^{x_F} = \frac{\sqrt{\frac{45}{8}} g_{\mathrm{DoF}} \, m_\chi \, M_{\mathrm{Pl}} \, c(c+2) \langle \sigma v \rangle}{2\pi^3 g_\star^{1/2} \sqrt{x_F}} , \qquad (1.8)$$

where $x \equiv m_\chi / T$ and the subscript F refers to the freeze-out time, g_{DoF} is the number of spin degrees of freedom of χ (2 for a spin-$1/2$ field), and $g_\star$ is defined to be the number of relativistic degrees of freedom at a given temperature (see [27] for a more detailed discussion about this point). With this information, one can calculate the relic abundance,

$$\Omega_{\mathrm{DM}} h^2 = \frac{1.04 \times 10^9 \, \mathrm{GeV}^{-1} m_{\mathrm{DM}}}{M_{\mathrm{Pl}} \int_{T_0}^{T_F} g_\star^{1/2} \langle \sigma v \rangle \mathrm{d}T} , \qquad (1.9)$$

where an overall factor of 2 should be added in the case of non self-conjugate DM, if the sum $\Omega_\chi h^2 + \Omega_{\bar{\chi}} h^2$ is considered. In order to obtain the right relic abundance, the cross section has to be of the order $\langle \sigma v \rangle \sim 3 \times 10^{-26} \, \mathrm{cm}^3 \, \mathrm{s}^{-1}$ (with the caveat that resonant annihilation or co-annihilation with other particles with similar mass would change drastically the story [28, 29]).

When numerical values are plugged in Eq. (1.9), a great coincidence appear: if we assume that χ interacts with the typical weak interaction scale, then its cross section would match the thermal one with a remarkable precision. Here is the origin of the term "WIMP", which we will use throughout this thesis.

This result is of course very exciting, especially in view of the fact that, as we are going to discuss in the next section, a number of beyond the standard model theories predict the existence of stable particles with this properties, and it is the reason why, before LHC found no hint of the existence of Supersymmetry, the word "dark matter" was almost confused with "WIMP" and with "neutralino". Anyway, it should be remembered that weak scale WIMPS are not the whole story. Indeed, appropriately recasting Eqs. 1.8 and 1.9 one gets that, in order to obtain the correct relic abundance, it is sufficient to impose [30]:

- $m_\chi \cdot \sigma \cdot M_{Pl} \gg 1$ in order to have a cold thermal relic;
- $\sigma \sim 10^{-8}\,\mathrm{GeV}^{-2}$ in order to match the correct relic abundance.

Assuming that the cross section scales roughly as $\sigma \sim g^4/m_\chi^2$ where g is a coupling constant one gets

$$g^2 \sim \frac{m_\chi}{10\,\mathrm{TeV}} \quad \text{and} \quad m_\chi \gg \frac{10^8\,\mathrm{GeV}^2}{M_{Pl}} \sim 0.1\,\mathrm{eV}. \tag{1.10}$$

This lower bound can be made stronger if one assumes that the WIMP annihilates through the standard weak interaction. In this case, the cross section will scale as $\sigma \sim G_F^2\, m_\chi^2$, and imposing that the particle χ have a relic abundance equal or lower than the measured one we get

$$\Omega_{DM} h^2 \sim 0.1 \left(\frac{10\,\mathrm{GeV}}{m_\chi}\right)^2 \lesssim 0.1 \quad \Rightarrow \quad m_\chi \gtrsim 10\,\mathrm{GeV}. \tag{1.11}$$

This bound is known as the Lee–Weinberg limit [31].

Correspondingly, an upper bound on the WIMP mass can be set by exploiting perturbative unitarity. Assuming for the cross section a scaling $\sim 1/m_\chi^2$, then unitarity imposes $\sigma \lesssim 4\pi/m_\chi^2$. Plugging this results into the expression for the relic abundance one gets $m_\chi \lesssim 120\,\mathrm{TeV}$ [32]. Notice that both these bounds were weaker in the original proposals, due to the assumption $\Omega_{DM} \approx 1$.

An interesting variation of the standard freeze-out mechanism is the one in which the DM particles are non interacting and their abundance is generated by the decay of parent WIMP particles. This is what happens for example in the so-called superWIMP scenario, for super-weakly interacting massive particles [33, 34]. An example of superWIMP is the supersymmetric gravitino, which interacts only gravitationally with the SM, in models in which the reheating temperature is low ($\lesssim 10^{10}\,\mathrm{GeV}$) so that their initial abundance is negligible. The next-to-lightest supersymmetric particle (NLSP) then freezes out as in the simplest scenario, but eventually decays to the gravitino, which is the lightest supersymmetric particle (LSP) and is stable if R-parity is assumed (see discussion in Sect. 1.3.4).

1.2.2 Freeze-In

The somehow opposite scenario with respect to the standard freeze-out one is that of freeze-in, originally proposed in [35] and then independently in [36] with an extended analysis. The mechanism works as follows. The dark matter χ is assumed to be quite heavier than the EW scale, and interact with SM particles through some renormalizable operator (so that the production cross section is always IR dominated) with very tiny coupling (because of this, χ is referred to as a FIMP, for feebly interacting massive particle). The initial abundance at very high temperature is assumed to be negligible, as a consequence of an inefficient production mechanism during reheating. As the temperature lowers, χ particles start to be produced, and the production rate has its maximum for $T \sim m_\chi$ (since later energetic enough particles become rare). The final yield n/s of χ particles is given by

$$Y_{\mathrm{FI}} \sim \lambda^2 \left(\frac{M_{\mathrm{Pl}}}{m_\chi} \right) \qquad (1.12)$$

Notice that the final abundance of DM is proportional to λ^2, and not to its inverse as in the freeze-out mechanism. The reason is that while in the latter the initial abundance is fixed by thermal equilibrium and the final one depends on how many DM particles annihilate before annihilations become too rare due to cosmic expansion, here the abundance depends on how many χ particles can be produced in the restricted window in which the temperature is not too high and the expansion rate is not too large. The distinction between the two mechanisms depend crucially on the size of the coupling.

1.2.3 Supermassive Dark Matter

An intriguing possibility is that DM is made up of supermassive particles produced non thermally after inflation or during its last stages. These go under the name of WIMPZILLAs, as originally proposed in [37–40]. The possible production mechanisms are diverse: they can be produced during reheating (even with a low reheating temperature), during preheating, at the end of inflation as a result of the rapid gravitational expansion, or finally in bubble collisions after a first order phase transitions. Two necessary conditions must hold in order for WIMPZILLAs to be a reliable DM candidate. First, they have to be stable, or have a lifetime larger than the age of the Universe. Second, they have to be produced non thermally, otherwise their present relic abundance would be greater than 1.

1.2.4 Asymmetric Dark Matter

A completely different mechanism from what discussed above is the one of asymmetric DM. The idea originates from the observation that the present densities of DM and baryons differ by a factor of $\Omega_{\rm DM}/\Omega_b \approx 5$, which is remarkably close to 1 if one considers the fact that the two densities are fixed at very different epochs and different phenomena. While this can be just an accident, or it can be attributed to anthropic reasons, it can as well have a deep physical motivation.

The baryon abundance in the Universe is related to the so-called "baryon asymmetry", i.e. the fact that baryons and anti-baryons are present in the Universe with different abundances. The correct measure of this asymmetry is given by the ratio $\eta_B = (n_B - n_{\bar{B}})/n_\gamma \approx 6 \times 10^{-10}$. Three necessary conditions (known as the Sakharov conditions [41]) must be fulfilled in order for a baryon asymmetry to be generated:

- The theory has to be B violating
- C and CP symmetries have to be violated
- Departure of thermal equilibrium

Departure from thermal equilibrium may happen in various phases during the evolution of the early Universe. For example, thermal equilibrium is lost if heavy particles are present which decay copiously, or right after the EW phase transition if it is of the first order. B violation is realised in the SM only at the quantum level. The baryon number is indeed anomalous (as well as the lepton number), and vacua with different B number are connected by the so-called "sphaleron" solutions. C and CP symmetries are violated in the SM, but to smaller extent than what is needed to justify the measured value of η_B. Some kind of new physics is therefore needed in order to explain the present baryon asymmetry. One possibility is connected to neutrino masses. If a Majorana mass term is added, then the lepton number is strongly violated. Non-perturbative phenomena, that violate B and L but respect B-L, can then transfer the lepton asymmetry to baryons.

A similar process can be envisaged adding DM to the game. Some CP violating process (which violates also B, L, or the quantum number that makes DM stable) produces an initial asymmetry in the visible or the dark sector, which is then communicated to the other sector by some other mechanism. Once the asymmetry is established the symmetric component is annihilated out, leaving only one between the particle and the anti-particle as a relic.

The idea that the DM and baryon asymmetries might be related to each other is quite old [42–44]. The initial motivation for the DM asymmetry was to solve the solar neutrino problem by accumulating DM that affects heat transport in the Sun [45]. The many possible mechanisms leading to a DM asymmetry are reviewed in [46]. An interesting example was introduced in [47], in which DM is a composite state of a dark QCD, linked to the usual one. Interestingly, this model can be testable at the LHC by looking for an "emerging jet" signature.

A consequence of the asymmetric DM paradigm is that the annihilation process in the present universe is forbidden, and indirect detection is therefore ineffective (unless its fractional asymmetry is very small, see e.g. [48]).

We have listed here only the main mechanisms proposed so far for the generation of the relic DM density. Other possibilities can be introduced, in particular in the framework of non standard cosmologies. For a review see [49].

1.3 Dark Matter Candidates

In this section we would like to give a short overview of the most studied DM candidates. Our starting point will be a discussion of the astrophysical candidates. We will then turn our attention to the only SM particles which could in principle be a viable DM candidate: neutrinos. Finally, we will talk about WIMP and non-WIMP candidates in BSM theories. We will conclude by mentioning the modified gravity approach as an alternative to the particle DM one.

1.3.1 MACHOs: Massive Compact Halo Objects

The simplest solution to the DM problem that one can think of is that the amount of missing mass in astrophysical structures is provided by astrophysical objects: dark or very faint bodies are indeed very hard to detect, and in principle a large population of this kind may fill the galaxy. The most attractive candidate is that of a population of planet-sized objects that can be baryonic in nature and fit within our current understanding of stellar evolution (called, in opposition to WIMPs, MACHOs: massive compact halo objects [50]).

The main tool we have to constrain the properties of these objects, and in particular the fraction of missing mass which they constitute, is that of microlensing. Gravitational lensing is the phenomenon, described by General Relativity, thanks to which light is deviated by the presence of a gravitational field. In the case of MACHOs, the idea is to observe eclipse-like events in which a MACHO passes through the line of sight of a distant star, distorting its image. The effect is tiny (hence the name microlensing) but the simultaneous observation of $\mathcal{O}(10^7)$ stars can lead to very strong constraints. Experiments like the MACHO Project [51], EROS [52], OGLE [53, 54] and POINT-AGAPE [55] observed millions of stars for 5–10 years, finding $\mathcal{O}(10)$ events each, while the expected number was 3–5 times larger if all the missing mass were given by MACHOs, constraining them to constitute no more than 20% of the missing mass of galaxies.

1.3.2 PBH: Primordial Black Holes

The idea that primordial black holes can constitute a large portion (if not all) of DM is quite old [56–58]. Black holes formed before the era of nucleosynthesis (therefore

called "primordial") evade the bound on the total amount of baryonic matter, and can
constitute a good DM candidate. Many mechanisms were proposed for the formation
of this kind of objects (see for example [59] and references therein). This scenario is
quite well constrained by MACHOs searches and considerations about their Hawking
radiation:

- If PBH are lighter than 5×10^{14} g they would have evaporated thanks to Hawking
  radiation [60].
- For masses $10^{15} - 10^{16}$ g Hawking radiation would produce a flux of $E \sim 100$ MeV
  photons [61] which are not seen in the extragalactic background [62, 63].
- In the range $10^{16} - 3 \times 10^{22}$ g PBH are excluded as the primary DM candidate by
  the possibility of the process of PBH capture in stars during their formation phase:
  PBH would accrete inside the star and destroy the star's compact remnants. The
  observation of neutron stars or white dwarfs in a DM rich environment poses a
  limit on the fraction of PBH DM [64].
- Similarly, the capture process of PBH by already formed neutron stars constrains
  PBH DM in the range $3 \times 10^{18} - 10^{24}$ g [65].
- In the range $5 \times 10^{18} - 10^{20}$ g PHB DM can be excluded by searches for femto-
  lensing effects[1] of gamma ray bursts [68].
- Masses $10^{26} - 10^{34}$ g are excluded by the Kepler planetary survey and the MACHO
  and EROS-2 microlensing surveys [51, 52, 69].
- PBH with mass $10^{33} - 10^{42}$ g can be constrained with CMB data, exploiting the
  fact that they are massive enough to accrete emitting X-rays that would modify
  cosmic reionization [70].
- Masses larger than $\sim 10^{34} - 10^{35}$ g are constrained to form a small fraction of
  the DM density by searches for perturbations in the orbits of halo binaries due to
  the presence of MACHOs [71] (the precise number depending on the subset of
  binaries chosen for the analysis).
- Very massive MACHOs heavier than 10^{40} g are finally ruled out by the dynamic
  of the galactic disk [72].

The idea of PBH as DM has raised new attention in recent months, after the LIGO
collaboration announced the first observation of a gravitational wave signal from the
merger of two black holes with mass $\sim 30 M_\odot$, corresponding to $\sim 6 \times 10^{34}$ g [73].
On the one hand, this mass is quite unusual for astrophysical black holes, which are
expected to be lighter (if they form in the collapse of old and heavy stars) or heavier
(as the supermassive black holes at the centre of galaxies). On the other hand, is
was noticed in [74] that if PBH are the DM, the rate of merger processes is in the
right ballpark to explain LIGO observations. PBH in this mass range are claimed
to be excluded by CMB observations [70]. Nevertheless, this point is still debated:
calculations are not yet fully under control and order of magnitude uncertainties could

---

[1]Femto-lensing effects are diffraction effects due to gravitational lensing from massive objects
in the line of sight of the luminous one. Objects with mass $\sim 10^{17} - 10^{20}$ g corresponding to
$10^{-16} - 10^{-13} M_\odot$ can produce fringes in the gamma ray burst emission at ~ 1 MeV with an
angular separation of $\sim 10^{-15}$ arc sec, hence the name "femto-lensing" [66, 67].

allow PBH to constitute a portion of DM large enough to be detectable by LIGO [75]; on the other hand, there are claims that new constraints from compact stellar systems in ultra-faint dwarf galaxies could definitely rule out that mass window [76].

This scenario is for sure interesting, and will be tested in the coming years in many ways. Two promising directions are gravitational waves detections (if the spatial distribution of mergers can be measured, it will be possible to compare it with the DM galactic halo) and the detection of fast radio bursts signals strongly lensed by the passing close to a $\sim 30 M_\odot$ black hole in the galaxy [77]. In the future, gravitational waves will also be a very useful tool to constrain the properties of other exotic objects formed by new dark sector particles, being or not these related to DM [78].

1.3.3 Standard Model Neutrinos

Neutrinos have been considered for long time a good DM candidate. With respect to all other possibilities, they have the considerable advantage of being known to exist. Despite of this, there are two reasons which prove that DM is not made of neutrinos (if not for a tiny contribution). The first reason is related to their abundance. In the early universe, neutrinos were in thermal equilibrium with other SM particles, until they decoupled when the temperature was $\mathcal{O}(1-10\,\mathrm{MeV})$. Their decoupling followed exactly the freeze-out mechanism described in Sect. 1.2.1 for DM. Their relic abundance can be easily computed to be

$$\Omega_\nu h^2 = \sum_{i=1}^{3} \frac{m_{\nu_i}}{93\,\mathrm{eV}} \,. \tag{1.13}$$

The current best laboratory upper limit on the neutrino masses is $m_\nu < 2.05\,\mathrm{eV}$ [79], which apply to all the three neutrino flavours (since their squared mass differences are constrained to be much lower from oscillation experiments), while even lower limits $\sum m_\nu < 0.17\,\mathrm{eV}$ may be set with cosmological probes [24]. These values are much lower than the one needed in order for neutrinos to constitute 100% of the dark matter, and this explanation has therefore to be disregarded.

Another argument against neutrinos as DM comes from our understanding of the formation of large scale structures. Being relativistic at the moment of their decoupling, neutrinos would act as hot DM, erasing fluctuations as scales lower than their free-streaming length $40\,\mathrm{Mpc} \times (m_\nu/30\,\mathrm{eV})$ [80]. This would imply a top-down formation history, in which large structures form first, at odds with the fact that our galaxy appears to be older than the Local Group and with the observation of galaxies up to redshifts $z > 4$.

1.3.4 WIMPs: Weakly Interacting Massive Particles

The by far most studied DM candidates are the ones that fall in the category of WIMPs. Historically, this was partly due to the theoretical prejudice that new physics had to appear at the EW scale in order to solve the so-called naturalness problem of the standard model. Briefly, the naturalness problem has to do with the mass of the Higgs boson. For dimensional reasons, the Higgs boson's mass should receive corrections from every layer of new physics to which it couples. Taking for granted the existence of new physics at high enough energy (gravity, at the very least), it is quite unlikely that very large corrections sum up to give such a light result. Stated in another, more general way, the hierarchy problem can be seen as the problem of understanding why the EW scale is so much smaller that the only other known fundamental scale, the Planck mass and, in particular, the stability of this scale separation under quantum corrections. Of course, it is not guaranteed that this line of thought is correct. It may well be that nature is just accidentally fine tuned, or that a certain amount of fine tuning is necessary for the development of life, so that we observe the only universe in which observers can exist, or, finally, that fine tuning is just an artifact of a complex dynamics in a non fine tuned universe [81]. On the other hand, the naturalness argument in favour of new physics at a "not too far" scale still looks compelling, and if nothing new will be seen in the near future a serious reconsideration of our understanding of particle physics will be necessary.

Generically speaking, new physics solutions to the naturalness problem require the existence of a new layer of physics at a scale close to the EW one. There are three main classes of models that can solve the naturalness problem by cancelling the large corrections to the Higgs mass:

- In supersymmetric theories, corrections are cancelled by the presence, for each particle, of a super-partner with different spin, which contributes to δm^2 with an opposite sign correction.
- In composite models, the Higgs is a composite particle, and it is not part of the spectrum of the theory above its compositeness scale. This class is intersected with that of Little Higgs theories, in which the Higgs boson is the pseudo-Nambu-Goldstone boson of some spontaneously broken global symmetry, and is therefore light.
- In models with extra dimensions, the separation between the Planck mass and the EW scale is cancelled by making fields propagate outside our 4 dimensional manifold.

Quite remarkably, in each of these frameworks there is room for a good DM candidate. This is one of the reasons why the WIMP paradigm become so well established.

Supersymmetric WIMPs

Supersymmetry (SUSY) is one of the best-motivated proposals for physics beyond the Standard Model. Leaving the hierarchy problem apart for the moment, there are many good (albeit speculative) reasons to believe in supersymmetry, most notably its ability to link matter and force carrier fields, the possibility of relating gravity to other fundamental interactions and its role in string theory.

Supersymmetry is also interesting from a phenomenological point of view. Firstly, low-scale (around 1 TeV) supersymmetry help stabilize the mass of the Higgs boson, thus solving the Higgs naturalness problem. Secondly, in models where R-parity is conserved, the lightest supersymmetric particle (LSP) is stable and provides a perfect cold DM candidate (such that for long years the identification DM = neutralino was almost taken for granted). Thirdly, supersymmetry plays an essential role in grand-unified theories (GUT), constructed over the observation that solving the renormalization group equations the SM gauge couplings including supersymmetric particles (*sparticles*) have the same value at a scale $\sim 10^{16}$ GeV.

The only viable supersymmetric theories are, from the phenomenological point of view, those with only one charge, indicated as $N = 1$ supersymmetry. The building blocks of such models are the chiral supermultiplets, consisting of a Weyl fermion and a complex scalar, and gauge supermultiplets, consisting of a gauge field and a "gaugino" fermion. Fields interact via the usual gauge interactions and through the introduction of a "superpotential", which contains bilinear and trilinear couplings of the chiral supermultiplets and gives rise to mass and Yukawa terms.

The so-called minimal supersymmetric extension of the standard model (MSSM) [82, 83] is characterized by the addition to the SM of only the minimal number of new fields, one per each SM one, and a second Higgs supermultiplet which is needed for anomaly cancellation. The addition of these fields is necessary, since SM fields do not have the correct quantum numbers to be combined in supermultiplets. The coupling of supersymmetry to gravity would also require the introduction of a spin 3/2 gravitino in a supermultiplet with the spin 2 graviton. The minimal superpotential is

$$W = \epsilon_{ij} \left(y_e H_1^j L^i e^c + y_d H_1^j Q^i d^c + y_u H_2^i Q^j u^c \right) + \epsilon_{ij} \mu H_1^i H_2^j . \tag{1.14}$$

Yukawa couplings are 3×3 matrices in flavour space. The presence of the second Higgs doublet (as opposed to the situation in the SM) makes it possible for the potential to be a holomorfic function of the fields, which is a necessary condition for the consistency of the theory.

Supersymmetry is spontaneously broken by a mechanism which is not known a priori. The breaking can be parametrized by an explicitly SUSY breaking Lagrangian, with the condition that the breaking is soft, i.e. such that the resulting theory is still free of quadratic divergences. Gauge bosons acquire masses with the usual mechanism of spontaneous breaking of the gauge symmetry, while fermion masses are given by the Yukawa terms after symmetry breaking.

Let us now look back to the superpotential of Eq. (1.14). Other terms may be added to this minimal version:

$$W_R = \frac{1}{2} \lambda \epsilon_{ij} L^i L^j e^c + \lambda' \epsilon_{ij} L^i Q^j d^c + \frac{1}{2} \lambda'' u^c d^c d^c + \mu' L^i H_2^i . \tag{1.15}$$

Each of these terms violate one between the baryon and lepton number, and can therefore be strongly constrained by the non observation of exotic phenomena such as proton decays. The presence of these terms can be avoided if a discrete symmetry, called R-parity, is imposed: to each field is associated an R-parity eigenvalue of $(-1)^{3B+L+2s}$, where s is the spin of the particle, so that SM fields have $R = +1$ and their superpartners have $R = -1$. R-parity is conserved multiplicatively, meaning that the supersymmetric partners can be produced from SM ones only in pairs, and that heavy supersymmetric particles can only decay to lighter SUSY states. This have an extremely important consequences: if R-parity is conserved, the LSP is absolutely stable, making it a perfect DM candidate.

To conclude this section, let us discuss what are the possible DM candidates of the MSSM. Electrically charged or coloured LSP are excluded as DM candidates. The remaining possibilities are therefore a sneutrino (which would have a sizeable scattering cross section on nuclei and is therefore excluded by direct detection, as far as a possible mixing between the left- and right-handed sneutrinos is not considered), the gravitino (with spin 3/2) and the lightest neutralino (with spin 1/2), the latter being the most studied option. There are four neutralinos in the MSSM: the supersymmetric partners of the two Higgs doublets (the Higgsinos), the partner of W^3 (the neutral wino) and the partner of the B (the bino). Four mass eigenstates are formed as linear combinations of these fields, and depending on the choice of parameters each of them can be the LSP.

Kaluza–Klein WIMPs

Studies of models in which the usual 4 dimensional space-time is embedded in a larger dimensional one started very long ago, with the pioneering works of T. Kaluza in 1921 [84] and O. Klein in 1926 [85]. The original idea of Kaluza was to unify gravity and electromagnetism by adding a fifth space-like dimension to general relativity and identifying the additional degrees of freedom of the metric with the photon. This is possible provided that the new dimension is compactified on a circle of radius R, very small in order to be unobservable, and if a $\mathbb{Z}_2$ identification is operated in order to avoid the presence of an additional unwanted scalar particle. Extra dimensions became then popular in the context of string theory, where the existence of 10–11 dimensions is typically assumed.

More recently, it was realised that extra dimension theories can help in solving the hierarchy problem, as in the very famous Arkani-Hamed, Dimopoulus and Dvali (ADD) 1998 proposal [86–90] or the Randall-Sundrum (RS) one [91, 92]. The ADD model postulates the existence of $n \geq 2$ extra dimensions in which only gravity can propagate, while the SM fields are confined on a 3+1-dimensional manifold (often referred to as the "brane", as opposed to the "bulk" 3+n+1-dimensional space). At energies above the TeV, SM fields start propagating in the new dimensions, in which gravity has an effect comparable to those of other forces. The largeness of the Planck mass with respect to the weak scale is then an artefact of the largeness of new compact dimensions with respect to the weak length scale. Indicating by $M_{\rm Pl}$ the usual 4-dimensional Planck mass and by $M_{\rm Pl(4+n)}$ the fundamental scale one obtains [86]

$$M_{\rm Pl} \sim M_{\rm Pl(4+n)}^{2+n} R^n \tag{1.16}$$

where n is the number of new compact dimensions and R is their size. Imposing that $M_{\rm Pl(4+n)} \sim m_{\rm EW}$ and demanding that R reproduces the observed value of $M_{\rm Pl}$ one gets

$$R \sim 10^{\frac{30}{n}-17} \, {\rm cm} \times \left(\frac{1 \, {\rm TeV}}{m_{\rm EW}}\right) . \tag{1.17}$$

For $n = 1$ the model is not viable, since Newtonian gravity would be modified from a $1/r^2$ to a $1/r^4$ power law at the scale of the Solar System $R \sim 10^{13}$ cm. For $n \geq 2$ instead one gets $R \sim 1$ mm or lower, interestingly in the range which is testable with present experiments on gravitation ($n = 2$ is indeed already excluded).

The RS model is a step further in the sense that it eliminates the residual hierarchy between the EW scale and the compactification scale $1/R$. In order to do so, a setup is proposed in which two 4-dimensional branes, the visible one in which we live and a hidden one, are embedded in a 5-dimensional space-time, whose additional dimension is not factorizable in the metric. Solution to the Einstein equations in this scenario lead to a metric of the form

$$ds^2 = e^{-2k R \phi} dx_\mu dx^\mu + R^2 d\phi^2 , \tag{1.18}$$

where x^μ are the usual 4-dimensional coordinates, the additional one is parametrised by $0 \leq \phi \leq \pi$, and k is a scale of the order $M_{\rm Pl}$. The very large hierarchy between the Planck and the EW scale is generated by the exponential factor in the metric, which does not necessarily require a very large value of the radius R, as in the ADD model. On the other hand, R can be sent to infinity, resulting in a model in which the extra dimension is not compact anymore, still with a viable phenomenology [92].

Extra dimensional theories prove to be very interesting for DM physics because for each field that propagates in the compact extra dimensions there exist a tower of excitations (called "Kaluza–Klein" particles, or KK), among which the lightest has mass $\sim R^{-1}$ and is absolutely stable [93, 94]. The most studied case is the one of universal extra dimensions (UED) [95], in which not only gravity but also SM fields propagate in the additional compact dimensions. Assuming that no 4-dimensional brane is present, translational invariance in the extra dimension leads to the conservation of the KK number, i.e. the number associated with KK excitations, which is broken by quantum effects to the conservation of KK parity. In the very same way in which R-parity makes the LSP stable in the MSSM, the lightest Kaluza–Klein particle (LKP) is stable, and a good DM candidate. The LKP is typically the first KK excitation of the photon, which must have a mass in the TeV range in order to reproduce the correct relic abundance via the standard freeze-out mechanism [96].

Other "natural" Scenarios

The list of WIMP particles we presented was far from be complete. Many other candidates and scenarios are described in the literature, often less studied but definitely not less interesting. In the brief introduction to WIMPs, we mentioned the composite

Higgs scenario. Dark matter candidates exist in this framework in the form of composite states of new fermions, made stable by the conservation of some new quantum number. Scalar models in which the Higgs serves as a portal are also interesting.

Finally, models in which the DM has no SM gauge interaction, and its relic abundance is obtained by some dynamics in a hidden sector, are called WIMPless scenarios. In this case, DM direct and indirect detection experiments may be totally ineffective, as far as some connection between the visible and the hidden sector is not introduced.

Minimal DM

An orthogonal approach to WIMPs is the one of Minimal DM [97]. Instead of looking for DM in an existing model designed to solve the naturalness problem, the idea of minimal DM is to add to the SM the minimal number of degrees of freedom (only one scalar or fermionic SU(2) multiplet, interacting with the SM only through gauge interactions in such a way that the mass m_χ is the only free parameter) and systematically look for the ideal candidate. The condition that such particle must respect are:

1. The lightest component of the multiplet (which represents the DM candidate) is stable on cosmological time-scales.
2. The mass splitting generated by quantum corrections are such that the lightest component is neutral.
3. The DM candidate is allowed by DM searches.

With these simple requirements, a SU(2)-quintuplet with $Y = 0$ and $m_\chi \sim 10\,\text{TeV}$ is identified a viable DM candidate, which survives all tests from DM searches and has the correct relic density. Interestingly, the PAMELA positron excess (which we will describe in details in Chap. 3) can be explained in terms of DM annihilation in this minimal model [97–103].

1.3.5 Sterile Neutrinos

Sterile neutrinos are proposed new fermions with no SM gauge interactions (hence the name "sterile") which play an important role in the see-saw mechanism of neutrino mass generation. The only allowed interaction terms of these new fermions are a Majorana mass term and a Yukawa coupling with the SM Higgs and the lepton doublets. The SM lagrangian is modified as

$$\mathcal{L} = \mathcal{L}_{\text{SM}} + \bar{N}_j i \partial_\mu \gamma^\mu N_j - f_{ij} \bar{L}_i N_j \widetilde{H} - \frac{M_j}{2} \bar{N}_j^c N_j + h.c. , \qquad (1.19)$$

where N_j are the sterile neutrinos, L_i are the SM lepton doublets and H is the Higgs doublet. After EW symmetry breaking, ordinary neutrinos get a mass from mixing with the additional ones. An interesting scenario is the one in which the additional neutrinos have a mass of the order of the EW scale. In this case, among other phenomenological implications, a sterile neutrino DM candidate with mass $\mathcal{O}(10\,\text{keV})$ is naturally provided. Interestingly enough, sterile neutrinos in this range are a candidate for warm dark matter, which could help in solving some tensions in the comparison of numerical simulations of the dynamics of cold DM halos with observations (see Sect. 1.1.3). For a more thorough discussion of sterile neutrinos in the see-saw mechanism and their role as DM candidates, see [104] and references therein.

1.3.6 Axions

The axion is a particle proposed by Peccei and Quinn [105, 106] to solve the so-called "strong CP problem". Before going to describe axions as DM candidates let us review what the strong CP problem is. Consider the Lagrangian of QCD with n generations of massive quarks:

$$\mathcal{L}_{\text{QCD}} = -\frac{1}{4}G^a_{\mu\nu}G^{a\,\mu\nu} + \sum_{j=1}^{n}\left[\bar{q}_j\gamma^\mu i D_\mu q_j - (m_j\bar{q}_{Lj}q_{Rj} + h.c.)\right] + \theta\frac{g^2}{32\pi^2}G^a_{\mu\nu}\widetilde{G}^{a\,\mu\nu}.$$

$$(1.20)$$

The last term is a quadri-divergence, and does not contribute in perturbation theory. Nevertheless, non trivial vacuum configurations exist in QCD, and that term has a physical meaning. Taking into account the possibility of redefining the quark phases, thanks to the Adler–Bell–Jackiw anomaly [107, 108] it turns out that the actual physical quantity is

$$\bar{\theta} = \theta - \arg(\det M),$$

$$(1.21)$$

where M is the quark mass matrix. The two terms on the right hand side have a very different origin: one is related to non perturbative QCD effects, while the other is originated by EWSB. It would therefore be natural to have $\bar{\theta}$ of order 1.

Since the θ-term in the QCD lagrangian is P and CP violating, the non observation of CP violation in strong interactions allow to set a very stringent limit on $\bar{\theta}$. The best upper limit is currently $\bar{\theta} \lesssim 10^{-9}$, coming from bounds on the electric dipole moment of neutrons $|d_n| < 3 \times 10^{-26}e$ cm.

A possible explanation for the smallness of $\bar{\theta}$ is the presence of a spontaneously broken global symmetry $U(1)_{\text{PQ}}$. This symmetry is assumed as a global symmetry of the lagrangian, which is then broken explicitly by non perturbative effects and finally spontaneously broken. In this way, the physical phase $\bar{\theta}$ becomes

$$\bar{\theta} = \theta - \arg(\det M) - \frac{a(x)}{f_a},$$

$$(1.22)$$

where $a(x)$ (the *axion* field) is the pseudo-Nambu-Goldstone boson associated to the breaking of the $U(1)_{\text{PQ}}$ symmetry, and f_a is called the axion decay constant.

Non perturbative effects also produce an effective potential $V(\bar{\theta})$ with a minimum at $\bar{\theta} = 0$, which solves the strong CP problem.

The mass of the axion is given by

$$m_a \simeq 0.6\,\mathrm{eV} \frac{10^7\,\mathrm{GeV}}{f_a} . \tag{1.23}$$

Depending on the value of f_a, observable axion masses span roughly 12 orders of magnitudes: from $\sim 10^3$ eV constrained by the CAST experiment down to 10^{-11} eV with the CASPEr experiment [109]. Axion searches typically exploit the axion coupling to two photons

$$\mathcal{L}_{a\gamma\gamma} = -g_\gamma \frac{\alpha}{\pi} \frac{a(x)}{f_a} \vec{E} \cdot \vec{B} \tag{1.24}$$

in a number of different searches: light shining through walls, microwave cavities, use of magnetic-resonance techniques, etc.

After their first proposal in the late 70s, axion-like particles showed up in different realizations, in particular in the context of string theory. The axion mechanism also provide a good DM candidate. Axions are produced both thermally and non thermally in the early universe. The thermal population of such a low mass particle is a hot DM component, and it is therefore forced to have only a minor abundance. The non thermal part is instead more interesting, because it acts as cold DM. A review of the role of axions in cosmology can be found in [110].

1.3.7 Modified Gravity as an Alternative to Dark Matter

Evidences of the existence of dark matter have been accumulated for more than 80 years, starting with the works of Oort and Zwicky. Nevertheless, it was not until the '80s that the majority of the physics community started to think of DM as some kind of new particle, instead of some exotic astrophysical object. Despite the many efforts over the last 40 years, we still have no real clue about the particle nature of DM, nor we have ever observed any non gravitational effect. It is therefore legitimate to ask whether the DM is really made of new particles. It could well be that, starting at galactic scales, Newtonian gravity differs from what observed at shorter scales, in such a way that the effect of a new particle population is mimicked. This was the idea behind the original proposal of "modified Newtonian dynamics" (MOND) in 1983 [111]. In this model, Newtonian gravity responsible for the rotation curves of galaxies, is modified in such a way that a test particle in a Newtonian gravitational potential Φ_N generated by the baryon distribution alone undergo an acceleration

$$\tilde{\mu}\left(\frac{|\vec{a}|}{a_0}\right)\vec{a} = -\vec{\nabla}\Phi_N\,, \qquad \text{where} \quad \begin{cases} \tilde{\mu}(x) \sim 1 & \text{for } x \gg 1 \\ \tilde{\mu}(x) \sim x & \text{for } x \ll 1 . \end{cases} \tag{1.25}$$

With this relation, one is able to reproduce very well the galactic rotation curves and the so-called Tully–Fisher relation for spiral galaxies that relates the total baryonic mass to the asymptotic rotation velocity.

On the one hand, Eq. (1.25) can be seen as a smart parametrization of the effect of non baryonic DM on the dynamics of galaxies, and the question is to understand how the acceleration scale a_0 and the function $\widetilde{\mu}$ are generated. On the other hand, (1.25) can be taken more seriously and viewed as a modification of inertia in the regime of low accelerations. This direction is not an easy one. Equation (1.25) can be derived from a variational principle with the so-called aquadratic Lagrangian (AQUAL) [112], and embedded in a relativistic formulation in the tensor-vector-scalar theory (TeVeS) [113]. A review of these approaches may be found in [114] (also included in [115]).

There are three main problems that MOND theories face when competing with the standard DM paradigm [116]:

- The difficulty when dealing with the dynamics of galaxy clusters (and particularly the bullet cluster)
- Problems in reproducing the path of peaks in the CMB power spectrum, in particular the third one (which is predicted to be very low in baryon dominated cosmologies)
- The last and most important problem has to do with the shape of the matter power spectrum. In the DM picture, baryons fall in the potential wells produced by DM, partially loosing memory of the baryon acoustic oscillations which they shared with radiation (and are imprinted in the CMB). On the contrary, in a no DM universe these fluctuations are not suppressed, leading to a totally different shape of the power spectrum.

References

1. F. Zwicky, Die Rotverschiebung von extragalaktischen Nebeln. Helv. Phys. Acta **6**, 110–127 (1933)
2. G. Bertone, D. Hooper, *A History of Dark Matter*, arXiv:1605.04909
3. K.G. Begeman, A.H. Broeils, R.H. Sanders, Extended rotation curves of spiral galaxies: dark haloes and modified dynamics. Mon. Not. Roy. Astron. Soc. **249**, 523 (1991)
4. K.C. Freeman, On the disks of spiral and S0 galaxies. Astrophys. J. **160**, 811 (1970)
5. J. Einasto, A. Kaasik, E. Saar, Dynamic evidence on massive coronas of galaxies, Nature **250**, 309–310 (1974)
6. J.P. Ostriker, P.J.E. Peebles, A. Yahil, The size and mass of galaxies, and the mass of the universe, Astrophys. J. Lett. **193**, L1–L4 (1974)
7. P.M.W. Kalberla, L. Dedes, J. Kerp, U. Haud, Dark matter in the Milky Way, II. the HI gas distribution as a tracer of the gravitational potential. Astron. Astrophys. **469**, 511–527 (2007), arXiv:0704.3925
8. J. Fan, A. Katz, L. Randall, M. Reece, Double-Disk Dark Matter. Phys. Dark Univ. **2**, 139–156 (2013), arXiv:1303.1521
9. J. Fan, A. Katz, L. Randall, M. Reece, Dark-Disk Universe. Phys. Rev. Lett. **110**(21) 211302 (2013), arXiv:1303.3271

10. L. Randall, M. Reece, Dark Matter as a Trigger for Periodic Comet Impacts. Phys. Rev. Lett. **112**, 161301 (2014), arXiv:1403.0576
11. N.J. Shaviv, The Paleoclimatic evidence for Strongly Interacting Dark Matter Present in the Galactic Disk, arXiv:1606.02851
12. D. Clowe, M. Bradac, A.H. Gonzalez, M. Markevitch, S.W. Randall, C. Jones, D. Zaritsky, A direct empirical proof of the existence of dark matter. Astrophys. J. **648**, L109–L113 (2006), arXiv:astro-ph/0608407
13. M. Markevitch, A.H. Gonzalez, L. David, A. Vikhlinin, S. Murray, W. Forman, C. Jones, W. Tucker, A Textbook example of a bow shock in the merging galaxy cluster 1E0657-56. Astrophys. J. **567**, L27 (2002), arXiv:astro-ph/0110468
14. D. Clowe, A. Gonzalez, M. Markevitch, Weak lensing mass reconstruction of the interacting cluster 1E0657-558: Direct evidence for the existence of dark matter. Astrophys. J. **604**, 596–603 (2004), arXiv:astro-ph/0312273
15. S.W. Randall, M. Markevitch, D. Clowe, A.H. Gonzalez, M. Bradac, Constraints on the Self-Interaction Cross-Section of Dark Matter from Numerical Simulations of the Merging Galaxy Cluster 1E 0657–56. Astrophys. J. **679**, 1173–1180 (2008), arXiv:0704.0261
16. M.J. Jee et al., Discovery of a ringlike dark matter structure in the core of the galaxy cluster Cl 0024+17. Astrophys. J. **661**, 728–749 (2007), arXiv:0705.2171
17. A. Mahdavi, H.y. Hoekstra, A.y. Babul, D.y. Balam, P. Capak, A Dark Core in Abell 520. Astrophys. J. **668**, 806–814 (2007), arXiv:0706.3048
18. M. Bradač, S.W. Allen, T. Treu, H. Ebeling, R. Massey, R.G. Morris, A. von der Linden, D. Applegate, Revealing the properties of dark matter in the merging cluster MACSJ0025.4-1222, Astrophys. J. **687** 959 (2008), arXiv:0806.2320
19. R. Massey, T. Kitching, J. Richard, The dark matter of gravitational lensing. Rept. Prog. Phys. **73**, 086901 (2010), arXiv:1001.1739
20. **2DFGRS** Collaboration, M. Colless et al., The 2dF Galaxy Redshift Survey: Spectra and redshifts, Mon. Not. Roy. Astron. Soc. **328** (2001) 1039, arXiv:astro-ph/0106498
21. S.D.S.S. Collaboration, M. Tegmark et al., The 3-D power spectrum of galaxies from the SDSS. Astrophys. J. **606**, 702–740 (2004), arXiv:astro-ph/0310725
22. V. Springel et al., Simulating the joint evolution of quasars, galaxies and their large-scale distribution. Nature **435**, 629–636 (2005), arXiv:astro-ph/0504097
23. N. Menci, A. Grazian, M. Castellano, N.G. Sanchez, A Stringent Limit on the Warm Dark Matter Particle Masses from the Abundance of z=6 Galaxies in the Hubble Frontier Fields, Astrophys. J. **825**(1) L1 (2016), arXiv:1606.02530
24. **Planck** Collaboration, P.A.R. Ade et al., Planck 2015 results. XIII. Cosmological parameters, Astron. Astrophys. **594**, A13 (2016), arXiv:1502.01589
25. S. Dodelson, *Modern Cosmology* (Academic Press, Amsterdam, 2003)
26. J. Lesgourgues, Cosmological Perturbations, in *Proceedings, Theoretical Advanced Study Institute in Elementary Particle Physics: Searching for New Physics at Small and Large Scales (TASI 2012)*, pp. 29–97, 2013, arXiv:1302.4640
27. P. Gondolo, G. Gelmini, Cosmic abundances of stable particles: Improved analysis. Nucl. Phys. B **360**, 145–179 (1991)
28. K. Griest, D. Seckel, Three exceptions in the calculation of relic abundances. Phys. Rev. D **43**, 3191–3203 (1991)
29. P. Ciafaloni, D. Comelli, A. De Simone, E. Morgante, A. Riotto, A. Urbano, The Role of Electroweak Corrections for the Dark Matter Relic Abundance. JCAP **1310**, 031 (2013), arXiv:1305.6391
30. S. Profumo, Astrophysical Probes of Dark Matter, in *Proceedings, Theoretical Advanced Study Institute in Elementary Particle Physics: Searching for New Physics at Small and Large Scales (TASI 2012)*, pp. 143–189, 2013, arXiv:1301.0952
31. B.W. Lee, S. Weinberg, Cosmological lower bound on heavy neutrino masses. Phys. Rev. Lett. **39**, 165–168 (1977)
32. K. Griest, M. Kamionkowski, Unitarity limits on the mass and radius of dark matter particles. Phys. Rev. Lett. **64**, 615 (1990)

33. J.L. Feng, A. Rajaraman, F. Takayama, Superweakly interacting massive particles. Phys. Rev. Lett. **91**, 011302 (2003), arXiv:hep-ph/0302215
34. J.L. Feng, A. Rajaraman, F. Takayama, SuperWIMP dark matter signals from the early universe. Phys. Rev. **D68**, 063504 (2003), arXiv:hep-ph/0306024
35. J. McDonald, Thermally generated gauge singlet scalars as selfinteracting dark matter. Phys. Rev. Lett. **88**, 091304 (2002), arXiv:hep-ph/0106249
36. L.J. Hall, K. Jedamzik, J. March-Russell, S.M. West, Freeze-In Production of FIMP Dark Matter. JHEP **03**, 080 (2010), arXiv:0911.1120
37. D.J.H. Chung, E.W. Kolb, A. Riotto, Superheavy dark matter. Phys. Rev. **D59**, 023501 (1999), arXiv:hep-ph/9802238
38. D.J.H. Chung, E.W. Kolb, A. Riotto, Nonthermal supermassive dark matter. Phys. Rev. Lett. **81**, 4048–4051 (1998), arXiv:hep-ph/9805473
39. E.W. Kolb, D.J.H. Chung, A. Riotto, WIMPzillas!, in Trends in theoretical physics II. Proceedings, 2nd La Plata Meeting, Buenos Aires, Argentina, November 29-December 4, 1998, pp. 91–105, 1998, arXiv:hep-ph/9810361
40. D.J.H. Chung, E.W. Kolb, A. Riotto, Production of massive particles during reheating. Phys. Rev. D **60**, 063504 (1999), arXiv:hep-ph/9809453
41. A.D. Sakharov, Violation of CP Invariance, c Asymmetry, and Baryon Asymmetry of the Universe, Pisma Zh. Eksp. Teor. Fiz. **5** (1967) 32–35. [Usp. Fiz. Nauk 161, 61(1991)]
42. S. Nussinov, Technocosmology: could a technibaryon excess provide a 'natural' missing mass candidate? Phys. Lett. B **165**, 55–58 (1985)
43. G.B. Gelmini, L.J. Hall, M.J. Lin, What is the cosmion? Nucl. Phys. **B281**, 726 (1987)
44. D.B. Kaplan, A single explanation for both the baryon and dark matter densities. Phys. Rev. Lett. **68**, 741–743 (1992)
45. D.N. Spergel, W.H. Press, Effect of hypothetical, weakly interacting, massive particles on energy transport in the solar interior. Astrophys. J. **294**, 663–673 (1985)
46. K.M. Zurek, Asymmetric Dark Matter: Theories, Signatures, and Constraints. Phys. Rept. **537**, 91–121 (2014), arXiv:1308.0338
47. Y. Bai, P. Schwaller, Scale of dark QCD. Phys. Rev. **D89**(6) 063522 (2014), arXiv:1306.4676
48. K. Murase, I.M. Shoemaker, Detecting Asymmetric Dark Matter in the Sun with Neutrinos, Phys. Rev. **D94**(6) 063512 (2016), arXiv:1606.03087
49. G. Gelmini, P. Gondolo, DM Production Mechanisms in *Bertone, G. (ed.): Particle dark matter* pp. 99–117, arXiv:1009.3690
50. K. Griest, Galactic microlensing as a method of detecting massive compact halo objects. Astrophys. J. **366**, 412–421 (1991)
51. **MACHO** Collaboration, C. Alcock et al., The MACHO project: Microlensing results from 5.7 years of LMC observations, Astrophys. J. **542** 281–307 (2000), arXiv:astro-ph/0001272
52. **EROS-2** Collaboration, P. Tisserand et al., Limits on the macho content of the galactic halo from the EROS-2 survey of the magellanic clouds, astron. Astrophys. **469** 387–404 (2007), arXiv:astro-ph/0607207
53. S. Calchi Novati, L. Mancini, G. Scarpetta, L. Wyrzykowski, LMC self lensing for OGLE-II microlensing observations. Mon. Not. Roy. Astron. Soc. **400**, 1625 (2009), arXiv:0908.3836
54. L. Wyrzykowski et al., The OGLE View of Microlensing towards the Magellanic Clouds. I. A Trickle of Events in the OGLE-II LMC data. Mon. Not. Roy. Astron. Soc. **397**, 1228–1242 (2009), arXiv:0905.2044
55. **POINT-AGAPE** Collaboration, S. Calchi Novati et al., POINT-AGAPE pixel lensing survey of M31: Evidence for a MACHO contribution to galactic halos, Astron. Astrophys. **443**, 911 (2005), arXiv:astro-ph/0504188
56. B.J. Carr, S.W. Hawking, Black holes in the early Universe. Mon. Not. Roy. Astron. Soc. **168**, 399–415 (1974)
57. P. Meszaros, The behaviour of point masses in an expanding cosmological substratum. Astron. Astrophys. **37**, 225–228 (1974)
58. B.J. Carr, The Primordial black hole mass spectrum. Astrophys. J. **201**, 1–19 (1975)

59. S. Clesse, J. García-Bellido, Massive Primordial Black Holes from Hybrid Inflation as Dark Matter and the seeds of Galaxies. Phys. Rev. **D92**(2) 023524 (2015), arXiv:1501.07565
60. S.W. Hawking, Black hole explosions. Nature **248**, 30–31 (1974)
61. D.N. Page, S.W. Hawking, Gamma rays from primordial black holes. Astrophys. J. **206**, 1–7 (1976)
62. **EGRET** Collaboration, P. Sreekumar et al., EGRET observations of the extragalactic gamma-ray emission, Astrophys. J. **494** 523–534 (1998), arXiv:astro-ph/9709257
63. B.J. Carr, K. Kohri, Y. Sendouda, J. Yokoyama, New cosmological constraints on primordial black holes. Phys. Rev. **D81**, 104019 (2010), arXiv:0912.5297
64. F. Capela, M. Pshirkov, P. Tinyakov, Constraints on primordial black holes as dark matter candidates from star formation. Phys. Rev. **D87**(2) 023507 (2013), arXiv:1209.6021
65. F. Capela, M. Pshirkov, P. Tinyakov, Constraints on primordial black holes as dark matter candidates from capture by neutron stars. Phys. Rev. **D87**(12) 123524 (2013), arXiv:1301.4984
66. A. Gould, Astrophys. J. **386**, L5 (1992)
67. A. Ulmer, J. Goodman, Femtolensing: Beyond the semiclassical approximation. Astrophys. J. **442**, 67 (1995), arXiv:astro-ph/9406042
68. A. Barnacka, J.F. Glicenstein, R. Moderski, New constraints on primordial black holes abundance from femtolensing of gamma-ray bursts. Phys. Rev. **D86**, 043001 (2012), arXiv:1204.2056
69. K. Griest, A.M. Cieplak, M.J. Lehner, Experimental Limits on Primordial Black Hole Dark Matter from the First 2 yr of Kepler Data, Astrophys. J. **786**(2), 158 (2014), arXiv:1307.5798
70. M. Ricotti, J.P. Ostriker, K.J. Mack, Effect of primordial black holes on the cosmic microwave background and cosmological parameter estimates. Astrophys. J. **680**, 829 (2008), arXiv:0709.0524
71. M.A. Monroy-Rodríguez, C. Allen, The end of the MACHO era- revisited: new limits on MACHO masses from halo wide binaries, Astrophys. J. **790**(2), 159 (2014), arXiv:1406.5169
72. C.G. Lacey, J.P. Ostriker, Massive black holes in galactic halos?, Astrophys. J. **299**, 633–652 (1985)
73. **Virgo, LIGO Scientific** Collaboration, B.P. Abbott et al., Observation of gravitational waves from a binary black hole merger. Phys. Rev. Lett. **116**(6) 061102 (2016), arXiv:1602.03837
74. S. Bird, I. Cholis, J.B. Muñoz, Y. Ali-Haïmoud, M. Kamionkowski, E.D. Kovetz, A. Raccanelli, A.G. Riess, Did LIGO detect dark matter?. Phys. Rev. Lett. **116**(20) 201301 (2016), arXiv:1603.00464
75. M. Sasaki, T. Suyama, T. Tanaka, S. Yokoyama, Primordial black hole scenario for the gravitational wave event GW150914. Phys. Rev. Lett. **117**(6) 061101 (2016), arXiv:1603.08338
76. T.D. Brandt, Constraints on MACHO Dark Matter from Compact Stellar Systems in Ultra-Faint Dwarf Galaxies, Astrophys. J. **824**(2) L31 (2016), arXiv:1605.03665
77. J.B. Muñoz, E.D. Kovetz, L. Dai, M. Kamionkowski, Lensing of fast radio bursts as a probe of compact dark matter. Phys. Rev. Lett. **117**(9) 091301 (2016), arXiv:1605.00008
78. G.F. Giudice, M. McCullough, A. Urbano, Hunting for Dark Particles with Gravitational Waves. JCAP **1610**(10) 001 (2016), arXiv:1605.01209
79. **Troitsk** Collaboration, V.N. Aseev et al., An upper limit on electron antineutrino mass from Troitsk experiment. Phys. Rev. **D84** 112003 (2011), arXiv:1108.5034
80. G. Bertone, D. Hooper, J. Silk, Particle dark matter: Evidence, candidates and constraints. Phys. Rept. **405**, 279–390 (2005), arXiv:hep-ph/0404175
81. P.W. Graham, D.E. Kaplan, S. Rajendran, Cosmological Relaxation of the Electroweak Scale. Phys. Rev. Lett. **115**(22), 221801(2015), arXiv:1504.07551
82. H.E. Haber, G.L. Kane, The search for supersymmetry: probing physics beyond the standard model. Phys. Rept. **117**, 75–263 (1985)
83. S. Dimopoulos, H. Georgi, Softly broken supersymmetry and SU(5). Nucl. Phys. **B193**, 150–162 (1981)
84. T. Kaluza, On the Problem of Unity in Physics, Sitzungsber. Preuss. Akad. Wiss. Berlin (Math. Phys.) **1921** 966–972 (1921)

85. O. Klein, Quantum Theory and Five-Dimensional Theory of Relativity. (In German and English), Z. Phys. **37**, 895–906 (1926). [Surveys High Energ. Phys.5,241(1986)]
86. N. Arkani-Hamed, S. Dimopoulos, G.R. Dvali, The Hierarchy problem and new dimensions at a millimeter. Phys. Lett. **B429** 263–272 (1998), arXiv:hep-ph/9803315
87. I. Antoniadis, N. Arkani-Hamed, S. Dimopoulos, G.R. Dvali, New dimensions at a millimeter to a Fermi and superstrings at a TeV. Phys. Lett. **B436**, 257–263 (1998), arXiv:hep-ph/9804398
88. N. Arkani-Hamed, S. Dimopoulos, G.R. Dvali, J. March-Russell, Neutrino masses from large extra dimensions. Phys. Rev. **D65**, 024032 (2002), arXiv:hep-ph/9811448
89. N. Arkani-Hamed, S. Dimopoulos, G.R. Dvali, Phenomenology, astrophysics and cosmology of theories with submillimeter dimensions and TeV scale quantum gravity. Phys. Rev. **D59**, 086004 (1999), arXiv:hep-ph/9807344
90. N. Arkani-Hamed, S. Dimopoulos, G.R. Dvali, N. Kaloper, Infinitely large new dimensions. Phys. Rev. Lett. **84**, 586–589 (2000), arXiv:hep-th/9907209
91. L. Randall, R. Sundrum, A Large mass hierarchy from a small extra dimension. Phys. Rev. Lett. **83**, 3370–3373 (1999), arXiv:hep-ph/9905221
92. L. Randall, R. Sundrum, An Alternative to compactification. Phys. Rev. Lett. **83**, 4690–4693 (1999), arXiv:hep-th/9906064
93. E.W. Kolb, R. Slansky, Dimensional reduction in the early universe: where have the massive particles gone? Phys. Lett. B **135**, 378 (1984)
94. E.W. Kolb, M.S. Turner, The Early Universe. Front. Phys. **69**, 1–547 (1990)
95. T. Appelquist, H.-C. Cheng, B.A. Dobrescu, Bounds on universal extra dimensions. Phys. Rev. **D64**, 035002 (2001), arXiv:hep-ph/0012100
96. G. Servant, T.M.P. Tait, Is the lightest Kaluza-Klein particle a viable dark matter candidate? Nucl. Phys. **B650**, 391–419 (2003), arXiv:hep-ph/0206071
97. M. Cirelli, N. Fornengo, A. Strumia, Minimal dark matter. Nucl. Phys. **B753**, 178–194 (2006), arXiv:hep-ph/0512090
98. M. Cirelli, A. Strumia, M. Tamburini, Cosmology and Astrophysics of Minimal Dark Matter. Nucl. Phys. **B787**, 152–175 (2007), arXiv:0706.4071
99. M. Cirelli, R. Franceschini, A. Strumia, Minimal Dark Matter predictions for galactic positrons, anti-protons, photons. Nucl. Phys. **B800**, 204–220 (2008), arXiv:0802.3378
100. M. Cirelli, A. Strumia, Minimal Dark Matter predictions and the PAMELA positron excess, PoS **IDM2008** (2008) 089, arXiv:0808.3867
101. M. Cirelli, A. Strumia, Minimal Dark Matter: Model and results. New J. Phys. **11**, 105005 (2009), arXiv:0903.3381
102. M. Cirelli, F. Sala, M. Taoso, Wino-like Minimal Dark Matter and future colliders. JHEP **10**, 033 (2014), arXiv:1407.7058. [Erratum: JHEP01,041(2015)]
103. M. Cirelli, T. Hambye, P. Panci, F. Sala, M. Taoso, Gamma ray tests of Minimal Dark Matter, JCAP **1510**(10), 026 (2015), arXiv:1507.05519
104. M. Shaposhnikov, Sterile neutrinos, in In *Bertone, G. (ed.): Particle dark matter* 228-248, 2010
105. R.D. Peccei, H.R. Quinn, CP Conservation in the presence of instantons. Phys. Rev. Lett. **38**, 1440–1443 (1977)
106. R.D. Peccei, H.R. Quinn, Constraints imposed by CP conservation in the presence of instantons. Phys. Rev. **D16**, 1791–1797 (1977)
107. S.L. Adler, Axial vector vertex in spinor electrodynamics. Phys. Rev. **177**, 2426–2438 (1969)
108. J.S. Bell, R. Jackiw, A PCAC puzzle: pi0 -> gamma gamma in the sigma model. Nuovo Cim. **A60**, 47–61 (1969)
109. K.A. Olive et al., Review of Particle Physics. Chin. Phys. **C38**, 090001 (2014)
110. D.J.E. Marsh, Axion Cosmology. Phys. Rept. **643**, 1–79 (2016), arXiv:1510.07633
111. M. Milgrom, A Modification of the newtonian dynamics as a possible alternative to the hidden mass hypothesis. Astrophys. J. **270**, 365–370 (1983)
112. J. Bekenstein, M. Milgrom, Does the missing mass problem signal the breakdown of Newtonian gravity? Astrophys. J. **286**, 7–14 (1984)

113. J.D. Bekenstein, Relativistic gravitation theory for the MOND paradigm. Phys. Rev. **D70**, 083509 (2004), arXiv:astro-ph/0403694. [Erratum: Phys. Rev. **D71**, 069901 (2005)]
114. J.D. Bekenstein, Alternatives to Dark Matter: Modified Gravity as an Alternative to dark Matter in *Bertone, G. (ed.): Particle dark matter* pp. 99–117, arXiv:1001.3876
115. J. Silk et al., in *Bertone, G. (ed.): Particle dark matter* (2010)
116. S. Dodelson, The real problem with MOND. Int. J. Mod. Phys. **D20**, 2749–2753 (2011), arXiv:1112.1320

Part I
Direct and Indirect WIMP Searches

Chapter 2
Direct Detection of WIMPs

2.1 Experimental Strategies for WIMP Detection

In the next chapters we are going to give an overview of the main WIMP search strategies, which fall in three big categories:

- Direct searches probe the scattering cross section of WIMPs with nuclei, by looking for nuclear recoils in some target material in a low background environment. When a WIMP of the galactic halo passes through the target, it may scatter off a nucleus, whose recoil will be measured by suitable detectors. The main astrophysical uncertainty is due to the DM density in the Solar System and to its velocity distribution.
- Indirect searches look for the products of WIMP annihilations in a number of astrophysical objects, including the galaxy and his satellites, other galaxies and galaxy clusters, as well as the Sun and the Earth.
- In LHC searches, DM particles (and generically the "dark sector") are studied by producing them in proton collisions, in association with other SM particles (being very weakly interacting, WIMPs would otherwise be invisible at colliders, and their typical signature is missing energy in the transverse plane).

© Springer International Publishing AG 2017
E. Morgante, *Aspects of WIMP Dark Matter Searches at Colliders and Other Probes*, Springer Theses, https://doi.org/10.1007/978-3-319-67606-7_2

Pictorially, WIMP searches can be summarized with this diagram:

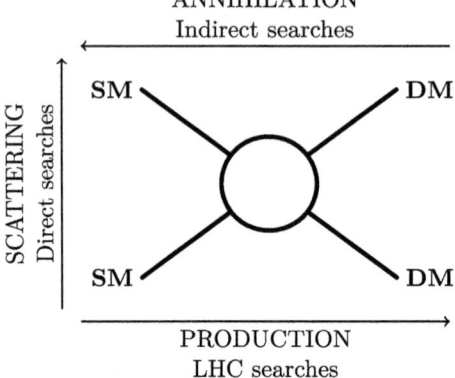

The simplicity of this representation is somehow misleading. The issue relies in the interpretation of experimental data in terms of a model. In direct searches, the momentum exchange is much lower than the other energy scales which enter the process, and a non-relativistic operators approach is the most suited one. On the contrary, the high energy reach of the LHC requires an interpretation in terms of models whose cut-off is larger than the LHC energy scales (which we will call UV complete models for simplicity), and a less model dependent EFT interpretation is only possible at the price of poorer bounds. It should also be noticed that the physical processes studied in these kind of searches involve very different energy scales, and the effect of running should be taken into account [1].[1] Finally, comparison with indirect searches is also model dependent, and is only meaningful when DM annihilation products include the main proton constituents (light quarks and gluons).

In this chapter we are going to discuss direct searches for WIMPs. The idea dates back to 1984, when Goodman and Witten [2] proposed to try to detect WIMPs by elastic scattering off nuclei in a terrestrial detector. The study was extended by Drukier, Freese and Spergel [3] to include a variety of cold dark matter candidates, as well as details of the detector and the halo model. They also showed that the Earth's motion around the Sun produces an annual modulation in the expected signal. We will start by reviewing the predictions for signal event rates and signatures; then we will turn to briefly discuss the backgrounds, and in particular the so called "neutrino floor"; finally we will describe the current status of direct detection and some of the planned future developments.

---

[1] The effect is nevertheless minor in the majority of cases, but could become important in view of future precise measurements of DM properties, when (and if) WIMP will ever be discovered.

2.2 Direct Detection: Prediction of Event Rates

Elastic scattering of WIMPs off nuclei has a rate per mass of target material

$$\frac{dR}{dE_R} = \frac{1}{m_{\mathcal{N}}} \frac{\rho_0}{m_\chi} \int_{\mathcal{V}} d^3\vec{v}\, f(\vec{v} + \vec{v}_E)\, v\, \frac{d\sigma}{dE_R}, \tag{2.1}$$

where E_R is the nuclear recoil energy, $m_{\mathcal{N}}$ is the mass of the target nuclei, m_χ is the WIMP mass, ρ_0 the local WIMP density in the galactic halo and $d\sigma/dE_R$ is the WIMP-nucleus differential cross section. $\vec{v}$ and $\vec{v}_E$ are the WIMP velocity with respect to the detector and the Earth velocity in the galactic rest frame (which can be assumed to be constant, as far as we are not interested in modulation effects and directional constraints). The function f is the WIMP velocity distribution in the galaxy. The integration domain is defined in such a way that the WIMP velocity with respect to the detector is larger than v_{min}, the minimal velocity of the incoming WIMP in order for the scattering event to be detected. The energy that is transferred to the recoiling nucleus is:

$$E_R = \frac{p^2}{2m_{\mathcal{N}}} = \frac{\mu_{\mathcal{N}}^2 v^2}{m_{\mathcal{N}}}(1 - \cos\theta), \tag{2.2}$$

where p is the momentum transfer, θ is the scattering angle in the WIMP-nucleus centre of mass frame, $m_{\mathcal{N}}$ is the nuclear mass and $\mu_{\mathcal{N}}$ is the WIMP-nucleus reduced mass:

$$\mu_{\mathcal{N}} = \frac{m_{\mathcal{N}} \cdot m_\chi}{m_{\mathcal{N}} + m_\chi}. \tag{2.3}$$

The minimum velocity is given by:

$$v_{\mathrm{min}} = \sqrt{\frac{m_{\mathcal{N}} E_{th}}{2\mu_{\mathcal{N}}^2}}, \tag{2.4}$$

where E_{th} is the energy threshold of the detector.

   In the simplest galactic model (the so-called standard halo model, which describes an isotropic, isothermal sphere of collisionless particles with density profile $\rho(r) \propto r^{-2}$) the velocity distribution of WIMPs is Maxwellian:

$$f(\vec{v}) = \frac{1}{\sqrt{2\pi}\sigma_v} \exp\left(-\frac{v^2}{2\sigma_v^2}\right). \tag{2.5}$$

The distribution is truncated (and normalized accordingly) at the local escape velocity, which depends on the position in the galaxy, such that $f(\vec{v}) = 0$ for $v \geq v_{\mathrm{esc}}$. The parameters used in the standard halo model are $\rho_0 = 0.3\,\mathrm{GeV cm^{-3}} = 5 \times 10^{-25}\,\mathrm{gcm^{-3}} = 8 \times 10^{-3} M_\odot\,\mathrm{pc^{-1}}$, $v_c = 220\,\mathrm{km s^{-1}}$ and a local escape velocity of $v_{esc} = 544\,kms^{-1}$ (see [4] for a recent discussion on the determination of these parameters and their uncertainties). The underlying assumption is that the phase-space distribution of the dark matter has reached a steady state and is smooth, which may not be the case for the Milky Way, in particular at small scales. Also, it is assumed that no DM disk is present in the Milky Way [5].

A rough numerical estimate of the scattering rate can be easily obtained [5]. Consider a WIMP scattering off a target nucleus, both with mass $\sim 100\,GeV$. The mean velocity of the WIMP relative to the target is roughly $\langle v \rangle \sim 220\,kms^{-1}$. The mean energy impinged on the nucleus is thus:

$$\langle E_R \rangle = \frac{1}{2} m_\chi \langle v \rangle^2 \sim 30\,keV. \tag{2.6}$$

Assuming a local dark matter density of $\rho_0 = 0.3\,GeVcm^{-3}$, the number density of WIMPs is $n_0 = \rho_0 m_\chi$, and their flux on Earth is:

$$\phi_0 = n_0 \times \langle v \rangle = \frac{\rho_0}{m_\chi} \times \langle v \rangle = 6.6 \times 10^4\,cm^{-2}\,s^{-1}. \tag{2.7}$$

An electroweak-scale interaction will have an elastic scattering cross section from the nucleus of $\sigma_{\chi N} \sim 10^{-38}\,cm^2$, leading to a rate for elastic scattering:

$$R \sim 0.1 \; \frac{events}{kg\,year} \left[\frac{100}{A} \times \frac{\sigma_{\chi N}}{10^{-38}\,cm^2} \times \frac{\langle v \rangle}{220\,km\,s^{-1}} \times \frac{\rho_0}{0.3\,GeV\,cm^{-3}} \right]. \tag{2.8}$$

where A is the atomic number of the target nucleus.

2.3   Scattering Cross Section

Being the WIMP velocity of the order of $220\,kms^{-1}$, the average momentum transfer may be estimated as

$$\langle p \rangle \simeq \mu_N \langle v \rangle, \tag{2.9}$$

which is approximately 6–70 MeV for m_χ between 10 GeV and 1 TeV. The scattering happens then in the extremely non relativistic regime, and will be isotropic in the centre of mass frame. The de Broglie wavelength corresponding to a momentum transfer of $p = 10\,MeV/c$ is:

$$\lambda = \frac{h}{p} \simeq 20\,fm > r_0 A^{1/3} = 1.25\,fm\ A^{1/3}. \tag{2.10}$$

This value is larger than the diameter of most nuclei, and therefore the scattering amplitudes on individual nucleons will, in first approximation, add coherently. Loss

Table 2.1 Non-relativistic operators for WIMP scattering on nuclei

$\mathcal{O}_1 = \mathbb{1}$	$\mathcal{O}_9 = i\vec{S}_\chi \cdot (\vec{S}_N \times \vec{q})$
$\mathcal{O}_2 = (v^\perp)^2$	$\mathcal{O}_{10} = i\vec{S}_N \cdot \vec{q}$
$\mathcal{O}_3 = i\vec{S}_N \cdot (\vec{q} \times \vec{v}^\perp)$	$\mathcal{O}_{11} = i\vec{S}_\chi \cdot \vec{q}$
$\mathcal{O}_4 = \vec{S}_\chi \cdot \vec{S}_N$	$\mathcal{O}_{10}\mathcal{O}_5$
$\mathcal{O}_5 = i\vec{S}_\chi \cdot (\vec{q} \times \vec{v}^\perp)$	$\mathcal{O}_{10}\mathcal{O}_8$
$\mathcal{O}_6 = (\vec{S}_\chi \cdot \vec{q})(\vec{S}_N \cdot \vec{q})$	$\mathcal{O}_{11}\mathcal{O}_3$
$\mathcal{O}_7 = \vec{S}_N \cdot \vec{v}^\perp$	$\mathcal{O}_{11}\mathcal{O}_7$
$\mathcal{O}_8 = \vec{S}_\chi \cdot \vec{v}^\perp$	

of coherence will affect only the tails of velocity distribution, and it is typically parametrized with a nuclear form factor.

The DM-nucleon interaction is determined by a set on non-relativistic operators. The general Lagrangian is

$$\mathcal{L}_{\text{int}} = \sum_{N=n,p} \sum_i c_i^{(N)} \mathcal{O}_i \chi^+ \chi^- N^+ N^-, \tag{2.11}$$

where $N^\pm$, $\phi^\pm$ are non-relativistic fields involving only creation or annihilation operators, i.e.

$$N^-(y) \equiv \int \frac{d^3k}{(2\pi)^3} \frac{1}{\sqrt{2m_N}} e^{-ik\cdot y} a_k^\dagger, \qquad N^+(y) \equiv (N^-(y))^\dagger. \tag{2.12}$$

The list of possible non-relativistic effective operators $\mathcal{O}_i$ is given in Table 2.1 [6] (here we limit ourselves to the operators which can arise from the exchange of spin-0 or spin-1 mediators). Given, in the language of usual relativistic QFT, the Lagrangian that controls the interaction of the WIMP with quarks and gluons, a "dictionary" to translate it into the corresponding non relativistic operators is given in [6, 7].

Generically speaking, operators which are proportional to the velocity of the DM particle or to the exchanged momentum result in a suppressed rate, since both quantities are typically small. In addition, a further suppression is associated to the nucleus spin. A special role in this discussion is played by the operators $\mathcal{O}_1, \mathcal{O}_4$. They arise as the non-relativistic limit of

$$\bar{\chi}\chi\bar{q}q \,, \quad \bar{\chi}\gamma_\mu\chi\bar{q}\gamma^\mu q \longrightarrow \mathcal{O}_1 \,, \qquad \bar{\chi}\gamma_\mu\gamma_5\chi\bar{q}\gamma^\mu\gamma_5 q \longrightarrow \mathcal{O}_4 \tag{2.13}$$

with appropriate coefficients. They are special in the sense that they are the only operators which are not suppressed by powers of the WIMP velocity or of the momentum transfer. For this reason, they are expected to dominate the scattering cross section, as far as the coefficient in front of the relativistic operator is not vanishing in a specific model (as in the models that will be considered in Chap. 9). These operators are commonly referred to respectively as *spin-independent* and *spin-dependent*, without further specifications, and limits by experimental collaborations are typically set in terms of the cross section obtained from them.

Assuming that only these two operators are present, the scattering cross section of a WIMP off a nucleus is given by

$$\frac{d\sigma}{dE_R} = \frac{m_{\mathcal{N}}}{2\mu_{\mathcal{N}}^2 v^2} \left[\sigma_{SI} F_{SI}^2(\mathcal{N}, E_R) + \sigma_{SD} F_{SD}^2(\mathcal{N}, E_R) \right], \qquad (2.14)$$

where F_{SI} and F_{SD} are the nuclear form factors, that depend on the recoil energy and σ_{SI} and σ_{SD} are the cross sections in the zero momentum transfer limits, which are given by

$$\sigma_{SI} = \frac{4\mu_{\mathcal{N}}^2}{\pi} \left[Z f_p + (A - Z) f_n \right]^2, \qquad (2.15)$$

$$\sigma_{SD} = \frac{32\mu_{\mathcal{N}}^2}{\pi} G_F^2 \frac{J+1}{J} \left[a_p \langle S_p \rangle + a_n \langle S_n \rangle \right]^2. \qquad (2.16)$$

Here J the total spin of the nucleus and f_p, f_n and a_p, a_n are the effective WIMP-couplings to neutrons and protons in the spin-independent and spin-dependent case, respectively. These can be calculated using the effective Lagrangian of the given theoretical model. They depend on the contributions of the light quarks to the mass of the nucleons and on the quark spin distribution within the nucleons, respectively, and on the composition of the dark matter particle. The terms in brackets $\langle S_{p,n} \rangle = \langle N | S_{p,n} | N \rangle$ are the expectation values of total proton and neutron spin operators in the limit of zero momentum transfer, and must be determined using detailed nuclear model calculations. We can immediately see that the spin-independent interaction cross section depends on the total number of nucleons, while the spin-dependent cross section is in general smaller, and only relevant for nuclei which have a non-zero spin in their ground state. The reason is that, for the typical DM momenta, its associated de Broglie wavelength is larger than the nuclear size, so the particle interacts coherently with all the nucleus seen as a collection of Z protons and A–Z neutrons. In the spin dependent case this effect does not lead to an enhancement of the cross section because, as it can be understood in terms of the nuclear shell model, nucleons tend to form pairs of zero spin, and the total spin is given only by the unpaired nucleons, whose number is typically not larger than one.

Let us now plug Eq. (2.14) into (2.1). If the scattering is dominated by its spin independent part (which is always the case as soon as a spin independent interaction occurs), and assuming for simplicity that $f_p = f_n$, we obtain

$$\frac{dR}{dE_R} = \frac{\rho_0 \sigma_{SI}^p}{2\mu_p^2 m_\chi} \times A^2 F_{SI}^2(\mathcal{N}, E_R) \times \mathcal{F}(v_{\min}(m_\chi, E_R, A); \vec{v}_E), \qquad (2.17)$$

where we have defined σ_{SI}^p to be the scattering cross section on nucleons, and μ_p the proton-WIMP reduced mass (which for practical purposes can be taken to be equal to the proton mass if the DM is heavier than a few GeV). The factor

$$\mathcal{F} = \int_{\mathcal{V}} \mathrm{d}^3 v \, \frac{f(\vec{v} + \vec{v}_E)}{v} \tag{2.18}$$

encodes the effect of the DM velocity distribution in the halo, and depends non trivially on m_χ, E_R and $m_\mathcal{N}$ via the minimal velocity $v_{\min}$. Written in this way Eq. (2.17) is very convenient, because it factorises the effect of the DM velocity distribution and the detector related quantities A and F_{SI} (even if, of course, also $\mathcal{F}$ depends on the target material).

Comparing the number of recoil events predicted by means of Eq. (2.17) with data, limits can be derived on σ_{SI}^p (or, analogously, on σ_{SD}^p). Two limits are relevant in order to understand the general feature of the bounds: when the DM particle is very light compared to the target nuclei ($m_\chi \ll m_\mathcal{N}$), the minimal velocity $v_{\min}$ in order for the recoil energy to be detectable becomes very large, and the integration domain of $\mathcal{F}$ closes up. The event rate then goes to zero, and limits get weaker. In order to test lighter and lighter WIMPs, correspondingly light target nuclei are needed. On the other hand, when $m_\chi \gg m_\mathcal{N}$ the minimal velocity does not depend on m_χ any more, and the scattering rate is sensitive to the ratio σ_{SI}^p / m_χ. For this reason, for large mass limits are expected to degrade precisely proportionally to m_χ. The transition between these two regimes lies at $m_\chi \sim m_\mathcal{N}$ (depending also on the threshold), and in this region limits are stronger. In the eventuality of the detection of an excess, in order to disentangle σ_{SI}^p and m_χ it would be necessary to obtain the same measure with different target nuclei [8].

2.4 Backgrounds

The main challenge in direct detection experiments is to reduce the background noise. The first source of background is constituted by environmental radioactivity and cosmic rays (and their secondary products). Environmental radioactivity is faced by shielding the detector and the target material with a combination of different Z materials, with a low intrinsic radioactivity. In particular, the main challenge is provided by neutrons, which can mimic a WIMP signal and constitute an irreducible background, which has to be estimated by means of numerical simulations. The hadronic component of cosmic rays is shielded by the many meters of rock above the detectors which are normally built underground. High energy muons can instead traverse the natural rock shield and produce a flux of energetic neutrons by interacting with nuclei in the rock. These neutrons are in turn slowed down by scattering in the rocks, and may scatter off nuclei in the detector with the typical WIMP energy. In order to reduce these backgrounds a cosmic ray veto is necessary. An additional complication results from the fact that the neutron flux has a seasonal variation which can mimic the WIMP signal modulation. Neutrons from cosmic rays muons will have an increasingly important role in future experiments.

Another threat to background rejection is posed by radioactivity in detector components, both because of radioactive isotopes intrinsically present in the material and because of long lived isotopes produced by cosmic rays during the exposure of the detector and target material at the Earth's surface. This background can only be reduced by appropriate choices of the material and by reducing its exposure time at the Earth's surface.

The most serious issue for future high-sensitivity experiment will be posed by the irreducible neutrino flux, the so-called "ultimate" background. Neutrinos come from a variety of sources. The most relevant for direct detection experiments are neutrinos from fusion reactions in the Sun, decay products of cosmic rays collisions with the atmosphere and relic supernovae neutrinos. Neutrinos coming from radioactive processes on Earth are subdominant [9].

There are two types of neutrino interactions to which dark matter experiments are potentially sensitive: the first is νe^- neutral current elastic scattering, and the second is coherent scattering of neutrinos with target nuclei.

Electron recoils that arise from solar neutrinos from pp reactions have low energies but a very high flux. Depending on the detector capability of discriminate electronic and nuclear recoil, they may become a relevant background at cross sections of 10^{-48} cm^2 or lower [5].

Coherent neutrino-nucleus scattering has never been observed, but it is theoretically a well understood phenomenon. Solar neutrinos (mostly from the decays of ^{8}B and ^{7}Be), atmospheric neutrinos and the ones from the diffuse supernovae background can undergo this kind of scattering. The typical recoil energy is quite low with respect to the one expected by WIMP events, which will make neutrino background even more problematic for future low threshold experiments. An indication of the importance of this background is given in [10]. In this paper, the effect of increased exposure is considered and it was shown that limits can not be pushed below a certain threshold, depending on the DM mass. For a DM particle of mass 10 GeV–10 TeV, the limit is at the level of 10^{-49}–10^{-48} cm^2, not far from the reach of next generation experiments (as we will discuss in Sect. 2.6).

Strictly speaking, a discovery limit exists only if the spectrum of nuclear recoil from WIMPs and from neutrinos coincide. This is the case, for example, of a 6 GeV DM particle with the background of ^{8}B neutrinos in a Xenon detector [10]. Otherwise, a precise determination of the energy spectrum of the events will help in discriminating between the two sources, together with the use of different target sources.

The other two tools that will help in discriminating the neutrino background are seasonal modulation and directional detection, which will be useful especially for light WIMP searches (since for heavy particles their kinetic energy may be large even if the WIMP velocity is directed parallel to the Earth's motion, resulting in a lower relative speed). The effect of directional detectors is carefully computed in [11]. Notice that the detection of neutrino scattering in next generation direct detection experiments can be also seen as a useful tool in order to precisely measure solar neutrino fluxes and test our knowledge of solar physics and of the Standard Model as well [12].

2.5 Specific WIMP Signatures

In order to detect a WIMP-induced signal in direct detection experiments, a specific signature unambiguously pointing towards a galactic halo particle would be needed. This signature might be provided by the Earth's motion around the Sun, which induces a seasonal modulation of the total event rate [3, 13]. The modulation has a purely kinematic origin. The Earth's velocity in the galactic rest frame is a superposition of the Earth's rotation around the Sun and the Sun's rotation around the galactic centre. As a consequence of this, the velocity distribution of the DM particles as viewed from Earth varies with a period of 1 year. The Earth's orbital speed is much smaller than the Sun's one, and therefore the seasonal variation is a small effect that can be parametrized as

$$
\frac{\mathrm{d}R}{\mathrm{d}E_R}(E_R, t) \simeq \frac{\mathrm{d}R}{\mathrm{d}E_R}(E_R) \left[1 + \Delta(E_R) \cos \frac{2\pi (t - t_0)}{T} \right] \tag{2.19}
$$

where, under the simplifying assumption that the DM velocity distribution in the galaxy is isotropic, $T = 1$ year and the phase is $t_0 = 150$ days [5]. The amplitude of the modulation $\Delta(E_R)$ becomes negative at small recoil energies, meaning that the differential event rates peaks in winter for small recoil energies, and in summer for larger recoils energies.

Seasonal modulated phenomena are unfortunately ubiquitous on Earth, and therefore many effects can mimic a WIMP signal. For this reason, directional signatures would highly improve the discrimination capability of direct search experiments [14, 15]. In particular, since the WIMP flux in the lab frame is peaked in the direction of motion of the Sun, namely towards the constellation Cygnus, the recoil spectrum is peaked in the opposite direction, leading to a very large forward-backward asymmetry of order $\mathcal{O}(v_\odot / \langle v \rangle) \approx 1$. Thus fewer events, namely a few tens to a few hundreds, depending on the halo model, are needed to discover a WIMP signal compared to the case where the seasonal modulation effect is exploited [15, 16]. The experimental challenge is to build massive detectors capable of detecting the direction of the incoming WIMP.

2.6 Present Status and Future Development

Present upper limits and future prospects on the spin independent WIMP-nucleon cross section are presented in Fig. 2.1. A large variety of techniques are employed in direct WIMP searches. Heat deposition, ionization, and scintillation are used in order to detect the recoil energy in a number of different materials. In the low WIMP mass region (roughly below 6 GeV) cryogenic detectors with a very low energy threshold and light target nuclei perform the best. Currently, the best limits come from the DAMIC [29], CRESST [18, 30], EDELWEISS [31] and SuperCDMS [19, 32] experiments, as well as the liquid Xenon experiments LUX [33] and XENON100 [34]. For

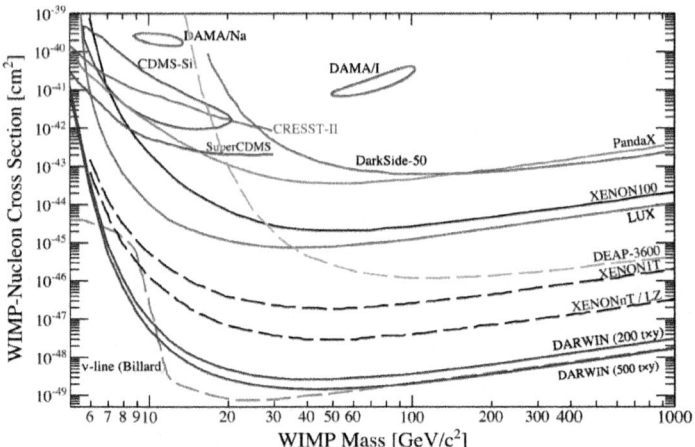

Fig. 2.1 Collection of bounds on σ_{SI}, from [17]. Upper limits are shown from CRESST-II [18], SuperCDMS [19], Panda-X [20], XENON-100 [21] and LUX [22], together with projections for DEAP-3600 [23], XENON1T [24], XENONnT [25], LZ [26] and DARWIN [27]. Updated results from LUX [28] improve current best upper limits by roughly a factor of 4 in the region $m_\chi \gtrsim 100\,\mathrm{GeV}$

larger WIMP masses, liquid Xenon experiments are able to place stronger bounds. The stronger ones come from the LUX experiment [22, 28, 33, 35], improving upon previous bounds from XENON100 [21].

When spin dependent interactions are considered, nuclei with a non-zero intrinsic angular momentum are needed. A particularly suitable one is ^{19}F, which is part of the target material of experiments as PICASSO [36], COUPP [37], SIMPLE [38] and PICO-2L [39]. The most stringent bounds for large DM mass come from the PICO-2L and LUX experiments, while for light DM PICASSO is stronger. The light spin $1/2$ nucleus ^{3}He could also be used for spin dependent WIMP interactions. The advantage is that it is technologically feasible to obtain large amounts of highly polarized material, which would help in rejecting the solar neutrino background [40], since the polarization modulated amplitude is large for neutrino-nucleus scattering (and can therefore cancel the polarization independent part if the polarization direction is chosen properly with respect to the direction of arrival from the Sun) but small for WIMPS due to velocity suppression.

In the past years, several anomalies were observed and later understood as due to background processes, in particular in the low WIMP mass region (as the one claimed by the CoGeNT experiment). Only two of them remain unexplained, by the CDMS-Si [41] and the DAMA/LIBRA [42, 43] experiments. The former have seen an excess of events over a very low background. The latter reports an annual variation of single-scatter event rate at low energy, compatible with WIMP interpretation. In both cases the relevant region in parameter space has been excluded by other searches, and the situation is still far from being clear. It can surely be said that, up to now, no settled WIMP signal has been seen by direct detection experiments.

Several new experiments are in data-taking, commissioning or planning phase, aiming towards improving the constrain capability both in the low and high WIMP mass region. Among them it is worth remember DEAP-3600 [23], XENON1T [24], XENONnT [25], LZ [26, 44] and DARWIN [27]. The latter, which will operate multi-tons of target material, will probably be the first experiment able to reach the region where the neutrino background becomes visible. To go beyond this point several technological advances will be needed. Directional detectors will probably be the next stage.

A complementary direction to the one of enlarging experiments is that of lowering the threshold by considering events different from nuclear recoils, in order to improve the exclusion potential for DM masses in the range keV-GeV. These include electronic recoils from ionization events [45–47], electronic excitation and de-excitation with the consequent photon emission [45, 48], molecular dissociation with the production of ions, and finally phonon or heat production [45], electron scattering in semiconductor materials [47, 49, 50], in superfluids [51, 52] and in superconductors [53, 54], exploiting bremsstrahlung in low recoil energy nuclear scatterings [55] or the high velocity of DM particles from evaporation in the Sun (a phenomenon which we will describe in details in Chap. 3) [56]. Interestingly, searches in the low mass range could have directional sensitivity if performed with 2D graphene targets, as recently proposed in [57]. The information on the incoming direction of the DM particle is retained by scattered electrons, and it is not lost in further electrons' scatterings inside the bulk material.

It should anyway be noticed that this discussion assumed that the interaction of the DM with target nuclei is spin independent. When the interaction is mediated by an operator different from $\mathcal{O}_1$ in Table 2.1 (*i.e.* when it is spin and/or velocity dependent) constraints are much weaker, and the neutrino background is not a concern. Models in which the spin independent part of the interaction is absent are non generic, but have very interesting phenomenology as we will discuss in the example of Chap. 9.

References

1. F. D'Eramo, B.J. Kavanagh, P. Panci, You can hide but you have to run: direct detection with vector mediators, JHEP **1608** (2016) 11, arXiv:1605.04917
2. M.W. Goodman, E. Witten, Detectability of certain dark matter candidates. Phys. Rev. D **31**, 3059 (1985)
3. A.K. Drukier, K. Freese, D.N. Spergel, Detecting cold dark matter candidates. Phys. Rev. D **33**, 3495–3508 (1986)
4. A.M. Green, Astrophysical uncertainties on direct detection experiments, Mod. Phys. Lett. A **27**, 1230004 (2012), arXiv:1112.0524
5. L. Baudis, Direct dark matter detection: the next decade, Phys. Dark Univ. **1**, 94–108 (2012) arXiv:1211.7222
6. A.L. Fitzpatrick, W. Haxton, E. Katz, N. Lubbers, Y. Xu, The effective field theory of dark matter direct detection, JCAP **1302** 004 (2013), arXiv:1203.3542
7. M. Cirelli, E. Del Nobile, P. Panci, Tools for model-independent bounds in direct dark matter searches, JCAP **1310**(019), (2013), arXiv:1307.5955

8. M. Pato, L. Baudis, G. Bertone, R.R. de Austri, L.E. Strigari, R. Trotta, Complementarity of dark matter direct detection targets, Phys. Rev. D **83** 083505, (2011), arXiv:1012.3458
9. J. Monroe, P. Fisher, Neutrino backgrounds to dark matter searches, Phys. Rev. D **76**, 033007 (2007), arXiv:0706.3019
10. J. Billard, L. Strigari, E. Figueroa-Feliciano, Implication of neutrino backgrounds on the reach of next generation dark matter direct detection experiments, Phys. Rev. D **89**(2), 023524 (2014), arXiv:1307.5458
11. P. Grothaus, M. Fairbairn, J. Monroe, Directional dark matter detection beyond the neutrino bound, Phys. Rev. D **90**(5), 055018 (2014), arXiv:1406.5047
12. D.G. Cerdeño, M. Fairbairn, T. Jubb, P.A.N. Machado, A.C. Vincent, and C. Bœhm, Physics from solar neutrinos in dark matter direct detection experiments, JHEP **05** 118, (2016), arXiv:1604.01025
13. K. Freese, J.A. Frieman, A. Gould, Signal modulation in cold dark matter detection. Phys. Rev. D **37**, 3388–3405 (1988)
14. D.N. Spergel, The motion of the earth and the detection of wimps. Phys. Rev. D **37**, 1353 (1988)
15. C.J. Copi, J. Heo, and L.M. Krauss, Directional sensitivity, WIMP detection, and the galactic halo, Phys. Lett. B **461** 43–48 (1999), arXiv:hep-ph/9904499
16. C.J. Copi L.M. Krauss, Angular signatures for galactic halo WIMP scattering in direct detectors: prospects and challenges, Phys. Rev. D **63**, 043507 (2001), arXiv:astro-ph/0009467
17. L. Baudis, Dark matter detection. J. Phys. G **43**(4), 044001 (2016)
18. CRESST-II Collaboration, G. Angloher et al., Results on low mass WIMPs using an upgraded CRESST-II detector, Eur. Phys. J. C **74**(12), 3184 (2014), arXiv:1407.3146
19. SuperCDMS Collaboration, R. Agnese et al., Search for low-mass weakly interacting massive particles with SuperCDMS, Phys. Rev. Lett. **112**(24), 241302 (2014), arXiv:1402.7137
20. PandaX Collaboration, M. Xiao et al., First dark matter search results from the PandaX-I experiment, Sci. China Phys. Mech. Astron. **57**, 2024–2030 (2014), arXiv:1408.5114
21. XENON100 Collaboration, E. Aprile et al., Dark matter results from 225 live days of XENON100 data, Phys. Rev. Lett. **109**, 181301 (2012), arXiv:1207.5988
22. LUX Collaboration, D.S. Akerib et al., First results from the LUX dark matter experiment at the Sanford underground research facility, Phys. Rev. Lett. **112**, 091303 (2014), arXiv:1310.8214
23. DEAP Collaboration, M.G. Boulay, DEAP-3600 dark matter search at SNOLAB, J. Phys. Conf. Ser. **375** 012027 (2012), arXiv:1203.0604
24. XENON1T Collaboration, E. Aprile, The XENON1T dark matter search experiment, Springer Proc. Phys. **148** 93–96 (2013), arXiv:1206.6288
25. XENON1T Collaboration, E. Aprile et al., Conceptual design and simulation of a water Cherenkov muon veto for the XENON1T experiment, JINST **9** 11006 (2014), arXiv:1406.2374
26. D.C. Malling et al., After LUX: the LZ program, arXiv:1110.0103
27. DARWIN Consortium Collaboration, L. Baudis, DARWIN: dark matter WIMP search with noble liquids, J. Phys. Conf. Ser. **375** 012028 (2012), arXiv:1201.2402
28. A. Manalaysay, L.U.X. the dark matter search, *Talk at IDM2016* (Sheffield, UK, 2016)
29. DAMIC Collaboration, J.R.T. de Mello Neto et al., The DAMIC dark matter experiment, in Proceedings, 34th International Cosmic Ray Conference (ICRC 2015), (2015), arXiv:1510.02126
30. CRESST Collaboration, G. Angloher et al., Results on light dark matter particles with a low-threshold CRESST-II detector, Eur. Phys. J. C 76(1), 25 (2016), arXiv:1509.01515
31. EDELWEISS Collaboration, E. Armengaud et al., Constraints on low-mass WIMPs from the EDELWEISS-III dark matter search, JCAP **1605**(5), 019 (2016), arXiv:1603.05120
32. SuperCDMS Collaboration, R. Agnese et al., New results from the search for low-mass weakly interacting massive particles with the CDMS low ionization threshold experiment, Phys. Rev. Lett. **116**(7), 071301 (2016), arXiv:1509.02448
33. LUX Collaboration, D. S. Akerib et al., Improved limits on scattering of weakly interacting massive particles from reanalysis of 2013 LUX data, Phys. Rev. Lett. **116**(16), 161301 (2016), arXiv:1512.03506

34. XENON100 Collaboration, E. Aprile et al., A low-mass dark matter search using ionization signals in XENON100, arXiv:1605.06262
35. LUX Collaboration, D.S. Akerib et al., The large underground xenon (LUX) Experiment, Nucl. Instrum. Meth. A **704**, 111–126 (2013), arXiv:1211.3788
36. PICASSO Collaboration, S. Archambault et al., Constraints on low-mass WIMP interactions on ^{19}F from PICASSO, Phys. Lett. B **711**, 153–161 (2012), arXiv:1202.1240
37. COUPP Collaboration, E. Behnke et al., First dark matter search results from a 4-kg CF_3I bubble chamber operated in a deep underground site, Phys. Rev. D **86**(5), 052001 (2012), arXiv:1204.3094. [Erratum: Phys. Rev.D90,no.7,079902(2014)]
38. M. Felizardo et al., Final analysis and results of the phase II simple dark matter search, Phys. Rev. Lett. **108**, 201302 (2012), arXiv:1106.3014
39. PICO Collaboration, C. Amole et al., Dark matter search results from the PICO-2L C_3F_8 bubble chamber, Phys. Rev. Lett. **114**(23), 231302 (2015), arXiv:1503.00008
40. T. Franarin, M. Fairbairn, Reducing the solar neutrino background using polarised Helium-3, Phys. Rev. D 94(5), 053004 (2016), arXiv:1605.08727
41. CDMS Collaboration, R. Agnese et al., Silicon detector dark matter results from the final exposure of CDMS II, Phys. Rev. Lett. **111**(25), 251301 (2013), arXiv:1304.4279
42. DAMA Collaboration, R. Bernabei et al., First results from DAMA/LIBRA and the combined results with DAMA/NaI, Eur. Phys. J. C **C56**, 333–355 (2008), arXiv:0804.2741
43. DAMA, LIBRA Collaboration, R. Bernabei et al., New results from DAMA/LIBRA, Eur. Phys. J. C **67**, 39–49 (2010), arXiv:1002.1028
44. LZ Collaboration, D.S. Akerib et al., LUX-ZEPLIN (LZ) Conceptual Design Report, arXiv:1509.02910
45. R. Essig, J. Mardon, T. Volansky, Direct detection of sub-GeV dark matter, Phys. Rev. D **85** 076007 (2012), arXiv:1108.5383
46. R. Essig, A. Manalaysay, J. Mardon, P. Sorensen, T. Volansky, First direct detection limits on sub-GeV dark matter from XENON10, Phys. Rev. Lett. **109**, 021301 (2012), arXiv:1206.2644
47. S.K. Lee, M. Lisanti, S. Mishra-Sharma, and B.R. Safdi, Modulation effects in dark matter-electron scattering experiments, Phys. Rev. D **92**(8), 083517 (2015), arXiv:1508.07361
48. S. Derenzo, R. Essig, A. Massari, A. Soto, T.-T. Yu, Direct detection of sub-gev dark matter with scintillating targets, Phys. Rev. D 96(1), 016026 (2017), arXiv:1607.01009
49. P.W. Graham, D.E. Kaplan, S. Rajendran, M.T. Walters, Semiconductor probes of light dark matter, Phys. Dark Univ. **1**, 32–49 (2012), arXiv:1203.2531
50. R. Essig, M. Fernandez-Serra, J. Mardon, A. Soto, T. Volansky, T.-T. Yu, Direct detection of sub-GeV dark matter with semiconductor targets, JHEP **05**, 046 (2016), arXiv:1509.01598
51. W. Guo, D.N. McKinsey, Concept for a dark matter detector using liquid helium-4, Phys. Rev. D **87**(11), 115001 (2013), arXiv:1302.0534
52. K. Schutz, K.M. Zurek, On the detectability of light dark matter with superfluid Helium, Phys. Rev. Lett. 117(12), 121302 (2016), arXiv:1604.08206
53. Y. Hochberg, Y. Zhao, K.M. Zurek, Superconducting detectors for superlight dark matter, Phys. Rev. Lett. **116**(1), 011301 (2016), arXiv:1504.07237
54. Y. Hochberg, M. Pyle, Y. Zhao, K.M. Zurek, Detecting superlight dark matter with fermi-degenerate materials, JHEP **1608**, 057 (2016), arXiv:1512.04533
55. C. Kouvaris, J. Pradler, Probing sub-GeV dark matter with conventional detectors, Phys. Rev. Lett. **118**(3), 031803 (2017), arXiv:1607.01789
56. C. Kouvaris, Probing light dark matter via evaporation from the sun, Phys. Rev. D **92**(7), 075001 (2015), arXiv:1506.04316
57. Y. Hochberg, Y. Kahn, M. Lisanti, C.G. Tully, K.M. Zurek, Directional detection of dark matter with 2D targets, Phys. Lett. B772, 239–246, (2017), arXiv:1606.08849

Chapter 3
Indirect Detection

An important corollary of the thermal relic production mechanism is that pairs of DM particles may annihilate into pairs of SM ones. If this happens, a flux of SM particles produced in DM annihilations should permeate our galaxy, and it would in principle be visible on Earth. A huge experimental effort is devoted to measure this flux. A large amount of results have been published by experiments looking for γ rays and antiparticles (positrons, antiprotons and antideuterium), and interesting lessons about DM have been learned (and still are). In this section we are going to review the main lines of research in this field, and give an overview of the current status.

3.1 DM Distribution in the Galactic Halo

The condition for the annihilation process to take place is that two DM particles encounter. It is then intuitive that the total annihilation rate will be roughly proportional to $n^2 \sigma_{\mathrm{ann}}$, where n is the local DM number density. It is therefore of primary importance to know the distribution of DM matter particles in the halo, in order to predict the flux of their annihilation products. Unfortunately, this distribution is far from being understood. A huge debate is ongoing on the shape of the DM profile with respect to the distance r from the galactic centre. While numerical simulations tend to favour peaked profiles, observations of galactic curves indicate cored profiles, as already seen in Sect. 1.1.3. The Navarro, Frenk and White (NFW) [1] profile (peaked as r^{-1} at the Galactic centre (GC)), the Einasto [2, 3] profile and the Moore [4] profile are the standard ones emerging from numerical simulations. The usually assumed cored profiles are the truncated Isothermal profile [5, 6] and the Burkert profile [7].

© Springer International Publishing AG 2017   43
E. Morgante, *Aspects of WIMP Dark Matter Searches at Colliders and Other Probes*, Springer Theses, https://doi.org/10.1007/978-3-319-67606-7_3

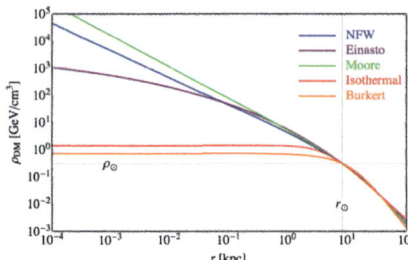

DM halo	α	r_s [kpc]	ρ_s [GeV/cm³]
NFW	–	24.42	0.184
Einasto	0.17	28.44	0.033
Isothermal	–	4.38	1.387
Burkert	–	12.67	0.712
Moore	–	30.28	0.105

Fig. 3.1 DM profiles and the corresponding parameters to be plugged in the functional forms of Eq. ((3.1)). Adapted from [14]

$$\text{NFW}: \quad \rho_{\text{NFW}}(r) = \rho_s \frac{r_s}{r}\left(1 + \frac{r}{r_s}\right)^{-2}$$

$$\text{Einasto}: \quad \rho_{\text{Ein}}(r) = \rho_s \exp\left\{-\frac{2}{\alpha}\left[\left(\frac{r}{r_s}\right)^{\alpha} - 1\right]\right\}$$

$$\text{Moore}: \quad \rho_{\text{Moo}}(r) = \rho_s \left(\frac{r_s}{r}\right)^{1.16}\left(1 + \frac{r}{r_s}\right)^{-1.84}$$

$$\text{Isothermal}: \quad \rho_{\text{Iso}}(r) = \frac{\rho_s}{1 + (r/r_s)^2}$$

$$\text{Burkert}: \quad \rho_{\text{Bur}}(r) = \frac{\rho_s}{(1 + r/r_s)(1 + (r/r_s)^2)} \tag{3.1}$$

The radius r_s and the density ρ_s are fixed in order to reproduce the local density at the Sun's position ($\rho_\odot = 0.3\,\text{GeV}/\text{cm}^3$ at a distance $r_\odot = 8.33\,\text{kpc}$ from the centre of the galaxy[1]) and the total mass contained within 60 kpc from the galactic centre ($M_{60} = 4.7 \times 10^{11} M_\odot$). The resulting choice of parameters is shown in Fig. 3.1. Varying the halo profile between the ones described in Eq. (3.1) introduces a roughly one order of magnitude uncertainty in the final flux of SM particles [12–15], depending on the galactic latitude (the uncertainty is larger when looking at the galactic centre, and becomes small at high latitude) and on the final state under study.

Two things should be noticed. Firstly, DM profiles have large differences close to the galactic centre, while they look similar for larger radii. As a consequence, the choice of the profile is expected to have a larger impact on γ rays from the galactic centre with respect to charged particles (and of course to γ rays from high latitude regions), because the trajectory of charged particles is dictated by their interactions with the interstellar medium (ISM) and galactic magnetic fields: their flux from a given direction is the sum of contributions from regions with different DM density, so that the differences in the choice of the profile are largely smeared out. Secondly, we are not considering the potential contribution from DM sub-halos, which are believed

---

[1]This value of $\rho_\odot$ is the typical one assumed in the literature. Recent determinations find higher central values, in the ballpark of 0.4 GeV/cm³ [8–11].

to form in a hierarchical picture. This is typically taken into account by means of effective multiplicative boost factors. The resolution of numerical simulations is typically far from the one needed to study this effect precisely, and the subject is still matter of debate (see [14] for a brief discussion).

3.2 Particle Production

Depending on the model, DM can annihilate into all the particles of the SM, with different branching ratios. The unstable final products (Higgs and heavy gauge bosons, heavy leptons, quarks) then decay/hadronize, and a final spectrum of stable particles is produced. Being the typical DM matter velocity in the galactic halo small ($\sim 10^{-3}c$), the DM annihilation process is typically parametrized by effective operators. A fundamental role is played by electro-weak radiative corrections [16], particularly in the case of heavy particles (roughly above $\mathcal{O}(100\,\text{GeV})$). Numerically, the importance of this effects comes from the soft/collinear divergence of processes where a gauge boson is emitted. Corrections appear in the form of multiplicative "splitting functions", and the cross section is enhanced by logarithms or double-logarithms of the ratio m_χ/m_W, which may become large for heavy DM. More over, EW corrections may lift the helicity suppression which arises when a Majorana DM particle annihilates into two fermions, leading to a much higher cross section and, as a consequence, a larger flux [17].

The second important aspect of EW corrections is that they open annihilation channels which, in general, may not be present in the model. As a simple - although unrealistic - example, let us consider a DM particle which couples only to electrons. When a couple of DM particles annihilate, an electron-positron pair is produced. If the initial energy is sufficient (*i.e.* if $2m_\chi > m_Z$) a Z boson can be emitted, which in turns decays hadronically into a $q\bar{q}$ pair. These quarks then hadronize, and the final flux of particles may include (anti-)protons and photons. The process is illustrated in Fig. 3.2.

Fig. 3.2 Example process in which a pair of DM particles annihilating into e^+e^- produce a flux of hadronic particles, photons and neutrinos in the final state thanks to electroweak corrections

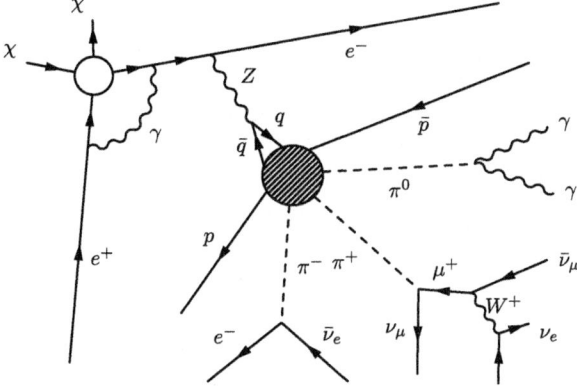

The final spectrum of SM particles, after applying EW corrections, has been computed in [16]. Automatized tools exists to compute this spectrum for any DM models. A very useful tool, which was used for some parts of this thesis, is `PPPC4DMID` [14], which provides fluxes and other useful functions related to cosmic rays' propagation, DM halo profiles and neutrino fluxes from DM annihilation in the Sun (which will be discussed in the following) in the form of handy tables, included in a `Mathematica` package [18].

3.3 Dark Matter Searches with Charged Cosmic Rays

In order to extract relevant informations from the measurement of the flux of charged cosmic rays possibly originated by DM annihilations, it is necessary to carefully consider their propagation in the galaxy. In the standard picture, cosmic rays are accelerated at the shock region of supernova remnants [19–21]. Charged particles ejected from supernovae explosions traverse repeatedly the shock surface and end up being accelerated through *diffusive shock acceleration*. Once injected in the interstellar medium, their motion through the galaxy is governed by a diffusion equation. The physical processes that come into play are [22–24]:

- Diffusion in the slowly varying component of the galactic magnetic field;
- Reacceleration in the turbulent component of galactic magnetic field;
- Energy loss due to bremsstrahlung, synchrotron radiation and inverse Compton scattering (only relevant for light particles, namely $e^{\pm}$);
- Interaction with the interstellar medium, which lead to annihilation and creation of nuclear species and, typically via pion decays, of electrons, positrons and photons;
- A possible galactic convective wind, pointing outside the galactic plane, that may influence the spectrum of nuclear species;
- Cosmic ray modulation due to the solar wind (only for low energy particles $E_{\text{kin}} \lesssim 10 \, \text{GeV}$).

The theory of cosmic rays' acceleration and propagation is reviewed in [21, 22, 24]. We will return on the details on the diffusion equation for antiprotons in Chap. 4.

A key role is played by the interactions with the interstellar medium. When *primary* cosmic rays (*i.e.* the ones accelerated by supernova remnants) interact with nuclei of the ISM, they annihilate and create a flux of *secondary* particles. It is in these interactions that stable antiparticles (e^{+}, $\bar{p}$, $\bar{d}$, *etc.*) are produced. Because of the intrinsic energy threshold of the production mechanism, positrons are more abundant than antiprotons (or even heavier anti-nuclei), roughly a factor 10^{-1} with respect to the corresponding particle, compared to 10^{-4} for antiprotons.

Antiparticles are produced as well in DM annihilation, in pair with the corresponding particle. For this reason, while particle production is DM annihilation is typically suppressed with respect to the cosmic ray background, antiparticles have a much higher signal/background ratio, and are more useful as a tool for indirect dark matter searches.

Uncertainties in the propagation parameters play a fundamental role. Very roughly, the value of these parameters are obtained by fitting the spectrum of different cosmic ray species and the ratio of secondary/primary channels. While scanning their allowed regions affect the final cosmic ray yield within the experimental uncertainties, the spectrum of antiparticles from DM annihilation, because of its different functional shape, may be shifted by orders of magnitudes. Following [13, 25], in order to quantify the uncertainty three benchmark choices of parameters called "MIN", "MAX" and "MED" can be assumed, which respectively minimise, maximise the final flux or give an intermediate yield. As a result, depending on the halo profile and on the energy, the fluxes have an uncertainty of roughly one order of magnitude due to uncertainties in the propagation parameters, on top of the one coming from our ignorance of the correct halo profile.

3.4 Dark Matter Searches with Photons

Propagation of γ rays in the galaxy is a much simpler issue. Indeed, absorption is negligible, and the total flux from a given direction is simply the integral of the contribution from all DM annihilations along the line of sight, which is proportional to the integral of the DM number density squared. On the other hand, the production mechanism of γ rays is much more varied than the one of charged particles:

- Prompt photons are produced in the annihilation of DM particles. They can be pair produced as primary annihilation channels of the annihilation process, they can be radiated from charged particles in the final state or from a charged mediators, or, finally, they can be the decay products of some intermediate metastable state. In the first case they would add a line to the background spectrum of γ rays, while in the other they will contribute to the continuum spectrum as an excess over the expected background. Being DM electrically neutral, annihilation into a γ pair is typically suppressed by loop factors, which give roughly a factor of 10^{-4}. The flux of γ rays from the centre of the galaxy (where most events are supposed to take place) is highly dependent on the DM halo profile, while the latter has little effect on the flux from external regions.
- Energetic electrons and positrons from DM annihilations may undergo inverse Compton scattering off photons of the CMB, the infrared background or from stellar light. The emitted photons may then reach Earth. In this case, propagation parameters do play a role, because the photons are originated by the $e^{\pm}$ flux. For the same reason, also the choice of the halo profile affect the final result, even if photons are not emitted in the galactic centre.
- Electromagnetic scattering of energetic electrons and positrons off the interstellar medium (both on charged HII and neutral HI, H_2 and He) lead to bremsstrahlung emission of gamma rays. Bremsstrahlung is the dominant source of energy loss for $e^{\pm}$, and it must be taken into account also in the computation of gamma rays from inverse Compton scattering, where it has the effect of reducing the final gamma

ray spectrum due to energy losses of the impinging $e^\pm$. The relative importance of photon emission from bremsstrahlung and inverse Compton scattering depends on the incoming electron's energy and on its position in the galaxy (and in particular on the strength of the galactic magnetic field, the density of the interstellar radiation field and the interstellar gas density). When this is convolved with the $e^\pm$ spectrum from DM, it turns out that bremsstrahlung is the dominant source of photons, together with prompt emission in the continuum, while inverse Compton scattering gives a smaller contribution [26].

- Strong magnetic fields in the galactic centre induce synchrotron radiation from $e^\pm$ produced in DM annihilations, with production of radio-waves (see e.g. [27, 28]).

While prompt photons are energetic (the typical energy scale is that of the mass of the annihilating particle, resulting in photons in the gamma spectrum), the ones emitted from electrons and positrons (both via scattering or synchrotron radiation) have a much broader spectrum, spanning from radio waves to gamma rays, and can therefore leave a signal in a number of different observations (the so-called "multi-wavelength" studies, see e.g. [29], also in [30]).

After having discussed the production mechanism of photons, it is important to understand what are the preferred target for the observations. In general, the most interesting regions are the ones in which the DM distribution is expected to peak, and in which the signal over noise ratio is the most favourable. The presence of a strong magnetic field is also necessary for observations of radio waves. In general, different regions are suited for different searches. Here we list the targets which experiments look at, remembering that new targets may always be added to this list:

- The first target is the Milky Way's galactic centre, used for prompt emission and radio-wave astronomy [31]. While the DM distribution is peaked in this region, it must be kept in mind that also the noise level is high.
- Small regions just outside the galactic centre have the advantage of reducing the background, for both prompt and secondary photon emission [32, 33]
- Wide regions of the galactic halo itself, several tens of degrees wide, typically centred at the galactic centre [34]. Those ase useful because they can include regions where inverse Compton scattering is efficient.
- Globular clusters, which are dense agglomerate of stars embedded in the Milky Way galactic halo. Even if they are supposed to be baryon dominated, they are interesting since some of them may have been originated by DM sub-halos, as summarised in [35], and prompt emission may be relevant.
- Sub-halos of the galactic DM halo [36].
- Dwarf spheroidal satellite galaxies, such as Sagittarius, Canis Major, Sculptor, Carina and many others. These are satellite galaxies of the Milky Way, with a very low star content and DM dominated. For this reason the signal/background ratio is expected to be large, even if it suffers from large uncertainties [37–44].
- Large scale structures such as the Virgo, Coma, Fornax and Perseus clusters in the nearby Universe [45–48].
- The "cosmological" gamma ray flux from DM annihilation is also interesting [49]. This is the total isotropic and redshifted flux from DM annihilations occurring in

all halos in the recent history of the Universe. An excess was reported in late years in the data of the ARCADE telescope [50], and an interpretation in terms of DM annihilating into leptons is possible [51, 52]. This is anyway in contrast with limits by the AMS-02 data [53].

- Other possibilities include satellite structures of the Milky Way made of DM only [54].

3.5 Summary of the Present Status

It is not easy to summarize in a few lines the present status of indirect searches. A large number of observations and data can be found in the literature, and a plethora of independent analysis of data provided by the experimental collaborations often lead do contrasting results. The most important highlights are the excess in the positron fraction by the PAMELA and AMS-02 experiments, which is a solid result but likely not related to DM; the galactic centre gamma rays excess, which is still a less established result but more promising from the point of view of DM; and finally, the putative gamma ray line at 3.5 keV, which has raised the interest in the last two years towards the sterile neutrino explanations of DM, but is still far from being a clear evidence of DM interactions. On the other hand, many other search channel have shown no deviation from the astrophysical background, and stringent limits have been derived on DM properties. A collection of up-to-date results is presented in Fig. 3.3. We are now going to describe in some details the current results of indirect searches, with a focus of the mentioned "DM hints".

3.5.1 Positrons

A lot of attention has been raised in the last years since the PAMELA satellite showed an increase in the energy spectrum of the positron fraction $e^+/(e^+ + e^-)$ for energies in the range 10–100 GeV [56]. These results, compatible with previous measurements by HEAT [57] and AMS-01 [58], where later confirmed and extended to roughly 400 GeV by measurements of the FERMI [59] satellite and AMS-02 [60, 61], a detector mounted on the International Space Station. AMS-02 data confirm the rise seen by PAMELA, and indicates a possible flattening above 300 GeV. Measurements of the total lepton flux $e^+ + e^-$ by FERMI [62] and AMS-02 [63] show no particular feature in the spectrum.

This measurements are extremely remarkable, because they indicate the existence of an additional source of positrons with respect to the standard astrophysical picture. An exciting possibility is that this excess is originated by DM annihilations (or decays). Generically, a spectrum of positrons originated by DM annihilations with a mass at the TeV scale would look similar to the measured one. With some dependence on the details of the model, it would present a "bump" shape with a sharp cut-off at

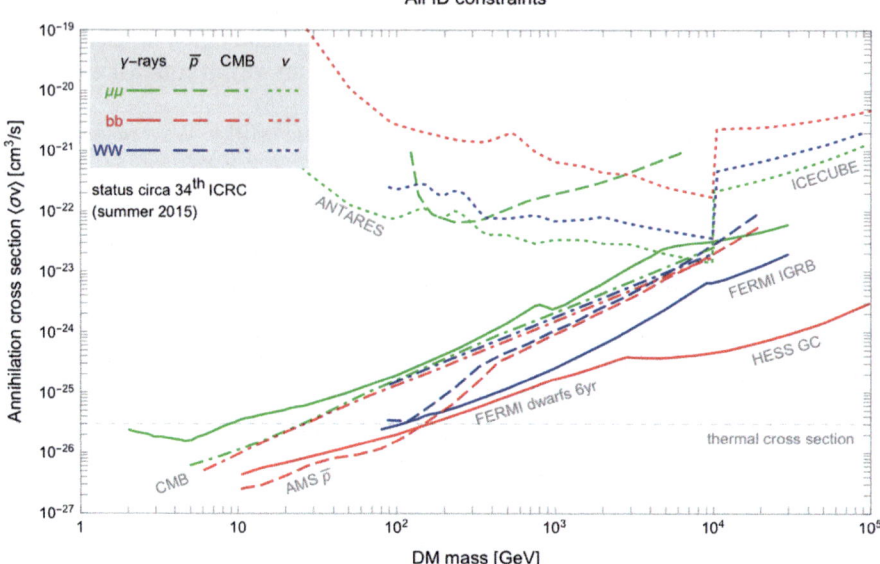

Fig. 3.3 Summary of bounds from indirect detection, rescaled in such a way to correspond to the Einasto profile. Neutrino bounds are not rescaled. (From [55])

high energies (corresponding to the DM mass) and a decreasing tail at low energies. A huge number of models have been proposed to explain the positron excess in terms of DM. To fit PAMELA data, the DM particle must have a mass of the order $2-3$ TeV, it must couple only to leptons (preferably $\tau^{\pm}$ [64]) in order to avoid tension with the antiproton data described below, and it must have an annihilation cross section $\langle \sigma v \rangle \sim 10^{-23}$, much larger than what naïvely expected for a thermal relic.[2]

The DM explanation has its drawbacks. Firstly, a certain tension is present between the positron fraction $e^{+}/(e^{+}+e^{-})$ data and the total lepton flux $e^{+}+e^{-}$ (see addendum to [67] and [55]). Secondly, a careful analysis of AMS-02 data points towards a much heavier DM ($10-40$ TeV) annihilating into $b\bar{b}$ and gauge/Higgs bosons, but this is at odds with the results from antiprotons and γ-rays searches and tends to disfavour the DM hypothesis [78]. Last, CMB constraints rule out the annihilation cross section necessary to explain PAMELA data, by noticing that overabundant DM annihilation at early times may inject energy that ionize the Universe at large redshift [79–81].

An important question to ask is what are the other possible sources of the positron excess, and whether they can be distinguished in data from the DM interpretation.

---

[2]In order to justify this discrepancy, one can rely on several possible, albeit ad-hoc, explanations: introduce a boost factor, possibly due to clumpiness of the dark matter halo [65, 66] or to the presence of a narrow resonance just below the threshold [67–69]; invoke non-perturbative effects operating at small velocities that can enhance the present day thermal cross section [70–77] or otherwise discard the standard thermal relic picture for DM particles.

The most famous alternative explanation is with pulsars. It has been known since a long time that a rise in the positron fraction can be due to the production of $e^{\pm}$ in pulsars [82]. In particular, a young nearby pulsar plus a diffuse background of mature pulsars can fit PAMELA positrons data [83, 84]. The intrinsic degeneracy between the pulsar and the DM interpretation of PAMELA and AMS-02 data cannot be broken by positron data alone [85]; nevertheless the two scenarios could be distinguished by a positive signal in the antiprotons channel since antiprotons are not expected to arise from pulsars.

A different possible mechanism involving pulsars is presented in [86, 87]. In these papers, it is shown that standard secondary emission of positrons, together with primary acceleration at nearby supernova remnants and primary emission from pulsar wind nebulae[3] can fit well the excess, even if the fit can be improved by adding a 50 GeV DM annihilating into pairs of $\mu^{\pm}$ with a cross section close to the thermal one [87].

Another possible mechanism is the acceleration of secondary $e^{\pm}$ at the shock region of supernova-remnants [88]. In the standard picture of cosmic ray propagation, particle (mostly protons) acceleration takes place at the shock region of supernova-remnants. Antiparticles are produced when cosmic rays interact with the nuclei of the interstellar medium. These interactions may as well take place at the acceleration region. In this case, antiparticles would be accelerated and give rise to a flux with a different power law behaviour from that originated by the standard mechanism, which can explain the PAMELA excess. A similar process can lead to antiproton production [89], as we will discuss in Chap. 4.

3.5.2 Antiprotons

After the announcement of the excess in the positron channel, the PAMELA collaboration released data in the antiproton channel [90, 91], showing no excess over the background up to an energy of 100–180 GeV. More recently, preliminary data were presented by AMS-02 [92], showing a spectrum of $\bar{p}/p$ flatter than what previously expected for energies up to $\sim$350 GeV. Contrarily to the case of the positron fraction, taking into account recent measurements of cosmic ray fluxes (in particular the B/C fraction) and their impact on the propagation parameters, the spectrum measured by

---

[3]The mechanism of pulsar wind nebulae is summarized in [86] and references therein. Pulsars are rapidly spinning neutron stars with a very strong surface magnetic field that, rotating, create an electric field which, in turn, can tear apart particles from the surface of the neutron stars. These particles are accelerated by the electromagnetic field and induce radiation, which subsequently produce particle/antiparticle pairs. Pulsars are located inside the remnant of the supernova explosion from which they originate. The interaction of the electromagnetic "wind" produced by the pulsar with the slow ejecta of the supernova produce a shock that propagates backward in the direction of the pulsar. Particles are accelerated by the shock, and accumulate in the region surrounding it where hot magnetized plasma is created. This is the so-called pulsar wind nebula (PWN). Particles are accumulated in the PWN until this is disrupted roughly 50 kyr after its creation, and are then released in the interstellar medium.

AMS-02 can be explained by secondary production alone [93–95]. In conclusion, even if future precise measurements could in principle favour a DM explanation, it is fair to say that presently no tantalizing hint of DM annihilation is seen in the antiproton channel.

3.5.3 Anti-deuterium and Anti-^{3}He

The idea to look for anti-deuterium produced in DM annihilation was first introduced in [96]. When DM annihilate producing anti-nucleons $\bar{p}$ and $\bar{n}$, under certain conditions these can merge to form an anti-nucleus. The process is typically described in the so-called "coalescence model". The condition under which this may happen is that the two anti-nucleons have parallel momenta and equal energies. This kinematical restriction makes the production rate of anti-deuterium much lower than that of other particles. On the other hand, its spectrum is predicted to be peaked at a fraction of GeV, with extremely low background from other astrophysical sources, such that it is commonly believed that the detection of a single $\bar{d}$ particle in the sub-GeV region would be a smoking gun of DM annihilation or decay. Current limits obtained with the BESS balloon [97] on the $\bar{d}$ flux are $1 - 2$ order of magnitudes above the level expected for realistic WIMP models. Searches with AMS-02 and GAPS (a specifically proposed balloon mission) are expected to have the sensitivity needed to exclude relevant parts of the parameter space [98, 99].

The same principle of anti-deuterium production can be applied to other anti-nuclei. Mass numbers $A > 2$ are suppressed essentially due to the reduced probability for a large number of nucleons to coalesce. For this reason, only anti-^{3}He seems to be a promising candidate [100, 101]. Interestingly enough, the AMS-02 collaboration has recently reported a tentative detection of few particles with $Z = -2$ and mass around the one of anti-^{3}He [102] (see also [103] for a discussion of the possible physical implications). Whether or not this should be taken seriously is still a matter of debate, since the signal-over-background ratio is so small that the collaboration itself warns that a sufficient instrumental understanding is still lacking, and it will take a few more years to establish the nature of this putative signal.

3.5.4 Gamma Rays from the Galactic Centre

The situation regarding gamma rays is intensively debated, and clear conclusions can not be drawn. The most debated issue concerns FERMI measurements of gamma rays from a region a few degrees around the galactic centre. Several authors reported, since 2009, the detection of a gamma ray signal, compatible with that obtained from DM annihilations (the first claim was made in [104], see [55] for additional references). In particular, the best fit is obtained for a 30–50 GeV DM particle, distributed according to a NFW profile, annihilating dominantly into $b\bar{b}$ with a cross section close to the thermal one, $\langle \sigma v \rangle \sim 10^{-26} \, \mathrm{cm^3/s}$ [105].

The presence of an excess was confirmed by the FERMI collaboration in [31], and it is robust against variations of the known astrophysical foregrounds [106–108]. Nevertheless, it is probably too early to claim with certainty that DM has been detected, since astrophysical explanations are still possible: a population of unresoved millisecond pulsars [109–115] or other point sources (even if this explanation is not universally accepted [116, 117]), Compton emission from cosmic ray outbursts in the past 10^6 years [118–120], allowing for space dependent propagation parameters [121] or finally reconsidered density of cosmic ray accelerators in the galactic centre (leading to an increased injection of cosmic rays) can yield the same signal [122, 123].

An important question that should be asked is whether the excess is compatible with constraints from other probes (indirect, direct and LHC) [124–128]. The answer is still not clear, and it seems to depend on the details of the analysis. A possible model that could fit the Fermi signal is one in which the Dark Matter is a fermion that couples to the Standard Model via a pseudo-scalar mediator, possibly embedded in a two Higgs doublet model in order to respect gauge invariance. The pseudo-scalar mediator would result in a spin-dependent interaction of DM with atomic nuclei, thus evading constraints from direct detection. A model of this kind could lead to interesting new signatures at the LHC, which could rule out or confirm the DM interpretation of the Fermi excess [129]. To conclude, the galactic centre excess is for sure an extremely intriguing puzzle, but, even if it would be great if the DM interpretation is confirmed, it is too early to draw definitive conclusions.

3.5.5 Gamma Ray Lines

Another interesting hint comes from the observation of a line at 3.5 keV in the spectrum of gamma rays from the Andromeda galaxy, the Perseus galaxy cluster [130] and in the stacked gamma ray spectrum of some galaxy clusters [131], with the XMM-Newton telescope. Again, these results are debated: while some groups do not find any evidence of the presence of the line in the XMM-Newton data [132], other groups relates the line with emission lines of Potassium and Chlorine [133, 134] (see also [135–137] for a discussion), and other use the morphology of the XMM-Newton data to exclude the DM interpretation [138].

In the past years a great excitement was raised by the observation of a line excess around 130 GeV in FERMI data [139, 140]. After a long and rich discussion in the literature, a reanalysis of FERMI data with an enlarged dataset [141] showed that the feature was due to a statistical effect. The presence of the line was excluded by subsequent HESS data [142].

3.5.6 *Neutrinos*

As any other SM particles, neutrinos may be produced in DM annihilations. Similarly
to photons, once they are produced neutrinos travel on a straight line without interacting with the galactic material, and may be detected at Earth with big Čerenkov
detectors located underground: when a neutrino interacts with the target material, secondary charged particles (in particular muons) may be produced with a speed larger
than that of light in the material, and traverse the experiment leaving a Čerenkov light
cone so that their energy and flight direction may be measured. The main background
are muons produced in cosmic rays when they interact with the atmosphere. For this
reason, neutrino detectors exploit the directional information to reject all the tracks
coming from above, using the Earth as a shield against these muons.

The targets for neutrino observation are the galactic centre (or the galactic halo),
satellite galaxies or cluster of galaxies and the Sun. Observations from the first two
targets are very similar to gamma rays observation, with the caveat that observations
of extragalactic neutrinos give typically weaker bounds. Observation of neutrinos
from the Sun are instead very peculiar, and we will return on them later on.

Limits from neutrino telescopes such as SUPERKAMIOKANDE, ICECUBE and
ANTARES fall in the ballpark of $\langle \sigma v \rangle \sim 10^{-22} - 10^{-20}$ cm^3 s^{-1} depending on the
annihilation channel, typically weaker than limits from gamma rays. Nevertheless,
they are less dependent on the DM mass, and become stronger than gamma ray limits
above few TeV. The reason is that, while a larger m_χ implies a smaller number density
and then a reduced flux, neutrinos from DM annihilations become more energetic,
and their cross section increases, with the net result of a reduced deterioration of the
limits.

3.6 Dark Matter Searches with Neutrinos from the Sun

Dark Matter searches with neutrinos from the Sun deserve a separate explanation. In
all the searches we have described so far, charged and neutral particles were produced
in DM annihilations/decays, and limits were set on the DM annihilation cross section
(or decay width). We are now going to discuss a very peculiar kind of search, in which
limits on the scattering cross section of DM with nucleons are obtained by looking at
DM annihilation products. In this sense, DM searches with neutrinos from the Sun
are in the middle between direct and indirect probes.

The mechanism was introduced in [143–147]. Dark matter particles of the galactic
halo can scatter off nuclei inside celestial bodies, in particular the Sun and the Earth.
When this happens, they may be slowed down to a velocity lower than the escape
velocity from that particular body, and being gravitationally captured. An overdensity
of DM particles may therefore originate in the core of these celestial objects, leading
to an enhanced rate of annihilations. Among the SM particles that are produced in
these interactions, an important role is played by neutrinos. Being weakly interacting,

they can traverse the body without interacting, and may be detected by Čerenkov detectors at Earth.

The number density of DM particles trapped inside the Sun (similarly for the Earth), and therefore their annihilation rate, is governed by three processes: gravitational capture, annihilation/decay and evaporation:

$$\frac{dN}{dt} = \Gamma_{capt} - 2\Gamma_{ann} - \Gamma_{evap} \tag{3.2}$$

Evaporation is the process in which DM particles thermalized by scattering off nuclei, gets accelerated to a velocity larger than the escape velocity and escape the Sun. It can safely be neglected for $m_\chi \gtrsim 4\,\text{GeV}$ [148]. The DM annihilation rate is proportional to the number of DM particles squared and to the annihilation cross section. It can be rewritten as $\frac{1}{2}C_{ann}N^2$, with $C_{ann} \propto \langle \sigma v \rangle$. On the other hand, the capture rate is constant and proportional to the scattering cross section of DM with nuclei. Equation (3.2) can then be solved to give

$$\Gamma_{ann} = \frac{\Gamma_{capt}}{2} \tanh^2\left(\frac{t}{\tau}\right) \xrightarrow{t \gg \tau} \frac{\Gamma_{capt}}{2} \tag{3.3}$$

where τ is is given by $1/\sqrt{\Gamma_{capt}C_{ann}}$. After a time larger than this equilibrium time scale, the total annihilation rate is fixed by the scattering cross section, and not by the annihilation one. If the annihilation cross section is increased, the number density decreases and viceversa, leaving the annihilation rate constant.

Before going on with the calculation of the capture and annihilation rates, let us comment briefly on this result. The fact that the annihilation rate is proportional to the scattering cross section makes possible a straight comparison with results from direct searches. Bounds obtained with neutrinos from the Sun are typically weaker than direct detection ones when spin independent scattering is considered, since the cross section is proportional to the total number of nucleons and typically the targets have large atomic number, whereas in the Sun scattering takes place primarily on hydrogen and helium. On the other hand, if the interaction is spin dependent the bounds are complementary to those from direct searches, and become stronger at large DM mass.

To conclude, let us notice that the fact that the annihilation rate is fixed by the scattering cross section does not mean that the details on the annihilation process are irrelevant for setting limits on a given model. On the contrary, the total neutrino flux depend on the branching ratios into all SM species separately, corrected to take into account electroweak corrections. This will be of crucial importance in the analysis of Chap. 9. The possible presence of a resonance plays instead a minor role, contrary to what happens in other indirect searches, because the resonant enhancement is factored out when computing branching ratios. When the DM mass is close to $M/2$, where M is the energy of the resonance, the branching ratios can change drastically, but this affects only a small region in m_χ as large as the resonance's width.

3.6.1 Calculation of the Annihilation and Capture Rates

Early calculations of the annihilation, capture and evaporation rates include [144, 149–155]. Here we will follow the derivation of [156]. The annihilation rate of DM in the Sun is given by

$$\Gamma_{\text{ann}} = \frac{1}{2} \int d^3 x \, n^2(\mathbf{x}) \langle \sigma v \rangle = \frac{1}{2} C_{\text{ann}} N^2 \tag{3.4}$$

where n is the DM number density and the $1/2$ is a symmetry factor for self-conjugate DM particles. After thermal equilibrium is reached, the DM distribution is given by

$$n(r) = n_0 \exp[-m_\chi \phi(r)/T_\odot], \tag{3.5}$$

where $\phi(r)$ is the gravitational potential at a given distance r from the centre and $T_\odot = 1.6 \times 10^7$ K is the temperature of the solar core. Taking for simplicity the density to be constant and equal to the central density $\rho_\odot = 151\,\text{g cm}^{-3}$ (with a necessary shift of notation with respect to the previous section), the integrals can be evaluated explicitly yielding [156]

$$n(r) = n_0 = e^{-r^2/r_{\text{DM}}^2}, \qquad r_{\text{DM}} = \left(\frac{3 T_\odot}{2\pi G_N \rho_\odot m_\chi} \right)^{1/2} \approx 0.01 R_\odot \sqrt{\frac{100\,\text{GeV}}{m_\chi}} \tag{3.6}$$

and

$$C_{\text{ann}} = \langle \sigma v \rangle \left(\frac{G_N \rho_\odot m_\chi}{3 T_\odot} \right)^{3/2}. \tag{3.7}$$

Here G_N is the Newton gravitational constant. The numerical approximation in Eq. (3.6) tells us that the DM particles are concentrated in the solar core, and is a consistency check of the assumptions of constant density and temperature.

The next ingredient is the calculation of the capture rate. As already discussed, this is proportional to the DM scattering cross section, summed over all nuclear species present in the solar core, weighted with their concentration. The result is:

$$\Gamma_{\text{capt}} = \frac{\rho_\chi}{m_\chi} \sum_i \sigma_i \int_0^{R_\odot} dr 4\pi r^2 n_i(r) \int_0^\infty dv 4\pi v^2 f_\odot(v) \frac{v^2 + v_{\odot\text{esc}}(r)^2}{v} \mathcal{P}_i(v, v_{\odot\text{esc}}(r)) \tag{3.8}$$

Let us discuss briefly the ingredients that enter this calculation: The first factor is just the DM number density (with a small shift of notation, $\rho_\chi = 0.3\,\text{GeV cm}^{-3}$ is the local DM density, which was called $\rho_\odot$ in the previous section). The cross section σ_i is summed over all nuclear species present in the solar core and multiplied by their individual concentration n_i, integrated over the entire volume. Note that the angular dependence of the cross section is neglected, and $v = 0$ limit can be assumed. The term $\mathcal{P}_i(v, v_{\odot esc}(r))$ gives the probability that a particle, once scattered off a nucleus at a distance r from the centre, is slowed down to a velocity lower than the escape velocity from the Sun at that position $v_{\odot esc}(r)$. In first approximation this is just given by the difference between the minimal energy loss required to have $v < v_{\odot esc}(r)$ after the scattering and the maximal energy loss in one scattering event, divided the maximal possible energy loss. In a more refined treatment, form factors must be introduced for each nuclear species in order to take into account the different response of nuclei depending on the energy [156]. The probability $\mathcal{P}_i$ is weighted with the DM velocity distribution in the solar reference frame, which is related to the one in the galactic rest frame by

$$f_\odot(v) = \frac{1}{2} \int_{-1}^{+1} d(\cos\theta) f\left(\sqrt{v^2 + v_\odot^2 + 2vv_\odot \cos\theta}\right). \tag{3.9}$$

Here $v_\odot \approx 233$ km/s is the solar velocity in the galactic rest frame and θ is the angle between v and $v_\odot$. The distribution of the DM velocity in the galaxy can be parametrized as

$$f(v) = N \left[\exp\left(\frac{v_{esc}^2 - v^2}{kv_0^2}\right) - 1\right]^k \theta(v_{esc} - v) \tag{3.10}$$

where N is a factor that normalizes f to 1, $v_{esc} \approx 450$–650 km/s is the escape velocity from the galaxy, $v_0 = 220$–270 km/s is a cut-off velocity and k is a factor in the range 1.5–3.5. The last ingredient in the calculation is the DM velocity which must be multiplied times the cross section to obtain the scattering rate. The effect of the solar gravitational field is taken into account by the escape velocity $v_{\odot esc}$, in such a way that the velocity of a DM particle falling towards the Sun from very far away with initial velocity v is $v^2 + v_{\odot esc}^2$.

With this in hand, the equilibrium time scale can be computed. Taking the age of the Solar System to be $t_\odot = t_\oplus = 4.5$ Gyr, the result is [157]

$$\frac{t_\odot}{\tau_\odot} = 1.9 \times 10^3 \left(\frac{\Gamma_{capt}}{s^{-1}}\right) \left(\frac{\langle\sigma v\rangle}{cm^3 \, s^{-1}}\right)^{1/2} \left(\frac{m_\chi}{100\,GeV}\right)^{3/4} \tag{3.11}$$

and

$$\frac{t_\oplus}{\tau_\oplus} = 1.1 \times 10^5 \left(\frac{\Gamma_{capt}^\oplus}{s^{-1}}\right) \left(\frac{\langle\sigma v\rangle}{cm^3 \, s^{-1}}\right)^{1/2} \left(\frac{m_\chi}{100\,GeV}\right)^{3/4}, \tag{3.12}$$

where the second equation is referred to WIMP capture by the Earth. While typically these numbers are larger than one, this is not guaranteed when the scattering cross section is spin and velocity suppressed (in particular it is not true in the case of the Earth, where only limits on Γ_{ann} can be set for spin dependent scattering [158]).

The next step will be to compute the neutrino production rate and its spectrum, taking into account neutrino propagation in the Sun, in the Solar System, and during Earth crossing.

3.6.2 Neutrino Spectra

Production of neutrinos in the Sun and their propagation to Earth is a complicate subject, involving many physical aspects. Here we will just recapitulate the building blocks of the calculation, that can be performed using different numerical tools. Early calculations of the neutrino spectrum were performed in [147, 159, 160]

The first ingredient is the calculation of the branching ratios into different SM channels. With this in hand, one can proceed to compute the final neutrino yield. Neutrinos are produced not only as primary annihilation channels, but also as a result of the hadronization process of quarks, in interactions of other annihilation products with the solar material and as a result of electroweak bremsstrahlung. Results of this calculation are provided in [156] and included in the Mathematica package PPPC4DMID [18]. We will use these results in Chap. 9. Hadronization of quarks and gluons from DM annihilation is performed using PYTHIA 8 [161], propagation of the stable and meta-stable hadronic states in the solar matter is simulated with GEANT4 [162, 163], and the effect of electroweak radiation is taken into account by means of fragmentation functions, as in [16].[4]

Once neutrinos are produced, they have to traverse the Sun, travel along the Solar System and, once reached Earth, traverse it towards the location of the experiment. The effects into play are neutrino oscillations (both in vacuum and in matter, where the MSW effect[5] has to be taken into account), neutral current scatterings (in which a neutrino excite a nucleus and is re-emitted with lower energy) and charged current scattering (in which a neutrino is removed from the flux and a charged lepton is created. If the lepton is a $\tau^{\pm}$ its decay produces back energetic neutrinos of all flavours). The appropriate formalism to deal with neutrino propagations is that of using two 3×3 density matrices (for ν and $\bar{\nu}$) evolved with an effective Hamiltonian [168, 169] (see also [170, 171] for earlier proposals). A complete calculation of neutrino fluxes was performed in [172], recently updated including the state-of-the-art knowledge of the neutrino mixing parameters [156]. Numerical tables are provided in PPPC4DMID [18].

---

[4]While the effect of electroweak corrections was not taken into account in PYTHIA 6, is taken into account in PYTHIA 8 [161, 164], so in principle there would be no need to use EW fragmentation functions any more. Anyway, this was not the approach of [156]. Even if PYTHIA 8 does not include all possible EW phenomena (e.g. $Z \rightarrow WW$, $W \rightarrow ZW$ splitting and γ emission from W are not included) results obtained with it are in agreement with those using the approach of fragmentation functions (compare for example [156, 165]).

[5]When neutrinos traverse matter, the effect of weak interactions with the surrounding material is to modify the phases of the three neutrino flavours, affecting the oscillation process. This is called MSW effect, from the names of Mikheev, Smirnov and Wolfenstein [166, 167].

3.6.3 Limits on the Scattering Cross Section

Neutrino fluxes described in the previous section can be compared with data from neutrino search experiment, and the non observation of any excess over the background allow to put bounds on the normalization of these spectra (the shape is fixed by the model and parameters such as the DM mass). The normalization depend on both the number of DM particles accumulated in the Sun (which is unknown, but calculable for a specific DM particle) and the annihilation cross section with neutrinos in the final state. For a given model, this can be related to the total annihilation cross section, in such a way that constraints are obtained on $\Gamma_{ann} \propto N^2 \langle \sigma v \rangle$. The last two quantities can not be disentangled, but if equilibrium is reached (3.3) allows to put bounds on Γ_{capt} and therefore on the DM-proton scattering cross section $\sigma_{\chi p}$.

Various experiments are currently putting bounds on the DM scattering cross section by looking for neutrinos from the Sun. A collection of latest results from the main experimental collaborations [173–176] is shown in Figs. 3.4 and 3.5 for spin independent and spin dependent scattering, respectively. Limits are shown for benchmark models in which DM annihilates into a single particle species. Typical choices are $b\bar{b}$ (as a reference for a channel with few, soft neutrinos) and $\tau\tau$ or W^+W^- (many hard neutrinos). Obtaining results for a given model with more annihilation channels is not a trivial task, even if of primary importance. An approximate procedure is described in Chap. 9 and in [177, 178]. In [176, 179], likelihood tables are provided to compute bounds in any DM models. Notice that, with respect to our results presented in Chap. 9 (as well as in [177, 178]), result presented by experimental collaborations do not include the effect of electroweak corrections yet.

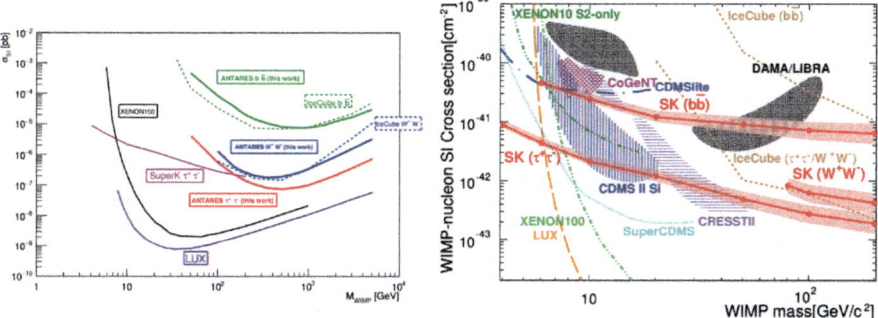

Fig. 3.4 Collection of limits on spin-independent scattering from DM annihilation in the Sun. *Top, left* ANTARES 2016 [175]. *Top, right* Super–Kamiokande 2015 [174]

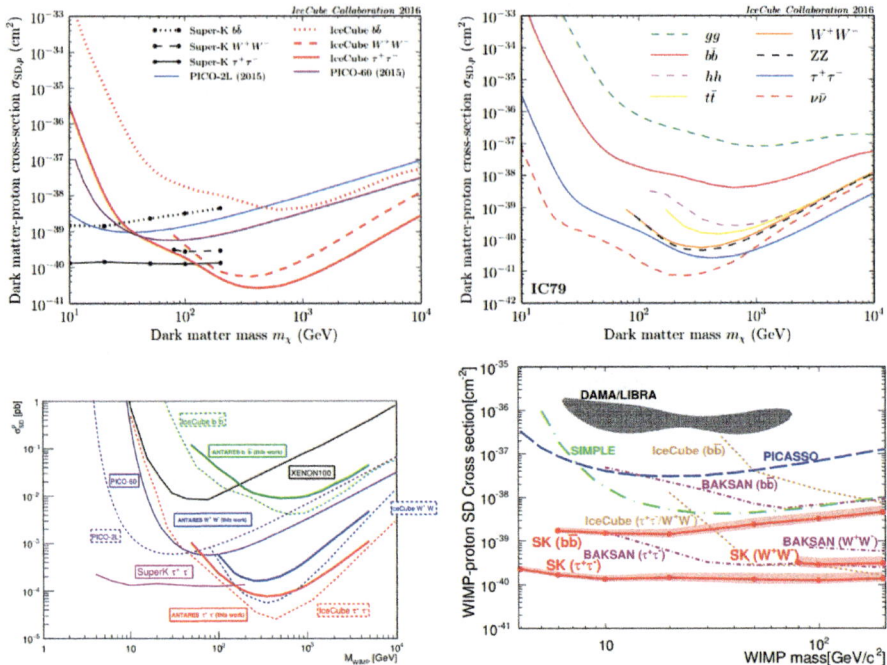

Fig. 3.5 Collection of limits on spin-dependent scattering from DM annihilation in the Sun. *Top* Ice-Cube 2016 [176]. *Centre, left* ANTARES 2016 [175]. *Centre, right* Super–Kamiokande 2015 [174]

References

1. J.F. Navarro, C.S. Frenk, S.D.M. White, The Structure of cold dark matter halos. Astrophys. J. **462**, 563–575 (1996), arXiv:astro-ph/9508025
2. A.W. Graham, D. Merritt, B. Moore, J. Diemand, B. Terzic, Empirical models for dark matter halos. I. Nonparametric construction of density profiles and comparison with parametric models. Astron. J. **132**, 2685–2700 (2006), arXiv:astro-ph/0509417
3. J.F. Navarro, A. Ludlow, V. Springel, J. Wang, M. Vogelsberger, S.D.M. White, A. Jenkins, C.S. Frenk, A. Helmi, The diversity and similarity of cold dark matter halos. Mon. Not. Roy. Astron. Soc. **402**, 21 (2010), arXiv:0810.1522
4. J. Diemand, B. Moore, J. Stadel, Convergence and scatter of cluster density profiles. Mon. Not. Roy. Astron. Soc. **353**, 624 (2004), arXiv:astro-ph/0402267
5. K.G. Begeman, A.H. Broeils, R.H. Sanders, Extended rotation curves of spiral galaxies: dark haloes and modified dynamics. Mon. Not. Roy. Astron. Soc. **249**, 523 (1991)
6. J.N. Bahcall, R.M. Soneira, The Universe at faint magnetidues. 2. Models for the predicted star counts. Astrophys. J. Suppl. **44**, 73–110 (1980)
7. A. Burkert, The Structure of dark matter halos in dwarf galaxies. IAU Symp. **171**, 175 (1996), arXiv:astro-ph/9504041. [Astrophys. J. 447, L25 (1995)]
8. R. Catena, P. Ullio, A novel determination of the local dark matter density. JCAP **1008**, 004 (2010), arXiv:0907.0018
9. M. Weber, W. de Boer, Determination of the local dark matter density in our galaxy. Astron. Astrophys. **509**, A25 (2010), arXiv:0910.4272
10. P. Salucci, F. Nesti, G. Gentile, C.F. Martins, The dark matter density at the Sun's location. Astron. Astrophys. **523**, A83 (2010), arXiv:1003.3101

11. W. de Boer, M. Weber, The dark matter density in the solar neighborhood reconsidered. JCAP **1104**, 002 (2011), arXiv:1011.6323
12. D. Maurin, R. Taillet, Spatial origin of galactic cosmic rays in diffusion models: ii- exotic primary cosmic rays. Astron. Astrophys. **404**, 949–958 (2003), arXiv:astro-ph/0212113
13. F. Donato, N. Fornengo, D. Maurin, P. Salati, Antiprotons in cosmic rays from neutralino annihilation. Phys. Rev. **D69**, 063501 (2004), arXiv:astro-ph/0306207
14. M. Cirelli, G. Corcella, A. Hektor, G. Hutsi, M. Kadastik, P. Panci, M. Raidal, F. Sala, A. Strumia, PPPC 4 DM ID: a poor particle Physicist cookbook for dark matter indirect detection. JCAP **1103**, 051 (2011), arXiv:1012.4515. [Erratum: JCAP 1210, E01 (2012)]
15. M. Cirelli, G. Giesen, Antiprotons from dark matter: current constraints and future sensitivities. JCAP **1304**, 015 (2013), arXiv:1301.7079
16. P. Ciafaloni, D. Comelli, A. Riotto, F. Sala, A. Strumia, A. Urbano, Weak corrections are relevant for dark matter indirect detection. JCAP **1103**, 019 (2011), arXiv:1009.0224
17. P. Ciafaloni, M. Cirelli, D. Comelli, A. De Simone, A. Riotto, A. Urbano, Initial state radiation in majorana dark matter annihilations. JCAP **1110**, 034 (2011), arXiv:1107.4453
18. Pppc 4 dm id - a poor particle physicist cookbook for dark matter indirect detection, http://www.marcocirelli.net/PPPC4DMID.html. Accessed 20 July 2015
19. A.R. Bell, The Acceleration of cosmic rays in shock fronts. I. Mon. Not. Roy. Astron. Soc. **182**, 147–156 (1978)
20. R. Blandford, D. Eichler, Particle acceleration at astrophysical shocks: a theory of cosmic ray origin. Phys. Rept. **154**, 1–75 (1987)
21. P. Blasi, The origin of galactic cosmic rays. Astron. Astrophys. Rev. **21**, 70 (2013), arXiv:1311.7346
22. A.W. Strong, I.V. Moskalenko, V.S. Ptuskin, Cosmic-ray propagation and interactions in the Galaxy. Ann. Rev. Nucl. Part. Sci. **57**, 285–327 (2007), arXiv:astro-ph/0701517
23. C. Evoli, D. Gaggero, D. Grasso, L. Maccione, Cosmic-ray Nuclei, antiprotons and Gamma-rays in the galaxy: a new diffusion model. JCAP **0810**, 018 (2008), arXiv:0807.4730. [Erratum: JCAP 1604(04), E01 (2016)]
24. M. Kachelriess, Lecture notes on high energy cosmic rays, arXiv:0801.4376
25. T. Delahaye, R. Lineros, F. Donato, N. Fornengo, P. Salati, Positrons from dark matter annihilation in the galactic halo: theoretical uncertainties. Phys. Rev. **D77**, 063527 (2008), arXiv:0712.2312
26. M. Cirelli, P.D. Serpico, G. Zaharijas, Bremsstrahlung gamma rays from light dark matter. JCAP **1311**, 035 (2013), arXiv:1307.7152
27. N. Fornengo, R.A. Lineros, M. Regis, M. Taoso, Galactic synchrotron emission from WIMPs at radio frequencies. JCAP **1201**, 005 (2012), arXiv:1110.4337
28. M. Cirelli, M. Taoso, Updated galactic radio constraints on dark matter, JCAP 1607(07), 041 (2016), arXiv:1604.06267
29. S. Profumo, P. Ullio, Multi-wavelength searches for particle dark matter, arXiv:1001.4086
30. S. Profumo, P. Ullio, Multi-wavelength studies, in *Particle Dark Matter*, ed by G. Bertone, pp. 547–564 (2010)
31. Fermi-LAT Collaboration, M. Ajello et al., Fermi-LAT observations of high-energy γ-ray emission toward the galactic center. Astrophys. J. **819**(1), 44 (2016), arXiv:1511.02938
32. H.E.S.S. Collaboration Collaboration, A. Abramowski et al., Search for a dark matter annihilation signal from the galactic center halo with H.E.S.S. Phys. Rev. Lett. **106**, 161301 (2011), arXiv:1103.3266
33. HESS Collaboration, V. Lefranc, E. Moulin, Dark matter search in the inner Galactic halo with H.E.S.S. I and H.E.S.S. II, in *Proceedings, 34th International Cosmic Ray Conference (ICRC 2015)* (2015), arXiv:1509.04123
34. Fermi-LAT Collaboration, M. Ackermann et al., Constraints on the galactic halo dark matter from Fermi-LAT diffuse measurements. Astrophys. J. **761**, 91 (2012), arXiv:1205.6474
35. HESS Collaboration, A. Abramowski et al., H.E.S.S. observations of the globular clusters NGC 6388 and M 15 and search for a Dark Matter signal. Astrophys. J. **735**, 12 (2011), arXiv:1104.2548

36. VERITAS Collaboration, D. Nieto, Hunting for dark matter subhalos among the Fermi-LAT sources with VERITAS, in *Proceedings, 34th International Cosmic Ray Conference (ICRC 2015)* (2015), arXiv:1509.00085

37. J. Aleksić et al., Optimized dark matter searches in deep observations of Segue 1 with MAGIC. JCAP **1402**, 008 (2014), arXiv:1312.1535

38. HESS Collaboration, A. Abramowski et al., Search for dark matter annihilation signatures in H.E.S.S. observations of Dwarf Spheroidal Galaxies. Phys. Rev. **D90**, 112012 (2014), arXiv:1410.2589

39. **Fermi-LAT** Collaboration, M. Ackermann et al., Searching for dark matter annihilation from milky way dwarf spheroidal galaxies with six years of fermi large area telescope data. Phys. Rev. Lett. **115**(23), 231301 (2015), arXiv:1503.02641

40. Fermi-LAT, MAGIC Collaboration, J. Rico, M. Wood, A. Drlica-Wagner, J. Aleksić, Limits to dark matter properties from a combined analysis of MAGIC and Fermi-LAT observations of dwarf satellite galaxies, in *Proceedings, 34th International Cosmic Ray Conference (ICRC 2015)* (2015), arXiv:1508.05827

41. DES, Fermi-LAT Collaboration, A. Drlica-Wagner et al., Search for Gamma-Ray emission from DES dwarf spheroidal galaxy candidates with Fermi-LAT data. Astrophys. J. **809**(1), L4 (2015), arXiv:1503.02632

42. HAWC Collaboration, J.P. Harding, B. Dingus, Dark matter annihilation and decay searches with the high altitude water cherenkov (HAWC) observatory, in *Proceedings, 34th International Cosmic Ray Conference (ICRC 2015)* (2015), arXiv:1508.04352

43. HAWC Collaboration, M.L. Proper, J.P. Harding, B. Dingus, First limits on the dark matter cross section with the HAWC Observatory, in *Proceedings, 34th International Cosmic Ray Conference (ICRC 2015)* (2015), arXiv:1508.04470

44. VERITAS Collaboration, B. Zitzer, Search for dark matter from dwarf galaxies using VER-ITAS, in *Proceedings, 34th International Cosmic Ray Conference (ICRC 2015)* (2015), arXiv:1509.01105

45. M. Ackermann et al., Constraints on dark matter annihilation in clusters of galaxies with the fermi large area telescope. JCAP **1005**, 025 (2010), arXiv:1002.2239

46. Fermi-LAT Collaboration, M. Ackermann et al., Search for extended gamma-ray emission from the Virgo galaxy cluster with Fermi-LAT. Astrophys. J. **812**(2), 159 (2015), arXiv:1510.00004

47. HESS Collaboration, A. Abramowski et al., Search for dark matter annihilation signals from the fornax galaxy cluster with H.E.S.S. Astrophys. J. **750**, 123 (2012), arXiv:1202.5494. [Erratum: Astrophys. J. 783, 63 (2014)]

48. VERITAS Collaboration, T. Arlen et al., Constraints on cosmic rays, magnetic fields, and dark matter from Gamma-Ray observations of the coma cluster of galaxies with VERITAS and fermi. Astrophys. J. **757**, 123 (2012), arXiv:1208.0676

49. Fermi-LAT Collaboration, M. Ackermann et al., Limits on dark matter annihilation signals from the fermi LAT 4-year measurement of the isotropic gamma-ray background. JCAP **1509**(09), 008 (2015), arXiv:1501.05464

50. D.J. Fixsen et al., ARCADE 2 measurement of the Extra-Galactic sky temperature at 3-90 GHz, Astrophys. J. 734, 1 (2011), arXiv:0901.0555

51. N. Fornengo, R. Lineros, M. Regis, M. Taoso, Possibility of a dark matter interpretation for the excess in isotropic radio emission reported by ARCADE. Phys. Rev. Lett. **107**, 271302 (2011), arXiv:1108.0569

52. N. Fornengo, R. Lineros, M. Regis, and M. Taoso, Cosmological radio emission induced by WIMP dark matter. JCAP **1203**, 033 (2012), arXiv:1112.4517

53. M. Fairbairn, P. Grothaus, Note on the dark matter explanation of the ARCADE excess and AMS data. Phys. Rev. **D90**(12), 127302 (2014), arXiv:1407.4849

54. Fermi-LAT Collaboration, M. Ackermann et al., Search for dark matter satellites using the FERMI-LAT. Astrophys. J. **747**, 121 (2012), arXiv:1201.2691

55. M. Cirelli, Status of indirect (and direct) dark matter searches, arXiv:1511.02031

56. PAMELA Collaboration, O. Adriani et al., An anomalous positron abundance in cosmic rays with energies 1.5-100 GeV. Nature **458**, 607–609 (2009), arXiv:0810.4995
57. HEAT Collaboration, S.W. Barwick et al., Measurements of the cosmic ray positron fraction from 1-GeV to 50-GeV, Astrophys. J. **482**, L191–L194 (1997), arXiv:astro-ph/9703192
58. AMS 01 Collaboration, M. Aguilar et al., Cosmic-ray positron fraction measurement from 1 to 30-GeV with AMS-01. Phys. Lett. **B646**, 145–154 (2007), arXiv:astro-ph/0703154]
59. Fermi-LAT Collaboration, M. Ackermann et al., Measurement of separate cosmic-ray electron and positron spectra with the fermi large area telescope. Phys. Rev. Lett. **108**, 011103 (2012), arXiv:1109.0521
60. A.M.S. Collaboration, M. Aguilar et al., First result from the alpha magnetic spectrometer on the international space station: precision measurement of the positron fraction in primary cosmic rays of 0.5-350 GeV. Phys. Rev. Lett. **110**, 141102 (2013)
61. A.M.S. Collaboration, L. Accardo et al., High statistics measurement of the positron fraction in primary cosmic rays of 0.5-500 GeV with the alpha magnetic spectrometer on the international space station. Phys. Rev. Lett. **113**, 121101 (2014)
62. **Fermi-LAT** Collaboration, A.A. Abdo et al., Measurement of the cosmic ray e+ plus e-spectrum from 20 GeV to 1 TeV with the fermi large area telescope, Phys. Rev. Lett. **102**, 181101 (2009), arXiv:0905.0025
63. A.M.S. Collaboration, M. Aguilar et al., Electron and positron fluxes in primary cosmic rays measured with the alpha magnetic spectrometer on the international space station. Phys. Rev. Lett. **113**, 121102 (2014)
64. P. Meade, M. Papucci, A. Strumia, T. Volansky, Dark matter interpretations of the e+- Excesses after FERMI. Nucl. Phys. **B831**, 178–203 (2010), arXiv:0905.0480
65. J. Lavalle, J. Pochon, P. Salati, R. Taillet, Clumpiness of dark matter and positron annihilation signal: computing the odds of the galactic lottery. Astron. Astrophys. **462**, 827–848 (2007), arXiv:astro-ph/0603796
66. J. Lavalle, Q. Yuan, D. Maurin, and X. J. Bi, Full calculation of clumpiness boost factors for antimatter cosmic rays in the light of Lambda-CDM N-body simulation results. Abandoning hope in clumpiness enhancement?. Astron. Astrophys. **479**, 427–452 (2008), arXiv:0709.3634
67. M. Cirelli, M. Kadastik, M. Raidal, A. Strumia, Model-independent implications of the e+-, anti-proton cosmic ray spectra on properties of dark matter. Nucl. Phys. **B813**, 1–21 (2009), arXiv:0809.2409. [Addendum: Nucl. Phys. B873, 530 (2013)]
68. M. Ibe, H. Murayama, T.T. Yanagida, Breit-wigner enhancement of dark matter annihilation. Phys. Rev. **D79**, 095009 (2009), arXiv:0812.0072
69. D. Feldman, Z. Liu, P. Nath, PAMELA positron excess as a signal from the hidden sector. Phys. Rev. **D79**, 063509 (2009), arXiv:0810.5762
70. A. Sommerfeld, Über die Beugung und Bremsung der Elektronen. Ann. der Phys. **403**(3), 257–330 (1931)
71. J. Hisano, S. Matsumoto, M.M. Nojiri, Explosive dark matter annihilation. Phys. Rev. Lett. **92**, 031303 (2004), arXiv:hep-ph/0307216
72. J. Hisano, S. Matsumoto, M.M. Nojiri, O. Saito, Non-perturbative effect on dark matter annihilation and gamma ray signature from galactic center. Phys. Rev. **D71**, 063528 (2005), arXiv:hep-ph/0412403
73. J. Hisano, S. Matsumoto, M. Nagai, O. Saito, M. Senami, Non-perturbative effect on thermal relic abundance of dark matter. Phys. Lett. **B646**, 34–38 (2007), arXiv:hep-ph/0610249
74. M. Cirelli, A. Strumia, M. Tamburini, Cosmology and astrophysics of minimal dark matter. Nucl. Phys. **B787**, 152–175 (2007), arXiv:0706.4071
75. M. Cirelli, R. Franceschini, A. Strumia, Minimal dark matter predictions for galactic positrons, anti-protons, photons. Nucl. Phys. **B800**, 204–220 (2008), arXiv:0802.3378
76. N. Arkani-Hamed, D.P. Finkbeiner, T.R. Slatyer, N. Weiner, A theory of dark matter. Phys. Rev. **D79**, 015014 (2009), arXiv:0810.0713
77. M. Lattanzi, J.I. Silk, Can the WIMP annihilation boost factor be boosted by the Sommerfeld enhancement?. Phys. Rev. **D79**, 083523 (2009), arXiv:0812.0360

78. M. Boudaud et al., A new look at the cosmic ray positron fraction. Astron. Astrophys. **575**, A67 (2015), arXiv:1410.3799

79. G. Giesen, J. Lesgourgues, B. Audren, Y. Ali-Haimoud, CMB photons shedding light on dark matter. JCAP **1212**, 008 (2012), arXiv:1209.0247

80. Planck Collaboration, P.A.R. Ade et al., Planck 2015 results. XIII. Cosmological parameters, Astron. Astrophys. 594, A13 (2016), arXiv:1502.01589

81. T.R. Slatyer, Indirect dark matter signatures in the cosmic dark ages. I. Generalizing the bound on s-wave dark matter annihilation from Planck results. Phys. Rev. **D93**(2), 023527 (2016), arXiv:1506.03811

82. F.A. Aharonian, A.M. Atoyan, H.J. Volk, High energy electrons and positrons in cosmic rays as an indicator of the existence of a nearby cosmic tevatron. Astron. Astrophys. **294**, L41–L44 (1995)

83. D. Hooper, P. Blasi, P.D. Serpico, Pulsars as the sources of high energy cosmic ray positrons. JCAP **0901**, 025 (2009), arXiv:0810.1527

84. Fermi-LAT Collaboration, D. Grasso et al., On possible interpretations of the high energy electron-positron spectrum measured by the fermi large area telescope. Astropart. Phys. **32**, 140–151 (2009), arXiv:0905.0636

85. M. Pato, M. Lattanzi, G. Bertone, Discriminating the source of high-energy positrons with AMS-02. JCAP **1012**, 020 (2010), arXiv:1010.5236

86. M. Di Mauro, F. Donato, N. Fornengo, R. Lineros, A. Vittino, Interpretation of AMS-02 electrons and positrons data. JCAP **1404**, 006 (2014), arXiv:1402.0321

87. M. Di Mauro, F. Donato, N. Fornengo, A. Vittino, Dark matter vs. astrophysics in the interpretation of AMS-02 electron and positron data. JCAP **1605**(05), 031 (2016), arXiv:1507.07001

88. P. Blasi, The origin of the positron excess in cosmic rays. Phys. Rev. Lett. **103**, 051104 (2009), arXiv:0903.2794

89. P. Blasi, P.D. Serpico, High-energy antiprotons from old supernova remnants. Phys. Rev. Lett. **103**, 081103 (2009), arXiv:0904.0871

90. O. Adriani et al., A new measurement of the antiproton-to-proton flux ratio up to 100 GeV in the cosmic radiation. Phys. Rev. Lett. **102**, 051101 (2009), arXiv:0810.4994

91. PAMELA Collaboration, O. Adriani et al., PAMELA results on the cosmic-ray antiproton flux from 60 MeV to 180 GeV in kinetic energy. Phys. Rev. Lett. **105**, 121101 (2010), arXiv:1007.0821

92. A. Kounine, Talk at AMS days at CERN (2015)

93. G. Giesen, M. Boudaud, Y. Génolini, V. Poulin, M. Cirelli, P. Salati, P.D. Serpico, AMS-02 antiprotons, at last! Secondary astrophysical component and immediate implications for Dark Matter. JCAP **1509**(09), 023 (2015), arXiv:1504.04276

94. C. Evoli, D. Gaggero, D. Grasso, Secondary antiprotons as a galactic dark matter probe. JCAP **1512**(12), 039 (2015), arXiv:1504.05175

95. R. Kappl, A. Reinert, M.W. Winkler, AMS-02 antiprotons reloaded. JCAP **1510**(10), 034 (2015), arXiv:1506.04145

96. F. Donato, N. Fornengo, P. Salati, Anti-deuterons as a signature of supersymmetric dark matter. Phys. Rev. **D62**, 043003 (2000), arXiv:hep-ph/9904481

97. H. Fuke et al., Search for cosmic-ray antideuterons. Phys. Rev. Lett. **95**, 081101 (2005), arXiv:astro-ph/0504361

98. P. von Doetinchem et al., Status of cosmic-ray antideuteron searches. PoS **ICRC2015**, 1218 (2015), arXiv:1507.02712

99. T. Aramaki et al., Review of the theoretical and experimental status of dark matter identification with cosmic-ray antideuterons. Phys. Rept. **618**, 1–37 (2016), arXiv:1505.07785

100. E. Carlson, A. Coogan, T. Linden, S. Profumo, A. Ibarra, S. Wild, Antihelium from dark matter. Phys. Rev. **D89**(7), 076005 (2014), arXiv:1401.2461

101. M. Cirelli, N. Fornengo, M. Taoso, A. Vittino, Anti-helium from dark matter annihilations. JHEP **08**, 009 (2014), arXiv:1401.4017

102. S. Ting, The first five years of the alpha magnetic spectrometer on the international space station, 8 December 2016, CERN, https://indico.cern.ch/event/592392/

103. A. Coogan, S. Profumo, Can AMS anti-helium events come from dark matter? Maybe!, arXiv:1705.09664
104. L. Goodenough, D. Hooper, Possible evidence for dark matter annihilation in the inner milky way from the fermi gamma ray space telescope, arXiv:0910.2998
105. T. Daylan, D.P. Finkbeiner, D. Hooper, T. Linden, S.K.N. Portillo, N.L. Rodd, T.R. Slatyer, The characterization of the gamma-ray signal from the central milky way: a case for annihilating dark matter. Phys. Dark Univ. **12**, 1–23 (2016), arXiv:1402.6703
106. B. Zhou, Y.-F. Liang, X. Huang, X. Li, Y.-Z. Fan, L. Feng, J. Chang, GeV excess in the milky way: the role of diffuse galactic gamma-ray emission templates. Phys. Rev. **D91**(12), 123010 (2015), arXiv:1406.6948
107. F. Calore, I. Cholis, C. Weniger, Background model systematics for the Fermi GeV excess. JCAP **1503**, 038 (2015), arXiv:1409.0042
108. F. Calore, I. Cholis, C. McCabe, C. Weniger, A tale of tails: dark matter interpretations of the fermi GeV excess in light of background model systematics. Phys. Rev. **D91**(6), 063003 (2015), arXiv:1411.4647
109. K.N. Abazajian, The Consistency of Fermi-LAT observations of the galactic center with a millisecond pulsar population in the central stellar cluster. JCAP **1103**, 010 (2011), arXiv:1011.4275
110. N. Mirabal, Dark matter versus pulsars: catching the impostor. Mon. Not. Roy. Astron. Soc. **436**, 2461 (2013), arXiv:1309.3428
111. Q. Yuan, B. Zhang, Millisecond pulsar interpretation of the Galactic center gamma-ray excess. JHEAp **3-4**, 1–8 (2014), arXiv:1404.2318
112. J. Petrović, P.D. Serpico, G. Zaharijas, Millisecond pulsars and the galactic center gamma-ray excess: the importance of luminosity function and secondary emission. JCAP **1502**(02), 023 (2015), arXiv:1411.2980
113. R. Bartels, S. Krishnamurthy, C. Weniger, Strong support for the millisecond pulsar origin of the Galactic center GeV excess. Phys. Rev. Lett. **116**(5), 051102 (2016), arXiv:1506.05104
114. S.K. Lee, M. Lisanti, B.R. Safdi, T.R. Slatyer, W. Xue, Evidence for unresolved γ-ray point sources in the inner galaxy. Phys. Rev. Lett. **116**(5), 051103 (2016), arXiv:1506.05124
115. R.M. O'Leary, M.D. Kistler, M. Kerr, J. Dexter, Young pulsars and the galactic center GeV gamma-ray excess, arXiv:1504.02477
116. D. Hooper, I. Cholis, T. Linden, J. Siegal-Gaskins, T. Slatyer, Pulsars cannot account for the inner Galaxy's GeV excess. Phys. Rev. **D88**, 083009 (2013), arXiv:1305.0830
117. I. Cholis, D. Hooper, T. Linden, Challenges in explaining the galactic center gamma-ray excess with millisecond pulsars. JCAP **1506**(06), 043 (2015), arXiv:1407.5625
118. E. Carlson, S. Profumo, Cosmic ray protons in the inner galaxy and the galactic center Gamma-Ray excess. Phys. Rev. **D90**(2), 023015 (2014), arXiv:1405.7685
119. J. Petrović P.D. Serpico, G. Zaharijas, Galactic center gamma-ray "excess" from an active past of the Galactic Centre? JCAP **1410**(10), 052 (2014), arXiv:1405.7928
120. I. Cholis, C. Evoli, F. Calore, T. Linden, C. Weniger, D. Hooper, The galactic center GeV excess from a series of leptonic cosmic-ray outbursts. JCAP **1512**(12), 005 (2015), arXiv:1506.05119
121. D. Gaggero, A. Urbano, M. Valli, P. Ullio, Gamma-ray sky points to radial gradients in cosmic-ray transport. Phys. Rev. **D91**(8), 083012 (2015), arXiv:1411.7623
122. D. Gaggero, M. Taoso, A. Urbano, M. Valli, P. Ullio, Towards a realistic astrophysical interpretation of the gamma-ray Galactic center excess. JCAP **1512**(12), 056 (2015), arXiv:1507.06129
123. E. Carlson, T. Linden, S. Profumo, Putting things back where they belong: tracing cosmic-ray injection with H2, Phys. Rev. Lett. **117**(11), 111101 (2016), arXiv:1510.04698
124. S. Ipek, D. McKeen, A.E. Nelson, A renormalizable model for the galactic center gamma ray excess from dark matter annihilation. Phys. Rev. **D90**(5), 055021 (2014), arXiv:1404.3716
125. T. Bringmann, M. Vollmann, C. Weniger, Updated cosmic-ray and radio constraints on light dark matter: implications for the GeV gamma-ray excess at the Galactic center. Phys. Rev. **D90**(12), 123001 (2014), arXiv:1406.6027

126. M. Cirelli, D. Gaggero, G. Giesen, M. Taoso, A. Urbano, Antiproton constraints on the GeV gamma-ray excess: a comprehensive analysis. JCAP **1412**(12), 045 (2014), arXiv:1407.2173

127. D. Hooper, T. Linden, P. Mertsch, What does the PAMELA antiproton spectrum tell us about dark matter? JCAP **1503**(03), 021 (2015), arXiv:1410.1527

128. K.N. Abazajian, R.E. Keeley, Bright gamma-ray galactic center excess and dark dwarfs: strong tension for dark matter annihilation despite milky way halo profile and diffuse emission uncertainties. Phys. Rev. **D93**(8), 083514 (2016), arXiv:1510.06424

129. P. Tunney, J.M. No, M. Fairbairn, A novel LHC dark matter search to dissect the galactic centre excess, arXiv:1705.09670

130. A. Boyarsky, O. Ruchayskiy, D. Iakubovskyi, J. Franse, Unidentified line in x-ray spectra of the andromeda galaxy and perseus galaxy cluster. Phys. Rev. Lett. **113**, 251301 (2014), arXiv:1402.4119

131. E. Bulbul, M. Markevitch, A. Foster, R.K. Smith, M. Loewenstein, S.W. Randall, Detection of an unidentified emission line in the stacked X-ray spectrum of galaxy clusters. Astrophys. J. **789**, 13 (2014), arXiv:1402.2301

132. M.E. Anderson, E. Churazov, J.N. Bregman, Non-detection of X-ray emission from sterile neutrinos in stacked galaxy spectra. Mon. Not. Roy. Astron. Soc. **452**(4), 3905–3923 (2015), arXiv:1408.4115

133. K.J.H. Phillips, B. Sylwester, J. Sylwester, The x-ray line feature at 3.5 KeV in galaxy cluster spectra. Astrophys. J. **809**, 50 (2015)

134. T.E. Jeltema, S. Profumo, Discovery of a 3.5 keV line in the galactic centre and a critical look at the origin of the line across astronomical targets. Mon. Not. Roy. Astron. Soc. **450**(2), 2143–2152 (2015), arXiv:1408.1699

135. A. Boyarsky, J. Franse, D. Iakubovskyi, O. Ruchayskiy, Comment on the paper "Dark matter searches going bananas: the contribution of Potassium (and Chlorine) to the 3.5 keV line" by T. Jeltema and S. Profumo, arXiv:1408.4388

136. E. Bulbul, M. Markevitch, A.R. Foster, R.K. Smith, M. Loewenstein, S.W. Randall, Comment on "Dark matter searches going bananas: the contribution of Potassium (and Chlorine) to the 3.5 keV line", arXiv:1409.4143

137. T. Jeltema, S. Profumo, Reply to two comments on "Dark matter searches going bananas the contribution of Potassium (and Chlorine) to the 3.5 keV line", arXiv:1411.1759

138. E. Carlson, T. Jeltema, S. Profumo, Where do the 3.5 keV photons come from? A morphological study of the Galactic Center and of Perseus. JCAP **1502**(02), 009 (2015), arXiv:1411.1758

139. T. Bringmann, X. Huang, A. Ibarra, S. Vogl, C. Weniger, Fermi LAT search for internal bremsstrahlung signatures from dark matter annihilation. JCAP **1207**, 054 (2012), arXiv:1203.1312

140. C. Weniger, A tentative gamma-ray line from dark matter annihilation at the fermi large area telescope. JCAP **1208**, 007 (2012), arXiv:1204.2797

141. Fermi-LAT Collaboration, M. Ackermann et al., Updated search for spectral lines from Galactic dark matter interactions with pass 8 data from the fermi large area telescope. Phys. Rev. **D91**(12), 122002 (2015), arXiv:1506.00013

142. HESS Collaboration, M. Kieffer, K.D. Mora, J. Conrad, C. Farnier, A. Jacholkowska, J. Veh, A. Viana, Search for Gamma-ray line signatures with H.E.S.S, in *Proceedings, 34th International Cosmic Ray Conference (ICRC 2015)* (2015), arXiv:1509.03514

143. J. Silk, K.A. Olive, M. Srednicki, The photino, the sun and high-energy neutrinos. Phys. Rev. Lett. **55**, 257–259 (1985)

144. L.M. Krauss, K. Freese, W. Press, D. Spergel, Cold dark matter candidates and the solar neutrino problem. Astrophys. J. **299**, 1001 (1985)

145. K. Freese, Can scalar neutrinos or massive Dirac neutrinos be the missing mass? Phys. Lett. B **167**, 295–300 (1986)

146. L.M. Krauss, M. Srednicki, F. Wilczek, Solar system constraints and signatures for dark matter candidates. Phys. Rev. D **33**, 2079–2083 (1986)

147. T.K. Gaisser, G. Steigman, S. Tilav, Limits on cold dark matter candidates from deep underground detectors. Phys. Rev. D **34**, 2206 (1986)

148. G. Busoni, A. De Simone, W.C. Huang, On the minimum dark matter mass testable by neutrinos from the sun. JCAP **1307**, 010 (2013), arXiv:1305.1817
149. K. Griest, D. Seckel, Cosmic asymmetry, neutrinos and the sun. Nucl. Phys. **B283**, 681 (1987). [Erratum: Nucl. Phys. B 296, 1034 (1988)]
150. A. Gould, WIMP distribution in and evaporation from the sun. Astrophys. J. **321**, 560 (1987)
151. A. Gould, Resonant enhancements in WIMP capture by the Earth. Astrophys. J. **321**, 571 (1987)
152. A. Gould, Direct and indirect capture of wimps by the Earth. Astrophys. J. **328**, 919–939 (1988)
153. A. Gould, Evaporation of WIMPs with arbitrary cross sections. Astrophys. J. **356**, 302–309 (1990)
154. A. Gould, Cosmological density of WIMPs from solar and terrestrial annihilations. Astrophys. J. **388**, 338–344 (1992)
155. A. Gould, Big bang archeology: WIMP capture by the earth at finite optical depth. Astrophys. J. **387**, 21 (1992)
156. P. Baratella, M. Cirelli, A. Hektor, J. Pata, M. Piibeleht, A. Strumia, PPPC 4 DMν: a poor particle physicist cookbook for neutrinos from dark matter annihilations in the sun. JCAP **1403**, 053 (2014), arXiv:1312.6408
157. G. Jungman, M. Kamionkowski, K. Griest, Supersymmetric dark matter. Phys. Rept. **267**, 195–373 (1996), arXiv:hep-ph/9506380
158. IceCube Collaboration, J. Kunnen, J. Lünemann, A search for dark matter in the centre of the Earth with the IceCube neutrino detector (2015), http://pos.sissa.it/archive/conferences/236/1205/ICRC2015_1205.pdf
159. S. Ritz, D. Seckel, Detailed neutrino spectra from cold dark matter annihilations in the sun. Nucl. Phys. B **304**, 877–908 (1988)
160. G. Jungman, M. Kamionkowski, Neutrinos from particle decay in the sun and earth. Phys. Rev. **D51**, 328–340 (1995), arXiv:hep-ph/9407351
161. T. Sjostrand, S. Mrenna, P.Z. Skands, A brief introduction to PYTHIA 8.1, Comput. Phys. Commun. **178**, 852–867 (2008), arXiv:0710.3820
162. GEANT4 Collaboration, S. Agostinelli et al., GEANT4: a simulation toolkit. Nucl. Instrum. Meth. **A506**, 250–303 (2003)
163. J. Allison et al., Geant4 developments and applications. IEEE Trans. Nucl. Sci. **53**, 270 (2006)
164. J.R. Christiansen, T. Sjöstrand, Weak gauge boson radiation in parton showers. JHEP **04**, 115 (2014), arXiv:1401.5238
165. A. Ibarra, M. Totzauer, S. Wild, Higher order dark matter annihilations in the Sun and implications for IceCube. JCAP **1404**, 012 (2014), arXiv:1402.4375
166. L. Wolfenstein, Neutrino oscillations in matter. Phys. Rev. D **17**, 2369–2374 (1978)
167. S.P. Mikheev, A. Yu, Smirnov, Resonance amplification of oscillations in matter and spectroscopy of solar neutrinos. Sov. J. Nucl. Phys. **42**, 913–917 (1985). [Yad. Fiz. 42, 1441 (1985)]
168. G. Raffelt, G. Sigl, L. Stodolsky, NonAbelian Boltzmann equation for mixing and decoherence. Phys. Rev. Lett. **70**, 2363–2366 (1993), arXiv:hep-ph/9209276. [Erratum: Phys. Rev. Lett. 98, 069902 (2007)]
169. G. Sigl, G. Raffelt, General kinetic description of relativistic mixed neutrinos. Nucl. Phys. B **406**, 423–451 (1993)
170. A.D. Dolgov, Neutrinos in the Early Universe. Sov. J. Nucl. Phys. **33**, 700–706 (1981). [Yad. Fiz. 33, 1309 (1981)]
171. R. Barbieri, A. Dolgov, Neutrino oscillations in the early universe. Nucl. Phys. B **349**, 743–753 (1991)
172. M. Cirelli, N. Fornengo, T. Montaruli, I.A. Sokalski, A. Strumia, F. Vissani, Spectra of neutrinos from dark matter annihilations. Nucl. Phys. **B727**, 99–138 (2005), arXiv:hep-ph/0506298. [Erratum: Nucl. Phys. B 790, 338 (2008)]
173. M.M. Boliev, S.V. Demidov, S.P. Mikheyev, O.V. Suvorova, Search for muon signal from dark matter annihilations inthe Sun with the Baksan Underground Scintillator Telescope for 24.12 years. JCAP **1309**, 019 (2013), arXiv:1301.1138

174. Super-Kamiokande Collaboration, K. Choi et al., Search for neutrinos from annihilation of captured low-mass dark matter particles in the sun by Super-Kamiokande. Phys. Rev. Lett. **114**(14), 141301 (2015), arXiv:1503.04858
175. ANTARES Collaboration, S. Adrian-Martinez et al., Limits on dark matter annihilation in the sun using the ANTARES neutrino telescope. Phys. Lett. **B759**, 69–74 (2016), arXiv:1603.02228
176. **IceCube** Collaboration, M.G. Aartsen et al., Improved limits on dark matter annihilation in the Sun with the 79-string IceCube detector and implications for supersymmetry, JCAP 1604(04), 022 (2016), arXiv:1601.00653
177. E. Morgante, D. Racco, M. Rameez, A. Riotto, The 750 GeV diphoton excess, dark matter and constraints from the IceCube experiment. JHEP **07**, 141 (2016), arXiv:1603.05592
178. T. Jacques, A. Katz, E. Morgante, D. Racco, M. Rameez, A. Riotto, Complementarity of DM searches in a consistent simplified model: the case of Z', JHEP 071, 1610 (2016), arXiv:1605.06513
179. P. Scott, C. Savage, nulike, http://nulike.hepforge.org/. Accessed 27 June 2016

Chapter 4
Focus on AMS-02 Anti-protons Results

4.1 Introduction

After the release of the PAMELA data showing a rise in the positron fraction at energies above 10GeV, interest rose around the measurement of the antiproton fraction $\bar{p}/p$. Indeed, DM annihilation could lead to a similar signal in antiprotons, where the degeneracy with a pulsar explanation is not present, since antiprotons are too heavy to be produced by pulsars' magnetic fields. As discussed in Sect. 3.5.2, in this channel no excess was seen by PAMELA, strongly constraining the DM interpretation of the positron effect. Indeed, the fact that no anomalous signal is seen in antiprotons in the same range of energies put severe constraints on DM properties [1] and tends to favour the so-called leptophylic models, in which DM only couples to leptons. In this scenario, antiprotons data can also be used to constrain DM properties [2, 3], since the positrons and antiprotons fluxes are correlated thanks to the electroweak corrections [4].

In this chapter, based on [5] (appeared before the release of preliminary antiproton data from the AMS-02 collaboration), we analyse the question whether a possible antiproton signal above the expected background would lead to a degeneracy problem between a possible DM origin and an astrophysical origin. The answer was, unfortunately, pessimistic: an excess in AMS-02 antiproton data could not be unambiguously attributed to DM if no other information is supplied. The analysis was carried on with mock AMS-02 data, which turned out not to be far from the actual measurements presented in [6]. Even if improved calculations of the backgrounds proved to be no evidence of an excess (as it will be discussed in Sect. 4.6), the point of [5] could be useful in view of new data releases by AMS-02 of possible future experiments.

The astrophysical source of antiprotons that we consider here is the one discussed in Ref. [7] to explain the rise of positrons and subsequently in Ref. [8] to predict the antiproton flux. The excess of positrons is due to secondary products of hadronic interactions inside the same SuperNova Remnants (SNR) that accelerate cosmic rays. Primary protons accelerated in shock regions of SNRs can undergo hadronic

© Springer International Publishing AG 2017
E. Morgante, *Aspects of WIMP Dark Matter Searches at Colliders and Other Probes*, Springer Theses, https://doi.org/10.1007/978-3-319-67606-7_4

interactions not only at late times after diffusion in the galaxy, but also when they are still in the acceleration region. These interactions will produce a flux of antiparticles that will in turn be accelerated by the same sources of the standard primary cosmic rays, resulting in an additional cosmic ray flux at Earth with a spectral shape different from that of standard secondaries. A generic prediction of the model is a flattening and eventually a weak rise of the antiparticle-over-particle ratio in both positrons and antiprotons channel [8]. The particular interest of this model is in the fact that it does not rely on any new source of antiparticles (since positrons and antiprotons are generated by the same primary protons that accelerate in SNR) and that it predicts similar signals both in positrons and in antiprotons, precisely as many DM model do. This leads to a possible degeneracy in the shape of signals of very different origin, weakening the discriminating power of AMS-02.

Using the projected sensitivity of AMS-02 for the antiproton channel, in [5] we studied this degeneracy under the assumption that the measurements of AMS-02 would show a significant antiproton excess above the background. First we assumed that the signal was generated by DM annihilation in the galactic halo, and we showed ho this coul be mimicked by SNR. Then we performed the opposite exercise, fitting a SNR signal with DM annihilation.

The chapter is organized as follows. In Sect. 4.2 we briefly review the mechanism for antiprotons production and acceleration from SNR and recall some results which will be used in the following. In Sect. 4.3, the calculation of the background of secondary antiprotons and their propagation is recalled, while the possible antiproton contribution from DM is discussed in Sect. 4.4. Then, in Sect. 4.5 we turn to investigate the degeneracies which may arise in the interpretation of a putative signal in antiprotons eventually measured by AMS-02. We first assume the signal is due to DM and we try to fit it with SNR, and subsequently we analyse briefly the possibility of a SNR signal intepreted as a DM. Finally, our conclusions are summarized in Sect. 4.7.

4.2 Antiprotons Accelerated in Supernova Remnants

Here we briefly recall the basics of the astrophysical mechanism leading to primary antiprotons and we refer to the original papers [7, 8] for further details. In particular, the analytical prescription for the ratio $\bar{p}/p$ that we will use for our analysis is derived in Ref. [8]. Simulations were also performed in [9].

Antiproton production inside the accelerator is described by the source function

$$Q_{\bar{p}}(E) = 2 \int_E^{E_{\max}} d\mathcal{E} N_{\mathrm{CR}}(\mathcal{E}) \sigma_{p\bar{p}}(\mathcal{E}, E) n_{\mathrm{gas}} c, \qquad (4.1)$$

where c is the speed of light, N_{CR} is the spectrum of protons inside the source, n_{gas} is the gas density in the shock region and $\sigma_{p\bar{p}}(\mathcal{E}, E)$ is the differential cross section for a proton of energy $\mathcal{E}$ to produce an antiproton of energy E in pp scattering, that we parametrize as in [10–12].

The energy E_{max} is the maximum energy of a proton accelerated in the SNR at the age relevant for this mechanism, which we will treat as a free parameter. The factor of 2 comes from the fact that, in pp collisions, an antineutron can be produced with equal probability than an antiproton (in the isospin symmetry limit); it will then decay into an antiproton, contributing equally to the final flux. For that, we are assuming that the characteristic size of the SNR is larger than the mean path travelled by a neutron before decay. Even if a careful treatment of this point could possibly lead to sizeable effects, this would not affect qualitatively our conclusions, which would still hold.

After being produced, the antiprotons undergo acceleration around the shock region. The $\bar{p}/p$ flux ratio at this stage is [8]

$$\left.\frac{J_{\bar{p}}(E)}{J_p(E)}\right|_{SNR} \sim 2\,n_1\,\epsilon\,c\,[\mathcal{A}(E) + \mathcal{B}(E)], \qquad (4.2)$$

where

$$\mathcal{A}(E) = \gamma\left(\frac{1}{\xi} + r^2\right)\int_m^E d\omega\,\omega^{\gamma-3}\frac{D_1(\omega)}{u_1^2}\int_\omega^{E_{max}} d\mathcal{E}\,\mathcal{E}^{2-\gamma}\sigma_{p\bar{p}}(\mathcal{E}, \omega) \qquad (4.3)$$

and

$$\mathcal{B}(E) = \frac{\tau_{SN}r}{2E^{2-\gamma}}\int_E^{E_{max}} d\mathcal{E}\,\mathcal{E}^{2-\gamma}\sigma_{p\bar{p}}(\mathcal{E}, E). \qquad (4.4)$$

The two terms $\mathcal{A}$ and $\mathcal{B}$ account for the antiparticles that are produced in the acceleration region and for the ones that are produced in the inner region of the SNR. In the above expressions, n_1 and u_1 are the background gas target density and the fluid velocity in the upstream region of the shock, fixed as in [8] to $2\,cm^{-3}$ and $0.5 \times 10^{-8}\,cm/s$, respectively.

The factor ξ in the $\mathcal{A}$ term gives the fraction of proton energy carried away by the produced secondary antiproton, which is here taken to be constant with energy. The validity of this assumption is discussed in [9]. In this work, we keep it as a constant and we consider it as a second free parameter for our analysis. Again, the assumpion of a constant value for ξ is not crucial for our conclusions, and a more detailed analysis would only change quantitatively our result.

Both $\mathcal{A}$ and $\mathcal{B}$ depend on the compression factor of the shock r, defined as the ratio of the fluid velocity upstream and downstream. The time parameter τ_{SN} is the typical SNR age. The index γ gives the slope of the spectrum in momentum space, and it is related to the shock compression factor by $\gamma = 3r/(r - 1)$. In order to compare the SNR $\bar{p}/p$ ratio with the ones generated by DM annihilation, we choose r to be consistent with the ones for the background antiproton spectrum and to satisfy

Table 4.1 Diffusion parameters used to propagate the secondary antiproton flux and the DM originated flux. No solar modulation is included

Model	z_t	δ	$D_0(10^{28}\text{cm}^2\text{s}^{-1})$	η	$v_A(\text{kms}^{-1})$	γ	v_c
KRA	4 kpc	0.50	2.64	−0.39	14.2	2.35	0
THK	10 kpc	0.50	4.75	−0.15	14.1	2.35	0

the relation $r = (2 + \gamma_{\text{pr}})/(\gamma_{\text{pr}} - 1)$, where $\gamma_{\text{pr}} = 2 - \gamma$ is the nuclei source spectral index for the Cosmic Ray (CR) propagation model, as defined in [13]; we then fix $r = 3.22$, which is consistent with $\gamma_{\text{pr}} = 2.35$ of both KRA and THK models of propagation (see Table 4.1).

The factor $\epsilon = 1.26$ in front of Eq. (4.2) accounts for the fact that $\bar{p}$ production happens not only in pp collisions, but also in collisions with heavier nuclei, depending on the chemical composition of the gas and it is fixed as in [8].

The diffusion coefficient upstream the shock D_1 is given by

$$D_1(E) = \left(\frac{\lambda_c c}{3\mathcal{F}}\right) \left(\frac{E}{eB\lambda_c}\right)^{2-\beta}, \tag{4.5}$$

where λ_c is the largest coherence scale of the turbulent component, $\mathcal{F} \sim (\Delta B/B)^2$ is the ratio of power in turbulent magnetic field over that in the ordered one, e is the unit charge, B is the magnetic field, and β is the index that characterizes the spectrum of B fluctuations. Following Ref. [8] we assume a Bohm-like diffusion index $\beta = 1$ and set $\mathcal{F} = 1/20$ and $B = 1\,\mu\text{G}$. The expression for D_1 symplifies to

$$D_1(E) \simeq 3.3 \times 20 \times 10^{22}\, E_{\text{GeV}}\text{cm}^2\text{s}^{-1}. \tag{4.6}$$

Note that this diffusion coefficient refers only to the acceleration region near the shock, and can be different from the one assumed in propagating particles through the galaxy. Diffusion in the galaxy affects in the same way both primary protons and antiprotons, so that the modifications in their spectra cancel out in the ratio. The flux ratio on Earth is then given by Eqs. (4.2), (4.3) and (4.4).

As for the parameters to vary in our following analysis, we have chosen E_{max} and ξ: we checked that they are the parameters having the largest impact on our estimate of the flux ratio. We have solved the equations above numerically in order to estimate the ratio (4.2) and we have checked that our results match the ones in [8] for the same choice of parameters. Note that for the analysis we have not used the expression for the background illustrated in [8], but rather the one obtained from the DRAGON [14] numerical code, as illustrated in the next section. Finally, we have neglected energy losses, which are not relevant for antiprotons, and solar modulation, which has negligible effect for $E \gtrsim 10\,\text{GeV}$, to which we restrict our analysis.

4.3 Secondary Antiprotons

As summarized in Sect. 3.3, the standard source of antiprotons in cosmic rays is the spallation of primary protons (i.e. protons accelerated in SNR) with nuclei of the interstellar medium (ISM). In a scenario in which the mechanism outlined in Sect. 4.2 is operative, the total antiproton flux ratio would be given by the secondary component computed in this section, plus the primary component given by Eq. (4.2).

The propagation of Cosmic Rays through the galaxy is regulated by the diffusion equation (see for instance [14])

$$\frac{\partial N_i}{\partial t} - \nabla \cdot (D\nabla - \mathbf{v}_c)N_i + \frac{\partial}{\partial p}\left(\dot{p} - \frac{p}{3}\nabla \cdot \mathbf{v}_c\right)N_i - \frac{\partial}{\partial p}p^2 D_{pp}\frac{\partial}{\partial p}\frac{N_i}{p^2} =$$

$$= Q_i(p, r, z) + \sum_{j>i} v\, n_{\text{gas}}(r, z)\sigma_{ij} N_j - v\, n_{\text{gas}}\sigma_i^{\text{in}}(E_k)N_i, \qquad (4.7)$$

where $N_i(p, \mathbf{x})$ is the number density of the i-th nuclear species, p is its momentum (not to be confused with the symbol for the proton) and v its velocity. D is the diffusion coefficient in the galaxy in real space, while D_{pp} is the diffusion coefficient in momentum space, that describes the diffusive reacceleration of CRs in the turbulent galactic magnetic field. The ISM gas density is given by n_{gas} and $\mathbf{v}_c$ is the convection velocity. The cross sections σ_i^{in} and σ_{ij} are the total inelastic cross section onto the ISM gas and the cross section for production of species i by fragmentation of species j, respectively. E_k is the kinetic energy of the particle under consideration. Finally, $Q_i(p, r, z)$ is the source function that describes the injection of primary CRs in the galaxy. The diffusion coefficients are parametrized as

$$D(\rho, R, z) = D_0 \left(\frac{v}{c}\right)^{\eta} e^{|z|/z_t} \left(\frac{\rho}{\rho_0}\right)^{\delta} \qquad (4.8)$$

and

$$D_{pp} = \frac{4}{3\delta(4 - \delta^2)(4 - \delta)} \frac{v_A^2 p^2}{D}, \qquad (4.9)$$

where (R, z) are the usual cylindrical coordinates, z_t is the half-height of the cylindrical diffusion box, $\rho = pv/(Ze)$ is the particle rigidity and v_A is the Alfvén velocity.

The spectrum of primary protons from SNR is parametrized as $Q_p \sim \rho^{-\gamma_{\text{pr}}}$. The diffusion Eq. (4.7) is solved numerically using the public avaiable DRAGON code [14].

We consider here two propagation models, namely KRA and THK, defined from the choice of propagation parameters and injection spectra illustrated in Table II of Ref. [13], found by looking for good fits to B/C data and PAMELA proton data. We report the values in Table 4.1 for convenience. We have not considered other propagation models here, as we expect different choices will not change dramatically our main conclusions.

We choose to restrict our analysis to antiproton ratio data with energy larger than 10 GeV. With this choice, solar modulation and the factors η and v_A in the propagation models do not play important role.

4.4   Antiprotons from DM

As discussed in Chap. 3, the three ingredients that control the production of CR's by DM annihilation are the density of DM particles in the galaxy, the details of the annihilation process (annihilation channel and fragmentation functions) and finally propagation to Earth. As a reference DM halo density profile, we have used the NFW and the Isothermal profiles of Eq. (3.1). The propagation of cosmic rays is still controlled by Eq. (4.7), with the source term $Q_{\bar{p}}$ now given by

$$Q_{\bar{p}}(\mathbf{r}, t, p) = \frac{1}{2} \left(\frac{\rho_{\mathrm{DM}}(\mathbf{r})}{m_{\mathrm{DM}}} \right)^2 \frac{\mathrm{d}N_{\bar{p}}}{\mathrm{d}E} \langle \sigma v \rangle, \qquad (4.10)$$

where $\langle \sigma v \rangle$ is the DM annihilation cross section and $\mathrm{d}N_{\bar{p}}/\mathrm{d}E$ is the number of antiprotons of a given energy E per DM annihilation. The diffusion parameters are still the ones given in Table 4.1. We have computed the antiproton flux at Earth using DRAGON [13] for various models of annihilating DM, as summarized in Table 4.2 and including electroweak corrections [4]. The models have been chosen so that they are not excluded by present antiproton data [15]. We use a DM mass of 3–4 TeV, depending on the model. While a lighter DM would produce a sizeable antiproton flux at lower energies which is excluded by data, a heavier DM would result in a smaller signal, therefore with an increased degeneracy with the standard astrophysical sources.

Table 4.2 DM annihilation models considered in this analysis

Name	Final state	Propagation model	DM mass (TeV)	σv_0 (cm^3/s)	Profile
bKN	$b\bar{b}$	KRA	3	7×10^{-25}	NFW
muKN	$\mu^+\mu^-$	KRA	4	8×10^{-23}	NFW
muKI	$\mu^+\mu^-$	KRA	4	1×10^{-22}	ISO
WKN	W^+W^-	KRA	3	7×10^{-25}	NFW
bTN	$b\bar{b}$	THK	3	7×10^{-25}	NFW
muTN	$\mu^+\mu^-$	THK	4	8×10^{-23}	NFW
muTI	$\mu^+\mu^-$	THK	4	1×10^{-22}	ISO
WTN	W^+W^-	THK	3	7×10^{-25}	NFW

4.5 Investigating the Degeneracies: Fit DM Signal Using SNR Model

Our aim is to test whether a putative signal in the ratio of $\bar{p}/p$ eventually observed by AMS-02 leads to degeneracies in the interpretation of its origin: DM or astrophysics? To this end, we produce a set of mock AMS-02 data through a set of benchmark DM models and ask if these data could be interpreted as due to SNR, based on the astrophysical mechanism described in Sect. 4.2 (and using the same propagation model).

As we mentioned already, we consider as free parameters in the SNR model the fraction of proton energy carried away by the antiproton ξ, and the energy cutoff E_{max}. In order to investigate possible degeneracies, we have performed the following steps:

- obtain the CR background expected for $\bar{p}/p$ using DRAGON, as described in Sect. 4.3;
- produce mock data for AMS, as described in the following;
- create a grid in the plane (E_{max}, ξ), in a range of values of 1 TeV $< E_{max} <$ 10 TeV and $0.1 < \xi < 0.5$ [7, 8];
- solve Eq. (4.2) numerically in order to get the ratio of $\bar{p}/p$ from SNR, as described in Sect. 4.2 on the grid, assuming the same cosmic ray background as the one used for DM models;
- calculate the χ^2, summed on each bin for a given mock dataset, between the DM mock flux and the SNR flux. We have performed this calculation on every point of the grid to get a function $\chi^2(E_{max}, \xi)$;
- estimate the minimum of the χ^2 for each mock dataset. Then, assuming a Gaussian distribution, the confidence contours in the plane (E_{max}, ξ) are plotted. The area within the contours will give us a measure of the degeneracy between DM and SNR interpretation of the mock data.

To create the mock data, we have considered a series of benchmark (fiducial) DM models and calculate the corresponding mock data for all of them, assuming a propagation method for Cosmic Rays (KRA or THK) and a DM halo profile. In particular, we have studied non-relativistic DM annihilating into two standard model (SM) fermions or gauge bosons with 100% branching ratio, such as $\chi\chi \rightarrow b\bar{b}$, $\chi\chi \rightarrow \mu^+\mu^-$, and $\chi\chi \rightarrow W^+W^-$. Their cross sections are chosen in such a way that they are consistent with the current PAMELA antiproton flux [16] and also not excluded by the other indirect detection observations: the positron fraction from PAMELA [17] and AMS-02 [18], Fermi LAT's gammay ray observation of dwarf galaxies [19] and diffuse background [20]. The DM benchmark models with different final states, annihilation cross section and density profiles are listed in Table 4.2.

To generate the AMS-02 mock data, we have first set the width of the energy bins based on the detector energy resolution to be [21]

$$\Delta E/E = (0.042(E/\text{GeV}) + 10)\,\%. \tag{4.11}$$

The mock data have as central value of $\bar{p}/p$ the one of the benchmark model in the centre of each bin. Uncertainties around each point have been calculated by summing up in quadrature systematic and statistical errors for the $\bar{p}/p$ ratio. The statistical error is approximately given by [15, 22]

$$\frac{\Delta(\bar{p}/p)^{\text{stat}}}{\bar{p}/p} \sim \frac{\Delta N_{\bar{p}}^{\text{stat}}}{N_{\bar{p}}} = \frac{1}{\sqrt{N_{\bar{p}}}}. \qquad (4.12)$$

We have fixed the relative systematic error to be $\Delta N_{\bar{p}}^{\text{syst}}/N_{\bar{p}} = 10\%$. Here $N_{\bar{p}}$ is the expected number of antiproton events per bin and is related to the specification parameters of the experiment via the relation $N_{\bar{p}} = \epsilon \, a_{\bar{p}} \, \Phi_i \Delta E \Delta t_i$. In particular, we have set the efficiency $\epsilon_i = 1$, the geometrical acceptance of the instrument $a_{\bar{p}} = 0.2 \, m^2$ sr and a reference operation time $\Delta t_i = 1$ yr. The flux Φ_i is the $\bar{p}$ flux in the centre of the bin i, while ΔE is the energy resolution for our binning, as found in Eq. (4.11). Mock data are plotted in Fig. 4.1 for KRA and Fig. 4.2 for THK propagation models. They extend up to $E_k \simeq 400$ GeV; having a higher energy reach would probably improve the discrimination between heavy DM and SNR models.

We are now able to quantify the capability of the SNR to reproduce possible antiproton fluxes generated by the DM models (as forecasted for the AMS-02). The SNR fluxes are calculated on the grid of values $(E_{\max}, \xi)$. Confidence contours in the plane $(E_{\max}, \xi)$ are shown in Figs. 4.3 and 4.4 for all benchmarks DM models in Table 4.2. Different colours represent 1σ to 5σ contours. We have assumed for

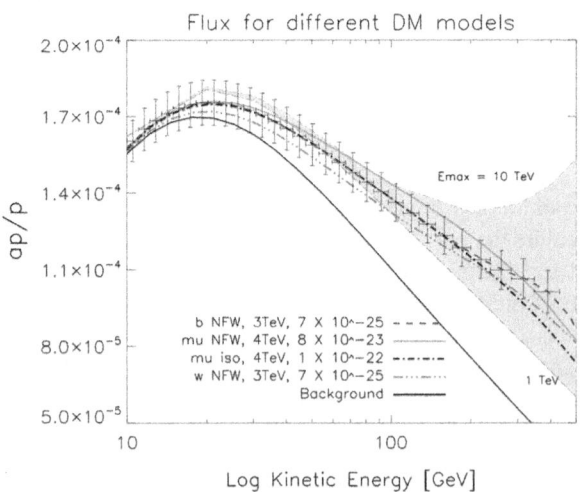

Fig. 4.1 The flux of $\bar{p}/p$ is plotted as a function of the kinetic energy for different DM models. The labels in the legend refer to annihilation channel, the DM halo profile, DM mass and annihilation cross section (in units of cm^3/s), respectively. The background from Cosmic Rays is shown in *solid black line*. For the first model we also overplot the corresponding mock data. The pink band corresponds to the region spanned by SNR when $\xi = 0.17$, as in [7] and 1 TeV $< E_{\max} < 10$ TeV. The propagation model used is KRA

Fig. 4.2 Same as Fig. 4.1 but with THK propagation model. In the lower panel we show PAMELA data [16] as compared to the same *background curve* as in Fig. 4.1 for KRA and to the *upper* panel of this Figure for THK. We keep the same range as in the other panel to facilitate the comparison

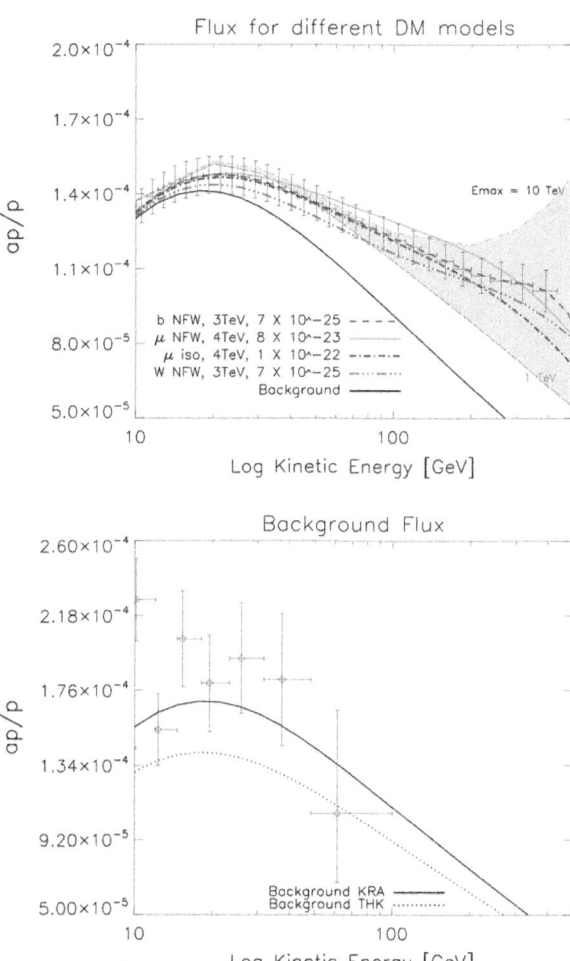

simplicity a Gaussian distribution. Figure 4.3 shows results for the four DM models in Table 4.2 whose propagation follows the KRA prescription. We see that for all annihilation channels (b, μ, W) there can be degeneracy between the corresponding DM model and SNR flux. A point in the grey region indicates that for those choice of ξ, E_{max} the SNR flux is compatible (and therefore degenerate) with mock data based on a DM hypothesis at 5σ. In particular, lower values of E_{max} allow for a larger degeneracy in all cases investigated here. The b- and W-channels seem to prefer larger values of ξ (with relative minimum at the edge of the grid) while the μ-channel has a minimum χ^2 for lower values of ξ. Notice though that the tendency towards lower values of ξ disappears when we change DM profile (Fig. 4.3, panel c) or when we change the propagation model, as in (Fig. 4.4, panel b). The values of the minimal χ^2 and number of degrees of freedom for all cases is shown in Table 4.3 for all models considered in the analysis.

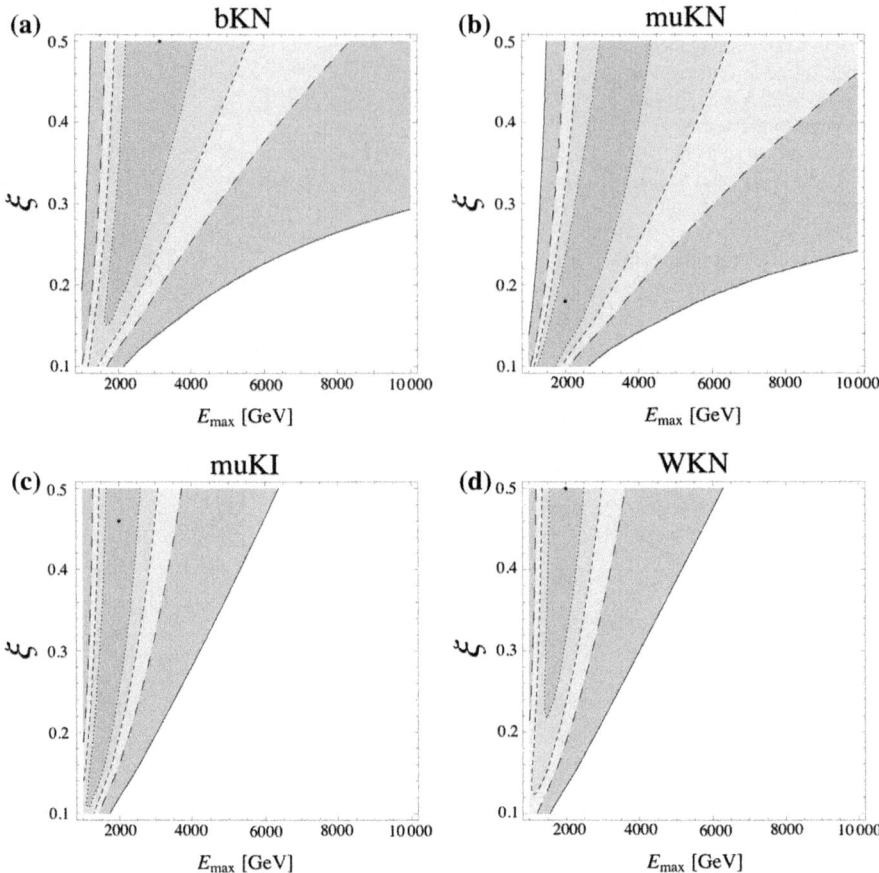

Fig. 4.3 Confidence contours for different DM models with propagation KRA. The names of the models refer to the ones given in Table 4.2. Colours indicate 1, 2, 3, 5 σ contours. The *black dot* corresponds to the minimum χ^2 value (relative minimum within the chosen grid)

There is indication that some portion of parameter space might be excluded by data on boron to carbon ration, as shown in [23]. However, we cannot make a direct comparison with the results of this paper because of a different choice of parameters. In particular our case corresponds indeed to $n_{gas} = 2\,\mathrm{cm}^{-3}$, $B = 1\,\mu G$ and $v = 0.5 \times 10^{-8}\,\mathrm{cm/s}$, which can be compared with Fig. 4.3 of their analysis (upper panel) for $K_B = 20$. We are however fixing r = 3.22 as explained in our Sect. 4.2 for consistency with the background spectrum. The paper [23] uses instead r = 4.

Finally, we have investigated the degeneracy following the inverse logic with respect to the analysis done so far; instead of assuming a DM benchmark model and test whether we can find a combination of (ξ, E_{max}) that fit our mock data, we reversed the procedure: we first produced a set of mock AMS-02 data through a benchmark SNR model and asked if these data could be interpreted as originated

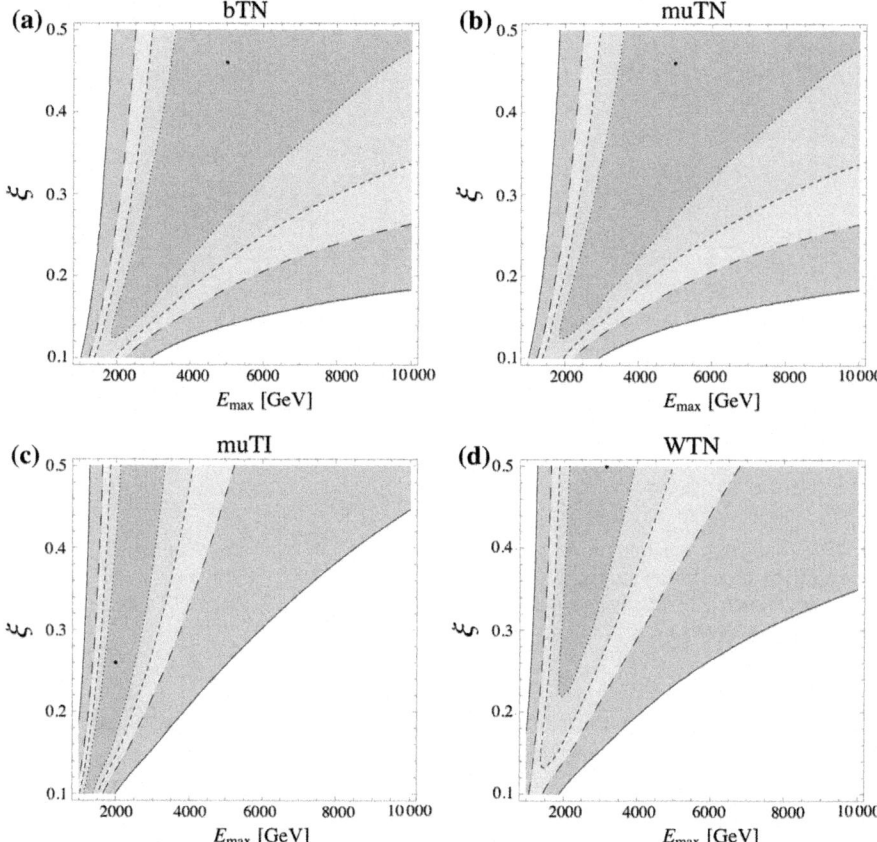

Fig. 4.4 Confidence contours for different DM models with propagation THK. The names of the models refer to the ones given in Table 4.2. Colours indicate 1, 2, 3, 5 σ contours. The *black dot* corresponds to the minimum χ^2 value (relative minimum within the chosen grid)

Table 4.3 χ^2 values for the models considered in this analysis. In all cases the number of degrees of freedom is N = 30 (data points) - 2 (parameters) = 28	Name	Minimum χ^2
	bKN	6.1
	muKN	6.3
	muKI	6.7
	WKN	21.0
	bTN	5.6
	muTN	5.6
	muTI	8.6
	WTN	8.6

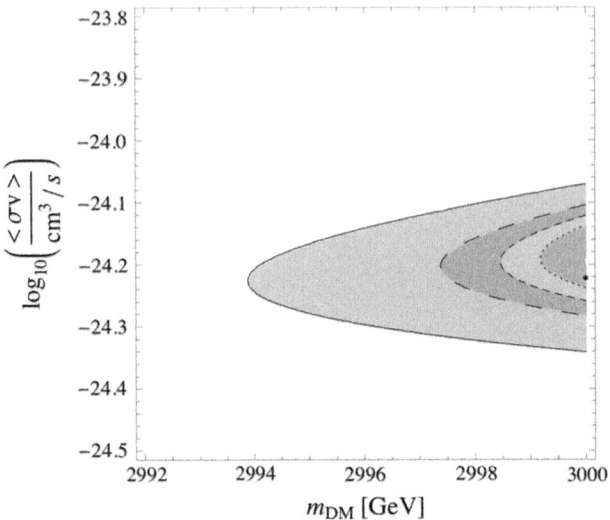

Fig. 4.5 Confidence contours in the parameter space ($\langle \sigma v \rangle$, M), for the $b\bar{b}$ annihilation channel and with KRA propagation model, as obtained fixing $E_{max} = 2500$ GeV and $\xi = 0.14$ in the SNR benchmark model. The (relative) minimum χ^2 within the grid for this case is 9.1 for 28 degrees of freedom. Colours indicate 1, 2, 3, 5 σ contours

from DM models (using the same propagation model). As expected, also in this case it is possible to find some degeneracy. In Fig. 4.5 we show an example of such a degeneracy, which, for the chosen SNR benchmark model and DM annihilation channel, peaks around a very small range in mass. This is in agreement with the value found in model bKN. The extension of the degeneracy does not vary much with the annihilation channels.

4.6   AMS-02 Antiprotons Data

Actual antiproton data from AMS-02 were presented first in preliminary form in [6], and then published in [24]. The very existence of an excess in these data is controversial. The analyses of [25–27], although different in the assumptions about parameters for the diffusion of cosmic rays in the galaxy, the estimates of the error of the primary slopes of the proton and α particles fluxes (used to calibrate the proton background) and on the effect of solar modulation on the expected flux, all agree on the fact that data are compatible with the background expectation within uncertainties (even if their plots all show some upward fluctuation). These analyses were performed comparing the preliminary data of [6] with the background obtained by fitting the updated AMS-02 data on the boron over carbon ratio.

More recently, Refs. [28–30] claimed the presence an excess in the antiproton data of [24], which could even be compatible with a DM explanation of the Fermi galactic centre excess discussed in Sect. 3.5.4. While the analysis in [28, 30] are not based on B/C data for estimating the backgrounds, therefore differing from [25–27], the one in [29] relies on B/C, but uses a different statistical treatment, which goes beyond the scope of this presentation.

To conclude, it is fair to say that a careful analysis of AMS-02 antiproton data is paramount is order to confirm or exclude the presence of an excess, whatever its nature is. As far as a consensus on this point is not reached, it is probably better to keep a conservative point of view, even if the possibility of a deviation from the expected flux is extremely exciting, whether this is due to DM or to new astrophysical phenomena.

4.7 Conclusions

Finding indirect signatures of DM is certainly one of the main targets of many current experimental efforts. Nevertheless, even in the optimistic case in which a signal above the expected background is found, the most pressing question is whether such a signal can be ascribed to DM annihilation (or decay) beyond any reasonable doubt. This is a legitimate question as there are astrophysical sources which can mimic a signal, the best example being pulsars which can generate a positron excess. In this chapter we have investigated this degeneracy problem focussing our attention on the antiproton signal. Indeed, antiprotons may be generated as secondaries accelerated in supernova remnants and we have shown that a potential signal from DM annihilation can be mimicked by such an astrophysical source.

References

1. M. Cirelli, Indirect searches for dark matter: a status review. Pramana **79**, 1021–1043 (2012), arXiv:1202.1454
2. A. De Simone, A. Riotto, W. Xue, Interpretation of AMS-02 results: correlations among dark matter signals, JCAP **1305**, 003 (2013), arXiv:1304.1336
3. N. Fornengo, L. Maccione, A. Vittino, Constraints on particle dark matter from cosmic-ray antiprotons, JCAP **1404**(04), 003 (2014), arXiv:1312.3579
4. P. Ciafaloni, D. Comelli, A. Riotto, F. Sala, A. Strumia, A. Urbano, Weak corrections are relevant for dark matter indirect detection, JCAP **1103**, 019 (2011), arXiv:1009.0224
5. V. Pettorino, G. Busoni, A. De Simone, E. Morgante, A. Riotto, W. Xue, Can AMS-02 discriminate the origin of an anti-proton signal?, JCAP **1410**(10), 078 (2014), arXiv:1406.5377
6. A. Kounine, Talk at AMS days at CERN, 2015, http://indico.cern.ch/event/381134/contributions/900587
7. P. Blasi, The origin of the positron excess in cosmic rays, Phys. Rev. Lett. **103** 051104, (2009), arXiv:0903.2794
8. P. Blasi, P.D. Serpico, High-energy antiprotons from old supernova remnants. Phys. Rev. Lett. **103**, 081103 (2009), arXiv:0904.0871

9. M. Kachelriess, S. Ostapchenko, R. Tomas, Antimatter production in supernova remnants, Astrophys. J. **733**, 119 (2011), arXiv:1103.5765
10. L.C. Tan, L.K. Ng, Parametrization of anti-p invariant cross-section in p p collisions using a new scaling variable. Phys. Rev. **D26**, 1179–1182 (1982)
11. L.C. Tan, L.K. Ng, Calculation of the equilibrium anti-proton spectrum. J. Phys. **G9**, 227–242 (1983)
12. T. Bringmann, P. Salati, The galactic antiproton spectrum at high energies: Background expectation vs. exotic contributions, Phys. Rev. **D75**, 083006 (2007), arXiv:0612514
13. C. Evoli, I. Cholis, D. Grasso, L. Maccione, P. Ullio, Antiprotons from dark matter annihilation in the Galaxy: astrophysical uncertainties, Phys. Rev. **D85** 123511(2012), arXiv:1108.0664
14. C. Evoli, D. Gaggero, D. Grasso, L. Maccione, Cosmic-ray nuclei, antiprotons and gamma-rays in the galaxy: a new diffusion model, JCAP **0810**, 018 (2008), arXiv:0807.4730. [Erratum: JCAP1604,no.04,E01(2016)]
15. M. Cirelli, G. Giesen, Antiprotons from dark matter: current constraints and future sensitivities, JCAP **1304**, 015(2013), arXiv:1301.7079
16. **PAMELA** Collaboration, O. Adriani et al., PAMELA results on the cosmic-ray antiproton flux from 60 MeV to 180 GeV in kinetic energy, Phys. Rev. Lett. **105**, 121101(2010), arXiv:1007.0821
17. **PAMELA** Collaboration, O. Adriani et al., An anomalous positron abundance in cosmic rays with energies 1.5–100 GeV, Nature **458**, 607–609(2009), arXiv:0810.4995
18. A.M.S. Collaboration, M. Aguilar et al., First result from the alpha magnetic spectrometer on the international space station: precision measurement of the positron fraction in primary cosmic rays of 0.5–350 GeV. Phys. Rev. Lett. **110**, 141102 (2013)
19. **Fermi-LAT** Collaboration, M. Ackermann et al., Dark matter constraints from observations of 25 milky way satellite galaxies with the fermi large area telescope, Phys. Rev. **D89**, 042001(2014), arXiv:1310.0828
20. **Fermi-LAT** Collaboration, M. Ackermann et al., Constraints on the galactic halo dark matter from fermi-lat diffuse measurements, Astrophys. J. **761**, 91(2012). arXiv:1205.6474
21. S. Ting, Slides of the talk at spacepart12, 5–7 november 2012, CERN. http://doi.org/indico.cern.ch/event/197799/page/1
22. M. Pato, D. Hooper, M. Simet, Pinpointing cosmic ray propagation with the AMS-02 experiment, JCAP **1006**, 022(2010), arXiv:1002.3341
23. I. Cholis, D. Hooper, Constraining the origin of the rising cosmic ray positron fraction with the boron-to-carbon ratio, Phys. Rev. **D89**(4), 043013 (2014), arXiv:1312.2952
24. **AMS** Collaboration, M. Aguilar et al., Antiproton flux, antiproton-to-proton flux ratio, and properties of elementary particle fluxes in primary cosmic rays measured with the alpha magnetic spectrometer on the international space station, Phys. Rev. Lett. **117**(9), 091103 (2016)
25. G. Giesen, M. Boudaud, Y.Génolini, V. Poulin, M. Cirelli, P. Salati, P.D. Serpico, AMS-02 antiprotons, at last! Secondary astrophysical component and immediate implications for dark matter, JCAP **1509**(09), 023 (2015), arXiv:1504.04276
26. C. Evoli, D. Gaggero, D. Grasso, Secondary antiprotons as a galactic dark matter probe, JCAP **1512**(12), 039(2015), arXiv:1504.05175
27. R. Kappl, A. Reinert, M.W. Winkler, AMS-02 antiprotons reloaded, JCAP **1510**(10), 034 (2015), arXiv:1506.04145
28. A. Cuoco, M. Krämer, M. Korsmeier, Novel dark matter constraints from antiprotons in light of AMS-02, Phys. Rev. Lett. **118**(19), 191102 (2017), arXiv:1610.03071
29. M.-Y. Cui, Q. Yuan, Y.-L. S. Tsai, Y.-Z. Fan, Possible dark matter annihilation signal in the AMS-02 antiproton data, Phys. Rev. Lett. **118**(19), 191101 (2017), arXiv:1610.03840
30. A. Cuoco, J. Heisig, M. Korsmeier, M. Krämer, Probing dark matter annihilation in the galaxy with antiprotons and gamma rays, arXiv:1704.08258

Part II
LHC Searches

Chapter 5
Dark Matter Searches at the LHC

5.1 Searches at a Hadron Collider

The Large Hadron Collider (LHC) is the world's largest and most powerful particle accelerator, located at CERN. It consists of a 27-km ring of superconducting magnets with a number of accelerating structures to boost the energy of the particles. Four main experiments are located along the tunnel, two of which (ATLAS and CMS) are general-purpose experiments built in order to find signals of new physics in the analysis of proton-proton and Pb-Pb collisions. In particular, one of the most important goals of the LHC is to shed new light on the DM puzzle, possibly with the long-awaited discovery of WIMP particles.

Protons accelerated by the LHC are currently colliding with a total energy in the centre of mass (c.o.m.) frame of 13 TeV. At these energies, protons can not be considered as elementary particles, but reveal their nature of composite states made of quarks and gluons. In first approximation, hadrons probed at large transferred momentum $\vec{p}$ behave as if they consist of "partons" of various types i, each with a probability $f_i(x)\mathrm{d}x$ of having a momentum between $x\vec{p}$ and $(x + \mathrm{d}x)\vec{p}$. This is the so-called "parton model", first introduced by Feynman in 1969 [1]. The parton index i runs over all particles present in the spectrum, and the functions f_i (called "parton distribution functions", or PDF) depend on x and on the momentum Q^2 exchanged in the interaction. Intuitively enough, gluons and u, d quarks are the ones with the highest PDF. In order of importance, after gluons and "valence" quarks we find the "sea" quarks, i.e. quarks and antiquarks that may be produced by SM interactions inside the proton. At the LHC energy scale, all other quarks but the top have non negligible PDFs. A plot of PDFs for two values of the exchanged momentum Q^2 is shown in Fig. 5.1.

Two important consequences of the parton picture of hadrons must be stressed. The first is that, when hadrons collide at high energies, a number of sub-interactions take place among partons. What typically happens is that a pair of partons undergo

© Springer International Publishing AG 2017
E. Morgante, *Aspects of WIMP Dark Matter Searches at Colliders and Other Probes*, Springer Theses, https://doi.org/10.1007/978-3-319-67606-7_5

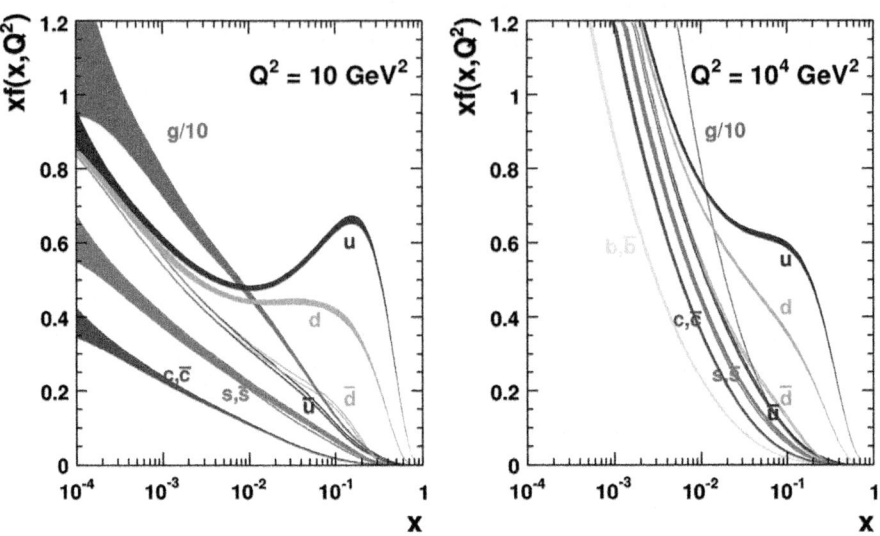

Fig. 5.1 Plots of $xf_i(x)$ for $Q^2 = 10, 10^4$ GeV^2 for all relevant partons from the MSTW 2008 PFS set [2]

a hard scattering, while the remaining ones produce a shower of hadronic particles collimated along the original axes of the incoming particles (in the c.o.m. frame), forming what is called the underlying event [3]. The second consequence is that the c.o.m. energy of the hard process and its momentum are not fixed, but depend on the fraction of the proton's energy carried by each of the two partons. This means that the energies of the particles produced in the hard scattering process (which are the ones one is typically interested in) do not sum to the initial energy. On the other hand, the transverse (with respect to the proton's velocity) component of the momentum carried by each parton is small and can be neglected.[1] As a consequence of this, the vectorial sum of the transverse momenta of the particles produced in the hard process sum up to 0. This consideration leads us to the definition of the *missing transverse energy* (denoted as $\not{E}_T$ or MET) which is one of the fundamental quantities for hadron colliders:

$$\not{E}_T = \left| \sum_{\text{visibles}} \vec{p}_T \right| \tag{5.1}$$

---

[1]The transverse momentum in boosted frame is equal to the one in the rest frame. This can be estimated by means of the uncertainty principle: for the proton to be a bound state with radius $\Delta x \sim$ 1 fm, the corresponding momentum is of the order of $\Delta p \sim \hbar/(2\Delta x) \sim 0.2$ GeV. Equivalently, one can require that the momentum does not exceed the scale Λ_{QCD}, in order for the protons to be confined.

where the sum is performed over all the particles produced in the hard process which are detected by means of their charged track or by an energy deposition in a calorimeter (visible particles). The sum do not include the invisible particles (which would simply be impossible from an operational point of view). This means that, when for example neutrinos are produced, the missing transverse energy is different from 0.

We can now start to discuss WIMP searches. At the LHC, a WIMP is not different from a neutrino, as far as its detection is concerned. Being only weakly interacting with SM particles, once produced WIMPs escape the detector without further scatterings, leaving as a signature a certain amount of missing energy. On its own, missing energy is not enough for an event to be registered by the detector, and at least one recoiling SM particle is necessary. The prototypical DM searches are then the so-called "mono-X plus MET" searches, which look at events in which a single SM object (a hadronic jet [4], a photon [5], a EW boson [6, 7], a heavy quark [8] or even a Higgs boson [9, 10]) recoils against a large amount of missing transverse energy. A different kind of signature is the one which involves the production of a "partner" particle, which may be the particle mediating the interaction of the DM with quarks and/or gluons, which then decays to DM and SM particle often leaving a rather complex signature. Extremely important in the context of simplified DM models (that we are going to discuss in the next chapter) are di-jet searches. A pair of hadronic jets may be produced in $pp \to V \to qq$ where the particle V exchanged in the s-channel is the one that mediates the DM-quarks interaction, or in a process like $pp \to \tilde{q}\tilde{q} \to q\chi q\chi$, where $\tilde{q}$ are t-channel coloured mediators that can be produced on shell and decay to quarks and DM particles. the first process typically leads to a resonant di-jet signal, while the second results in a di-jet + MET event.

Generically speaking, there are two categories of backgrounds for LHC DM searches. The first is a non-physical background due to a wrong determination of (because of uncertainties in the measurements, particles falling into non equipped areas of the detectors, etc.). This is called a reducible background, because its effect can be softened by, for example, a better knowledge of the detector's properties. There are then physical backgrounds, due to the production of neutrinos. This can be both reducible (as in the case of the process $W \to \ell\nu$, where ℓ is a charged lepton that can be used in order identify the process and distinguish it from a WIMP production event) or irreducible (as $Z \to \nu\nu$). Background estimation is a very important task, and requires a precise knowledge of the underlying SM processes and of the detector performance. We will shortly describe the estimation of the backgrounds in the example of mono-jet searches that discussed in Sect. 5.2.

To conclude this section, let us add that the best search strategy to constrain DM properties is not uniquely determined, but it depends on the properties of the model which is under examination. Let us take two extreme examples, to which we will return in Sect. 5.3. In the case in which DM is considered as a component of a more general theory, like various realizations of Supersymmetry, extremely refined analyses can be performed which exploit complex decay chains and *ad hoc* kinematic variables in order to reduce the background contamination to the minimum. On the other hand, when the point of view of being as model independent as possible is assumed and the DM-SM interaction is parametrized by simple effective operators, only a few search strategies remain possible, and the mono-X plus MET searches (especially mono-jet) give the best results. The experimental status of DM searches at the LHC after Run I is summarized in [11].

5.2 Mono-Jet Searches

As an example of how DM searches are performed at the LHC, we will here describe in some details a recent mono-jet analysis by the ATLAS collaboration [4], performed with $32\,\text{fb}^{-1}$ of data at $\sqrt{s} = 13\,\text{TeV}$. The ATLAS detector system consists of a inner tracking detector immersed in a $2\,\text{T}$ magnetic field, which reconstructs the curved track left by a charged particle that traverses it. Externally to the tracker there are an electromagnetic and a hadronic calorimeters, which measure the energy of $e^{\pm}$ and γ the first, and hadronic particles the second. Finally, the most external layer consists of muon chambers, again immersed in a large magnetic field, which reveal the track of muons produced in the interaction point (and serve as a veto for cosmic ray muons). A similar description also applies to CMS. As for the signature of SM particles, photons are detected by their energy deposition in the electronic calorimeter. The same happens for $e^{\pm}$, which being charged also leave a track in the inner tracker. Muons are also charged particles, but being massive are not absorbed by the electromagnetic calorimeter and escape the detector, leaving a trace in the inner tracker and in the muon chambers. Taus decay promptly, mostly in hadronic final states. The typical signature of a tau lepton is a slightly displaced well collimated jet, with low track multiplicity and a sizeable energy deposition in the electromagnetic calorimeter due to photons from π^0 decays. Light quarks (u, d, s, c) and gluons, being coloured, can not exist as asymptotic states. When they are produced they fragment into a shower of other coloured states, which then immediately hadronize forming colourless particles that in turn are detected in the form of jets. Jets include neutral and charged particles, both hadrons and leptons, and each component is seen with the suitable detector. In order for a particle to be associated with a jet, complex algorithms are used based on the definition of proper "distances" between their energy depositions, such as the anti-k_t one [12]. Heavy quarks are particular with respect to lighter ones. Bottom quarks form long-lived B mesons, which decay at a distance of a few millimetres from the interaction point, leaving displaced vertex tracks. This provides a tool to discriminate jets originated by b quarks (the so called "b-tagging procedure").[2] Top quarks decay into b quarks and W, and their typical

---

[2]Two are the characteristics of b-hadrons that are exploited in the so-called "b-tagging" algorithm: they are long lived, with a mean life of $\sim 1.6\,\text{ps}$, and that they are massive ($m_{B^{\pm}} \simeq 5.28\,\text{GeV}$). The first property allows them to travel a significant distance before decaying: for example, a b-quark with $p_T = 50\,GeV/c$ will fly on average almost half a centimeter. The mass affects the opening angle between the daughter particles: since it is large, the tracks associated with the charged decay products have a sizeable impact parameters with respect to the interaction point. Top quarks decay 100% into bW^+, and are therefore identified as a b-tagged jet plus lepton traces.

signature is therefore a b-tagged jet with leptons and missing energy from the decay $W \to \ell\nu$. Finally, neutrinos do not interact with the detector, and leave large amounts of missing energy, just as DM particles are supposed to do.

The correct choice of coordinates to describe outgoing particles in the detector is (ϕ, η, p_T). The azimuthal angle ϕ is defined around the beam pipe, and it only relevant when differences $\Delta\phi$ are measured between tracks (the absolute value is irrelevant). The pseudo-rapidity η is defined in terms of the polar angle with respect to the beam pipe as $\eta = -\ln[\tan(\theta/2)]$. Its importance relies on the fact that it is a good approximation of the rapidity of a particle $y = \ln[(E + p_z)/(E - p_z)]/2$, a quantity which is addictive under boosts along the axes direction z.

The first point to be addressed in a search is the simulation with Monte Carlo (MC) tools of the signal and of the background. The model considered in [4] is a simplified model in which a Dirac DM particle interacts with quarks via the exchange of a vector mediator, with axial couplings to both the DM and quarks. The resonance is assumed to have a Breit-Wigner shape, with the width fixed to its minimal value given by tree-level decays of the mediator into quarks and the DM. Coupling constants are fixed to $g_\chi = 1$ and $g_q = 0.25$, while the DM and the mediator mass are varied in the ranges $1\,\text{GeV} - 1\,\text{TeV}$ and $10\,\text{GeV} - 2\,\text{TeV}$ respectively. A detailed simulation of the signal requires NLO precision, as discussed in [13].

The main background for the $\chi\chi$ + jet signal comes predominantly from the processes $Z(\to \nu\bar{\nu})$ + jets, and W + jets (mostly $W(\to \tau\nu)$ + jets). Smaller contributions are given by $Z/\gamma^*(\to \ell^+\ell^-)$ + jets, multi-jet, $t\bar{t}$, single-top and diboson processes. These processes are simulated with MC tools, and their normalization is fixed by a fit with data.

Events are selected according to their MET, the nature of particles present and according the kinematic of these particles, in order to maximize the signal/background ratio. The criteria applied in [4] are:

• A primary vertex have to be reconstructed with at least two associated tracks with $p_T > 0.4\,\text{GeV}$.
• The missing transverse energy is required to be $\not{E}_T > 250\,\text{GeV}$. Events with large MET are less likely to be originated by mismeasurement of the jet energy. More over, a separation of $\Delta\phi > 0.4$ is required between the direction of the missing momentum $\vec{\not{p}}_T$ and the jet direction, in order to further reduce the multi-jet background in which one of the jets is badly measured.
• The leading jet (i.e. the one with the highest p_T) must have $p_T > 250\,\text{GeV}$ and $|\eta| < 2.4$ (corresponding to $\theta \gtrsim 10°$). Additional jets are allowed, up to a maximum of 4 jets with $p_T > 30\,\text{GeV}$ and $|\eta| < 2.8$. The requirement on η is necessary due to the poorer resolution in the very forward region, part of which can not be equipped because of the presence of the beam pipe.
• Additional constraints are imposed on the jets (tighter on the leading one) in order to reject all events in which a jet is not genuinely generated in proton-proton collisions but as a consequence of spurious effects in the detector, following [14, 15].

- Events in which muons with $p_T > 10\,\text{GeV}$ or electrons with $p_T > 20\,\text{GeV}$ are vetoed.

Since the efficiency of the search will depend in general on the p_T of the leading jet, events are grouped into "signal regions" according to that value. Signal and background estimation are done separately for each signal region. Both "inclusive" and "exclusive" signal regions are defined:

Inclusive signal region	IM1	IM2	IM3	IM4	IM5	IM6	IM7
$\not{E}_T$ (GeV)	> 250	> 300	> 350	> 400	> 500	> 600	> 700
Exclusive signal region	EM1	EM2	EM3	EM4	EM5	EM6	
$\not{E}_T$ (GeV)	250–300	300–350	350–400	400–500	500–600	600–700	

A large number of effects induce systematic uncertainties both in the background and in the signal estimations, including uncertainties in the energy scale of jets and $\not{E}_T$, variation of the renormalization, factorization and parton-shower matching scales, uncertainties induced by the background normalization fixing with respect to data, uncertainties in the top production cross sections, and uncertainty in the total luminosity. The total uncertainty in the background prediction goes up to $\mathcal{O}(10\%)$, depending on the signal region considered. The uncertainty in the signal and in the acceptance are somewhat larger: those related to the jet and $\not{E}_T$ scale and resolution vary between $1 - 3\%$; modelling of initial and final state radiation introduce a 20% uncertainty in the acceptance; the choice of different PDF sets results in a 20% uncertainty in the acceptance and 10% in the cross section; the choice of the normalization and factorization scales introduce another 3% uncertainty in the acceptance and 5% in the cross section; finally, a 5% uncertainty is due to the measure of the integrated luminosity. When summed up in quadrature, the total uncertainty in the signal estimation is $\sim 37\%$.

The good agreement found in data with SM predictions is translated into limits on the number of WIMP production events and, in turn, on the visible cross section for the process $pp \rightarrow \bar{\chi}\chi + \text{jets}$, defined as the product of the actual cross section times the acceptance times the efficiency $\sigma \times A \times \epsilon$ (Table 5.1). For WIMP pair production, the product $A \times \epsilon$ ranges from 25% for IM1 to 2% for IM7.

Figure 5.2 (left) shows the observed and expected 95% CL exclusion limits in the m_χ–m_A parameter plane, with $\pm 1\sigma$ theoretical uncertainties in the signal cross sections. For low m_χ the WIMP production cross section is resonantly enhanced, and mediator masses up to 1 TeV are excluded. For heavier DM the decay into a pair of WIMPs is kinematically suppressed, and the analysis loses its sensitivity. Perturbative unitarity is violated in the parameter region defined by $m_\chi > \sqrt{\pi/2}\, m_A$ [16]. The masses corresponding to the correct relic density as measured by the Planck and WMAP satellites [17, 18], in the absence of any interaction other than the one considered, are indicated in the figure as a line that crosses the excluded region at $m_A \sim 880$ and $m_\chi \sim 270\,\text{GeV}$. The region towards lower WIMP masses or higher mediator masses corresponds to dark matter overproduction.

Table 5.1 Observed and expected 95% CL upper limits on the number of signal events, S_{obs}^{95} and S_{exp}^{95}, and on the visible cross section, defined as the product of cross section, acceptance and efficiency, $\langle \alpha \varepsilon \rangle_{obs}^{95}$, for the IM1–IM7 selections

Signal channel	$\langle \alpha \varepsilon \rangle_{obs}^{95}$ [fb]	S_{obs}^{95}	S_{exp}^{95}
IM1	553	1773	1864_{-548}^{+829}
IM2	308	988	1178_{-348}^{+541}
IM3	196	630	694_{-204}^{+308}
IM4	153	491	401_{-113}^{+168}
IM5	61	196	164_{-45}^{+63}
IM6	23	75	84_{-23}^{+32}
IM7	19	61	48_{-13}^{+18}

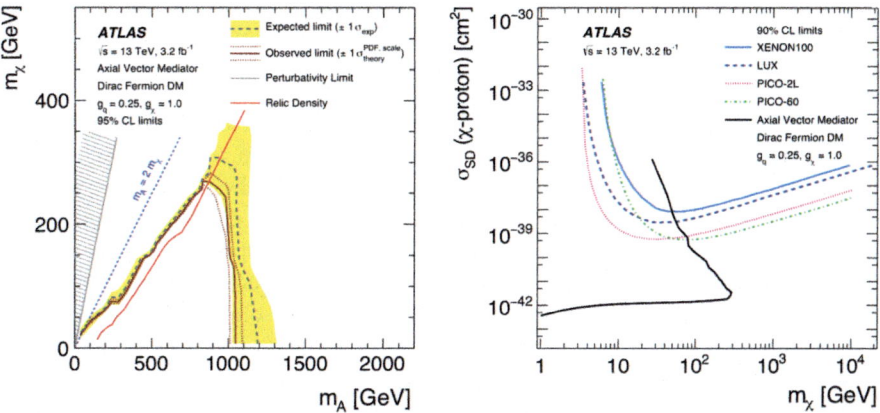

Fig. 5.2 *Left* 95% CL exclusion contours in the m_χ–m_A plane, from [4]. The solid (*dashed*) curve shows the median of the observed (expected) limit, while the bands indicate the $\pm 1\sigma$ theory uncertainties in the observed limit and $\pm 1\sigma$ range of the expected limit in the absence of a signal. The *red curve* corresponds to the expected relic density. The region excluded due to perturbativity, defined by $m_\chi > \sqrt{\pi/2}\, m_A$, is indicated by the hatched area. *Right* A comparison of the inferred limits to the constraints from direct detection experiments on the spin-dependent WIMP–proton scattering cross section in the context of the Z'-like simplified model with axial-vector couplings. Unlike in the m_χ–m_A parameter plane, the limits are shown at 90% CL. The results from this analysis, excluding the region to the *left* of the contour, are compared with limits from the XENON100 [21], LUX [22], and PICO [23, 24] experiments. The comparison is model-dependent and solely valid in the context of this model, assuming minimal mediator width and the coupling values $g_q = 1/4$ and $g_\chi = 1$

In Fig. 5.2 (right) the results are translated into 90% CL exclusion limits on the spin-dependent WIMP–proton scattering cross section as a function of the WIMP mass, following the prescriptions explained in Refs. [19, 20]. Results from direct-detection experiments XENON100 [21], LUX [22], and PICO [23, 24] are also shown for comparison. For WIMP masses up to $300\,\mathrm{GeV}$, limits on the scattering cross section of the order of $10^{-42}\mathrm{cm}^2$ can be inferred, roughly 3 orders of magnitude stronger than direct detection one, and extend to lower DM mass (below $10\,\mathrm{GeV}$). The loss of sensitivity in models where WIMPs are produced off-shell is expressed by the turn of the exclusion line, reaching back to low WIMP masses and intercepting the exclusion lines from the direct-detection experiments at around $m_\chi = 80\,\mathrm{GeV}$.

5.3   Theoretical Tools

In order to interpret the cross section limits obtained from the LHC $\not{E}_T$ searches, and to relate these bounds to the constraints that derive from direct and indirect detection, one needs a model of DM. There are a large number of qualitatively different possible models, that populate in principle the full "theory space" of the theories beyond the SM which include a particle that is a viable WIMP DM candidate. In the perspective of DM searches, the latter may be organized in three different classes, in order of increasing complication (and completeness):

1. The simplest class consists of theories where the DM is the only accessible state to our experiments. In such a case, the DM-SM interaction mediated by inaccessible particles is described in a universal way by a set of effective operators with mass dimension > 4. To obtain a Lagrangian term with the right dimensionality, a mass scale Λ must be introduced, and in this minimal approach this is the only new parameter apart from the DM mass. The EFT approach has proven to be very useful in the analysis of LHC Run I data [25–40] despite of its intrinsic limitations: in fact, effective theories are a valid description only when the relevant energy scale of the experiment is lower than the cut-off scale Λ. This poses a serious threat to the usage of EFT for DM searches at the LHC, as was shown in [27, 29, 31, 34, 35, 37, 39, 41–46]: Chap. 6 will be devoted to discuss this issue and the possible ways out.
2. A step beyond EFT is the class of DM simplified models. The idea of simplified models was firstly adopted in the context of Supersymmetry searches, as a way to grasp the most relevant features of different SUSY models which have similar signatures at colliders [47–49]. Such models are characterized by the most important state mediating the interaction of the DM particle with the SM, as well as the DM particle itself (see for example [33, 39, 50–53] for early proposals). Including the effect of the mediator's propagator allows to avoid the energy limitation of the EFT, and the simplified models are able to describe correctly the full kinematics of DM production at the LHC, at the price of an increased number of parameters. The effective scale Λ is traded in for the mass of the mediator and a handful of coupling constants, which poses important questions about the best way to constrain the parameter space and to present results. An additional advantage of simplified models over the EFT approach is that they allow to exploit the complementarity between different LHC searches, such as searches for nar-

row resonances in the di-jet channel or di-jet + MET searches. The approach of simplified models will be described in greater detail in Chap. 7.

3. On the other end of the spectrum, there are full BSM theories which include a DM candidate, the prototypical example being the Minimal Supersymmetric SM (MSSM). It is quite reductive to refer to those models as "DM models". Theories of this class have been built over the last decades inspired by the Naturalness problem of the SM (or the strong CP problem in the case of axion DM), and they may account for all phenomena up to a very high energy scale, typically much higher than what is or may be testable at present or future colliders. Reasonable phenomenological models of this kind have a large number of new parameters, leading to varied visions of DM, and it is typically hard to constrain the properties of DM itself without making any further assumption. On the other hand, symmetry-enforcing relations among couplings in this class of models result in important correlations between otherwise unrelated observables, which may have important phenomenological consequences.

From the point of view of LHC searches, the enlarged physical spectrum and parameter space of full new physics theories with respect to simplified models, and of the latter with respect to the EFT, lead to a greater variety of possible search channels. While within the EFT approach the mono-X searches give the best sensitivity, simplified models of DM can be constrained also with multi-jet + MET searches, with di-jet resonance searches and possibly others, depending on the degree of sophistication of the model. The possibilities are even richer with full theories. In SUSY searches, for example, cascade decays to the LSP with the production of SM particles lead to a huge variety of experimental signatures (see [54] for a review of the status of SUSY searches after Run I). Of course, the large number of parameters requires an even larger number of searches to be constrained and to avoid degeneracies [55].

To summarize, on the good side the additional degrees of freedom in going from the EFT to simplified DM models and to full theories allow to put limits on the DM properties by exploiting new search channels and the complementarity with other experimental searches; on the bad side, it involves more model dependence and requires care in the choice of the parameters and in the presentation of results.

All the three scenarios described above are interesting and must be considered as potential sources of new understanding on DM physics. It is extremely important that, in the future, all these tools keep being used in order to exploit the full LHC potential as a DM discovery machine.

In the following chapters, firstly we are going to discuss the EFT approach and the issues about its validity, then we will turn to a detailed description of simplified models, and finally we are going to highlight the role of the relic density as a tool to get information on the parameter space of simplified models.

References

1. R.P. Feynman, Very high-energy collisions of hadrons. Phys. Rev. Lett. **23**, 1415–1417 (1969)
2. A.D. Martin, W.J. Stirling, R.S. Thorne, G. Watt, Parton distributions for the LHC. Eur. Phys. J. C **63**, 189–285 (2009), arXiv:0901.0002
3. C.D.F. Collaboration, T. Affolder et al., Charged jet evolution and the underlying event in $p\bar{p}$ collisions at 1.8 TeV. Phys. Rev. D **65**, 092002 (2002)
4. ATLAS Collaboration, M. Aaboud et al., Search for new phenomena in final states with an energetic jet and large missing transverse momentum in pp collisions at $\sqrt{s} = 13$ TeV using the ATLAS detector, Phys. Rev. D **94**(3), 032005 (2016), arXiv:1604.07773
5. ATLAS Collaboration, M. Aaboud et al., Search for new phenomena in events with a photon and missing transverse momentum in pp collisions at $\sqrt{s} = 13$ TeV with the ATLAS detector, JHEP **06** 059 (2016), JHEP 059, 1606 (2016), arXiv:1604.01306
6. ATLAS Collaboration, G. Aad et al., Search for Dark Matter in events with a hadronically decaying W or Z boson and missing transverse momentum in pp collisions at $\sqrt{s} = 8$ TeV with the ATLAS detector, Phys. Rev. Lett. **112**(4) 041802 (2014), arXiv:1309.4017
7. ATLAS Collaboration, Search for Dark Matter produced in association with a hadronically decaying vector boson in pp collisions at $\sqrt{s} = 13$ TeV with the ATLAS detector at the LHC, ATLAS-CONF-2015-080
8. ATLAS Collaboration, G. Aad et al., Search for Dark Matter in events with heavy quarks and missing transverse momentum in pp collisions with the ATLAS detector, Eur. Phys. J. **C75**(2) 92 (2015), arXiv:1410.4031
9. ATLAS Collaboration, G. Aad et al., Search for Dark Matter produced in association with a Higgs boson decaying to two bottom quarks in pp collisions at $\sqrt{s} = 8$ TeV with the ATLAS detector, Phys. Rev. **D93**(7), 072007 (2016), arXiv:1510.06218
10. ATLAS Collaboration, Search for Dark Matter in association with a Higgs boson decaying to *b-quarks in pp* collisions at $\sqrt{s} = 13$ TeV with the ATLAS detector, ATLAS-CONF-2016-019
11. A. Askew, S. Chauhan, B. Penning, W. Shepherd, M. Tripathi, Searching for Dark Matter at hadron colliders. Int. J. Mod. Phys. A **29**, 1430041 (2014), arXiv:1406.5662
12. M. Cacciari, G.P. Salam, G. Soyez, The anti-k(t) jet clustering algorithm. JHEP **04**, 063 (2008), arXiv:0802.1189
13. U. Haisch, F. Kahlhoefer, E. Re, QCD effects in mono-jet searches for Dark Matter. JHEP **12**, 007 (2013), arXiv:1310.4491
14. ATLAS Collaboration, Selection of jets produced in 13TeV proton-proton collisions with the ATLAS detector, ATLAS-CONF-2015-029
15. ATLAS Collaboration, G. Aad et al., Characterisation and mitigation of beam-induced backgrounds observed in the ATLAS detector during the 2011 proton-proton run, JINST **8** P07004 (2013), arXiv:1303.0223
16. F. Kahlhoefer, K. Schmidt-Hoberg, T. Schwetz, S. Vogl, Implications of unitarity and gauge invariance for simplified Dark Matter models, JHEP **02**, 016 (2016), arXiv:1510.02110. [JHEP02,016(2016)]
17. Planck Collaboration, R. Adam et al., Planck 2015 results. I. Overview of products and scientific results, Astron. Astrophys. A1, **594**, (2016), arXiv:1502.01582
18. W.M.A.P. Collaboration, G. Hinshaw et al., Nine-year wilkinson microwave anisotropy probe (WMAP) observations: cosmological parameter results. Astrophys. J. Suppl. **208**, 19 (2013), arXiv:1212.5226
19. O. Buchmueller, M.J. Dolan, S.A. Malik, C. McCabe, Characterising Dark Matter searches at colliders and direct detection experiments: vector mediators, JHEP **037**, 1501 (2015), arXiv:1407.8257
20. G. Busoni et al., Recommendations on presenting LHC searches for missing transverse energy signals using simplified s-channel models of Dark Matter, arXiv:1603.04156
21. XENON100 Collaboration, E. Aprile et al., Limits on spin-dependent WIMP-nucleon cross sections from 225 live days of XENON100 data. Phys. Rev. Lett. **111**(2), 021301 (2013), arXiv:1301.6620

22. LUX Collaboration, D.S. Akerib et al., First spin-dependent WIMP-nucleon cross section limits from the LUX experiment, Phys. Rev. Lett. **116**(16), 161302 (2016), arXiv:1602.03489
23. PICO Collaboration, C. Amole et al., Dark Matter search results from the PICO-60 CF_3 I bubble chamber, Phys. Rev. D **93**(5), 052014 (2016), arXiv:1510.07754
24. PICO Collaboration, C. Amole et al., Improved Dark Matter search results from PICO-2L Run 2, Phys. Rev. **D93**(6), 061101 (2016), arXiv:1601.03729
25. M. Beltran, D. Hooper, E.W. Kolb, Z.A. Krusberg, T.M. Tait, Maverick Dark Matter at colliders. JHEP **1009**, 037 (2010), arXiv:1002.4137
26. J. Goodman, M. Ibe, A. Rajaraman, W. Shepherd, T.M. Tait et al., Constraints on Light Majorana Dark Matter from Colliders. Phys. Lett. B **695**, 185–188 (2011), arXiv:1005.1286
27. Y. Bai, P.J. Fox, R. Harnik, The tevatron at the frontier of Dark Matter direct detection. JHEP **1012**, 048 (2010), arXiv:1005.3797
28. J. Goodman, M. Ibe, A. Rajaraman, W. Shepherd, T.M. Tait et al., Constraints on Dark Matter from colliders. Phys. Rev. D **82**, 116010 (2010), arXiv:1008.1783
29. P.J. Fox, R. Harnik, J. Kopp, Y. Tsai, LEP shines light on Dark Matter. Phys. Rev. D **84**, 014028 (2011), arXiv:1103.0240
30. A. Rajaraman, W. Shepherd, T.M. Tait, A.M. Wijangco, LHC bounds on interactions of Dark Matter. Phys. Rev. D **84**, 095013 (2011), arXiv:1108.1196
31. P.J. Fox, R. Harnik, J. Kopp, Y. Tsai, Missing energy signatures of Dark Matter at the LHC. Phys. Rev. D **85**, 056011 (2012), arXiv:1109.4398
32. I.M. Shoemaker, L. Vecchi, Unitarity and monojet bounds on models for DAMA, CoGeNT, and CRESST-II. Phys. Rev. D **86**, 015023 (2012), arXiv:1112.5457
33. H. An, X. Ji, L.-T. Wang, Light Dark Matter and Z' dark force at colliders. JHEP **07**, 182 (2012), arXiv:1202.2894
34. R. Cotta, J. Hewett, M. Le, T. Rizzo, Bounds on Dark Matter interactions with electroweak gauge bosons. Phys. Rev. D **88**, 116009 (2013), arXiv:1210.0525
35. H. Dreiner, M. Huck, M. Krämer, D. Schmeier, J. Tattersall, Illuminating Dark Matter at the ILC. Phys. Rev. D **87**(7), 075015 (2013), arXiv:1211.2254
36. Y.J. Chae, M. Perelstein, Dark Matter search at a linear collider: effective operator approach. JHEP **1305**, 138 (2013), arXiv:1211.4008
37. P.J. Fox, C. Williams, Next-to-leading order predictions for Dark Matter production at hadron colliders. Phys. Rev. D **87**, 054030 (2013), arXiv:1211.6390
38. A. De Simone, A. Monin, A. Thamm, A. Urbano, On the effective operators for Dark Matter annihilations. JCAP **1302**, 039 (2013), arXiv:1301.1486
39. H. Dreiner, D. Schmeier, J. Tattersall, Contact interactions probe effective Dark Matter models at the LHC. Europhys. Lett. **102**, 51001 (2013), arXiv:1303.3348
40. J.-Y. Chen, E.W. Kolb, L.-T. Wang, Dark Matter coupling to electroweak gauge and Higgs bosons: an effective field theory approach, Phys. Dark. Univ. **2**, 200–218 (2013), arXiv:1305.0021
41. G. Busoni, A. De Simone, E. Morgante, A. Riotto, On the validity of the effective field theory for Dark Matter searches at the LHC. Phys. Lett. B **728**, 412–421 (2014), arXiv:1307.2253
42. O. Buchmueller, M.J. Dolan, C. McCabe, Beyond effective field theory for Dark Matter searches at the LHC. JHEP **1401**, 025 (2014), arXiv:1308.6799
43. G. Busoni, A. De Simone, J. Gramling, E. Morgante, A. Riotto, On the validity of the effective field theory for Dark Matter searches at the LHC, Part II: complete analysis for the s-channel, JCAP **060**, 1406 (2014), arXiv:1402.1275
44. A. Berlin, T. Lin, L.-T. Wang, Mono-Higgs detection of Dark Matter at the LHC. JHEP **06**, 078 (2014), arXiv:1402.7074
45. G. Busoni, A. De Simone, T. Jacques, E. Morgante, A. Riotto, On the validity of the effective field theory for Dark Matter searches at the LHC Part III: analysis for the t-channel, JCAP **022**, 1409 (2014), arXiv:1405.3101
46. D. Racco, A. Wulzer, F. Zwirner, Robust collider limits on heavy-mediator Dark Matter, JHEP **05**, 009 (2015), arXiv:1502.04701

47. N. Arkani-Hamed, P. Schuster, N. Toro, J. Thaler, L.-T. Wang, B. Knuteson, S. Mrenna, MAR-MOSET: The Path from LHC Data to the New Standard Model via On-Shell Effective Theories, arXiv:hep-ph/0703088
48. J. Alwall, P. Schuster, N. Toro, Simplified models for a first characterization of new physics at the LHC. Phys. Rev. D **79**, 075020 (2009), arXiv:0810.3921
49. LHC New Physics Working Group Collaboration, D. Alves et al., Simplified models for LHC new physics searches. J. Phys. **G39**, 105005 (2012), arXiv:1105.2838
50. E. Dudas, Y. Mambrini, S. Pokorski, A. Romagnoni, (In)visible Z-prime and Dark Matter. JHEP **08**, 014 (2009),arXiv:0904.1745
51. J. Goodman, W. Shepherd, LHC bounds on UV-complete models of Dark Matter, arXiv:1111.2359
52. M.T. Frandsen, F. Kahlhoefer, A. Preston, S. Sarkar, K. Schmidt-Hoberg, LHC and Tevatron bounds on the Dark Matter direct detection cross-section for vector mediators. JHEP **07**, 123 (2012), arXiv:1204.3839
53. R.C. Cotta, A. Rajaraman, T.M.P. Tait, A.M. Wijangco, Particle physics implications and constraints on Dark Matter interpretations of the CDMS signal. Phys. Rev. D **90**(1), 013020 (2014), arXiv:1305.6609
54. P. Bechtle, T. Plehn, C. Sander, Supersymmetry, in *The Large Hadron Collider: Harvest of Run 1*, ed. by T. Schörner-Sadenius (2015), pp. 421–462, arXiv:1506.03091
55. N. Arkani-Hamed, G.L. Kane, J. Thaler, L.-T. Wang, Supersymmetry and the LHC inverse problem. JHEP **08**, 070 (2006), arXiv:hep-ph/0512190

Chapter 6
The EFT Approach and Its Validity

6.1 Introduction

Given the plethora of particle physics model beyond the SM providing a WIMP candidate, it is highly desirable to study the signatures of this DM candidate in a model-independent way. In this and the following chapters, we are going to analyse the two main tools for such a model independent study, namely effective operators and simplified models.

Our starting point is the EFT approach where the interactions between the DM particle and the SM sector are parametrized by a definite set of effective (non-renormalizable) operators, generated after integrating out heavy mediators [1–16]. This approach is a very powerful and economical way to grasp the main features of a physical process, only in terms of the degrees of freedom which are excited at the scale of the process. EFT techniques are successfully applied in many branches of physics, and in particular they have become a standard way to present experimental results for DM searches.

As far as DM searches are concerned, if we consider a single effective operator and we extrapolate the EFT to high energies, the parameter space is reduced to a single energy scale, Λ (sometimes called M_* in the literature), in addition to the DM mass, and the potential number of WIMP models is reduced down to a relatively small basis set. Since direct and indirect detection of WIMPs, as well as WIMP production at the LHC, all require an interaction of the WIMPs with the SM particles, and such an interaction may be generated by the same operator, the EFT approach has the additional advantage of facilitating the analysis of the correlations between the various kinds of experiments.

The EFT description is only justified whenever there is a clear separation between the energy scale of the process to describe and the scale of the underlying microscopic interactions. In other contexts of DM searches, the energy scales involved are such that the EFT expansion is completely justified. For instance, for indirect DM searches the annihilations of non-relativistic DM particles in the galaxy occurs with momentum transfers of the order of the DM mass m_{DM}; in direct searches, the

© Springer International Publishing AG 2017

E. Morgante, *Aspects of WIMP Dark Matter Searches at Colliders and Other Probes*, Springer Theses, https://doi.org/10.1007/978-3-319-67606-7_6

momentum transfers involved in the scattering of DM particles with heavy nuclei are of the order of tens of keV. In all these cases, it is possible to carry out an effective description in terms of operators with an Ultra-Violet (UV) cutoff larger than the typical momentum transfer,[1] and reliable limits on the operator scale can be derived (see e.g. Refs. [17, 18]).

However, when collider searches are concerned, with the LHC being such a powerful machine, it is not guaranteed that the events used to constrain an effective interaction are not occurring at an energy scale larger than the cutoff scale of the effective description. In other words, some (or many) events of DM production may occur with such a high momentum transfer that the EFT is not a good description anymore [3, 5, 7, 10, 11, 13, 15, 19–24].

A possible way of using the EFT approach in a consistent way by considering only the fraction of events in which the typical momentum exchange is lower than the cutoff scale is described in Sect. 6.7 [21, 22, 24]. Here we will discuss in particular the technique introduced in [21], which is currently adopted by both the ATLAS and CMS collaborations in the presentation of their EFT results.

The rest of the present chapter is organized as follows. In Sect. 6.2 we discuss the use and the limitations of the EFT approach on a qualitative basis. Section 6.3 presents an estimate of the average momentum exchange in proton collisions at the LHC. In Sect. 6.4 we show how a simple effective operator compares a possible UV extension. In Sect. 6.5 we present and discuss the results of our analytical approach to assess the validity of EFT. In Sect. 6.6, the fully numerical approach is described and the results are compared with the analytical calculations. In Sect. 6.7 we analyze the impact of the limitation of the validity of the EFT for the limits from the LHC searches, and we introduce the rescaling procedure for the limits on the effective scale. Finally, we draw our conclusions in Sect. 6.8. The details of the analytical results can be found in the Appendix A.

6.2   General Considerations

An EFT is a powerful and economical way to describe physical processes occurring at a given energy scale in terms of a tower of interactions, involving only the degrees of freedom present at such scale. These interactions are generically non-renormalizable and with mass dimensions Λ^k, for some $k \geq 1$. For example, if one imagines that the UV theory contains a heavy particle of mass M, the low-energy effective theory at energies less than $\Lambda \sim M$ only contains the degrees of freedom lighter than Λ. The effects of the heavy field in the processes at low momentum transfer $Q_{tr} \ll \Lambda$ are encoded by a series of interactions, scaling as $(Q_{tr}/\Lambda)^k$ and whose coefficients

---

[1] Strictly speaking, the EFT approach is reliable in direct and indirect detection if m_χ is much smaller than the effective scale of the interaction, which in the simplest cases coincides with the scale at which the particle mediating the interaction goes on shell. This happens at $m_\chi \sim M_{med}/2$ for s-channel DM annihilation, or for $m_\chi \sim M_{med}$ in a χ-quark scattering process where the mediator is exchanged in the t-channel.

are matched to reproduce the UV theory at $Q_{tr} = \Lambda$. Therefore, the scale Λ sets the maximum energy at which the operator expansion in the EFT can be trusted. So generally speaking, the condition for the validity of an EFT is that the momentum transfer Q_{tr} in the relevant process one wants to describe must be less than the energy scale Λ.

In order to assess to what extent the effective description is valid, one has to compare the momentum transfer Q_{tr} of the process of interest (e.g. $pp \to \chi\chi + \text{jet}/\gamma$) to the energy scale Λ and impose that

$$\Lambda \gtrsim Q_{tr}. \tag{6.1}$$

Of course, there is some degree of arbitrariness in this choice, as one does not expect a sharp transition between a valid and an invalid EFT, but more precisely that the observables computed within the EFT are a less and less accurate approximation of the ones of the unknown UV theory as the cutoff scale Λ is approached. A more precise information on what Λ is can only come from knowing the details of the UV theory, the mass spectrum and the strength of the interactions.

The lower limits on Λ, extracted from interpreting the experimental data in terms of effective operators, should be considered together with the condition (6.1) on the validity itself of the effective approach. This means that one has to make sure that the lower limits on Λ obtained by the experiments satisfy to some extent the condition (6.1).

In order to clarify this point better, let us consider the simple example of a fermionic DM, whose interactions with quarks are mediated by a heavy scalar particle S through the Lagrangian

$$\mathscr{L}_{UV} \supset \frac{1}{2}M^2S^2 - g_q\bar{q}qS - g_\chi\bar{\chi}\chi S. \tag{6.2}$$

At energies much smaller than M the heavy mediator S can be integrated out, resulting in a tower of non-renormalizable operators for the fermionic DM interactions with quarks. The lowest-dimensional operator has dimension six

$$\mathscr{O}_S = \frac{1}{\Lambda^2}(\bar{\chi}\chi)(\bar{q}q), \tag{6.3}$$

and the matching condition implies

$$\frac{1}{\Lambda^2} = \frac{g_\chi g_q}{M^2}. \tag{6.4}$$

The Feynman diagrams for the processes under consideration are depicted in Figs. 6.1 and 6.2. The procedure of integrating out the heavy mediator and retaining the operator of lowest dimension can be viewed in terms of the expansion of the heavy particle propagator

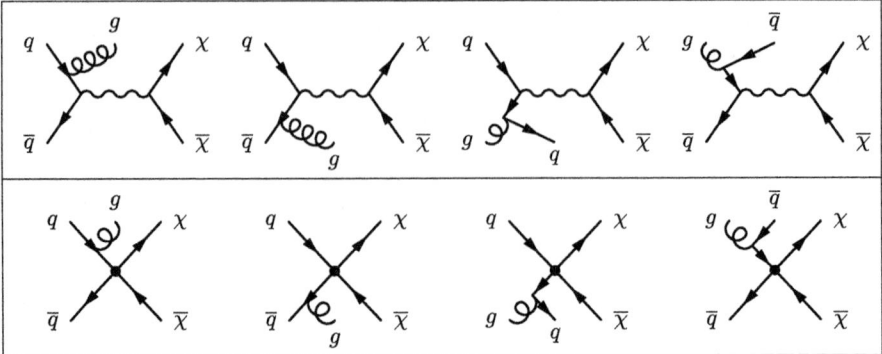

Fig. 6.1 The Feynman diagrams for DM pair production with ISR of a photon or jet, for a simplified model with mediator exchange (*top panel*) and its effective operator (*bottom panel*)

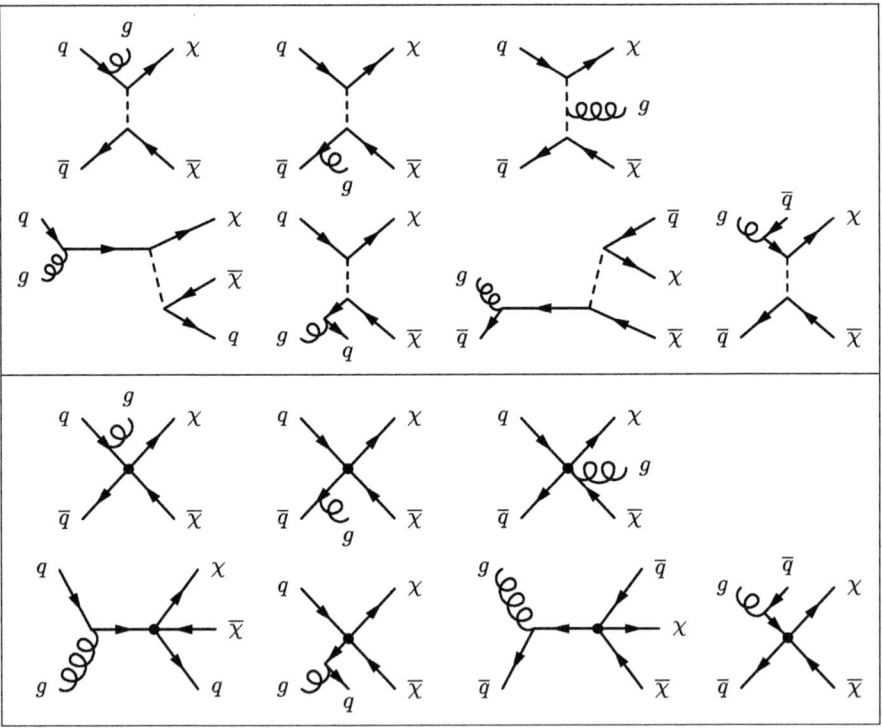

Fig. 6.2 Same as Fig. 6.1, for a *t*-channel mediator. In the *top panel*, the third diagram in the first row represents gluon emission from the coloured mediator. The corresponding process in the EFT limit (*bottom panel, first row, third diagram*) is suppressed by two additional powers of the effective scale Λ, and therefore can be ignored in our calculations

$$\frac{1}{Q_{\text{tr}}^2 - M^2} = -\frac{1}{M^2}\left(1 + \frac{Q_{\text{tr}}^2}{M^2} + \mathcal{O}\left(\frac{Q_{\text{tr}}^4}{M^4}\right)\right),\tag{6.5}$$

where only the leading term $1/M^2$ is kept. The higher-order terms in the expansion correspond to higher-dimensional operators. It is obvious that retaining only the lowest-dimensional operator is a good approximation as long as $Q_{\text{tr}}^2 \ll M^2 \sim \Lambda^2$. Thus, the parameter Q_{tr}/M characterizes the goodness of the truncation of the tower of effective operators to the lowest dimensional ones.

For the couplings to stay in the perturbative regime, one needs $g_q, g_\chi < 4\pi$. Also, we need a mediator heavier than the DM particle, that is $M > m_\chi$. So, Eq. (6.4) gives [4]

$$\Lambda \gtrsim \frac{m_\chi}{4\pi},\tag{6.6}$$

which depends linearly on the DM mass. This is a very minimal requirement on Λ and it is what, for instance, ATLAS uses in Ref. [25]. On top of this condition, the validity of the truncation to the lowest order in the expansion (6.5) requires that $Q_{\text{tr}} < M$, i.e. $Q_{\text{tr}} < \sqrt{g_q\, g_\chi}\Lambda < 4\pi\Lambda$, so that

$$\Lambda > \frac{Q_{\text{tr}}}{\sqrt{g_q g_\chi}} > \frac{Q_{\text{tr}}}{4\pi},\tag{6.7}$$

which depends on m_χ through Q_{tr} and refines the condition (6.1). However, the details of this condition depend case by case on the values of g_q, g_χ, and therefore on the details of the UV completion.

As an example of what the condition in Eq. (6.1) means, let us consider the hard scattering process of the production of two DM particles at the parton-level $q(p_1) + \bar{q}(p_2) \rightarrow \chi_1(p_3) + \chi_2(p_4)$. Since the two χ's are produced on-shell, the energy injected into the diagram needs to be $Q_{\text{tr}} \geq 2m_\chi$, so $\Lambda \gtrsim Q_{\text{tr}}$ implies $\Lambda \gtrsim 2m_\chi$, which would be much stronger than Eq. (6.6). However, the relevant process used for extracting lower limits on Λ at the LHC is when a single jet or photon is emitted from the initial state quark and this is the process we will consider in the next sections. We will proceed in two different ways. In Sect. 6.3 we will provide an estimate of the typical momentum transfer involved in the DM production associated to mono-jet or mono-photon, as a function of the transverse momentum and rapidity of the mono-jet or mono-photon. This calculation, even though not rigorous, will serve to provide an idea of the minimum value of Λ compatible with the EFT approach. Subsequently, in Sects. 6.4 and 6.5, we will asses the validity of the EFT more precisely by studying the effect of the condition $Q_{\text{tr}} < \Lambda$ on the cross sections for the production of DM plus mono-jet/photon, and by comparing the cross section in the EFT and in the UV theory where the mediator has not been integrated out.

6.2.1 Operators and Cross Sections

The starting point of our analysis is the list of the 18 operators reported in Table 6.1 which are commonly used in the literature [1]. We have considered not only the operators connecting the DM fermion to quarks (D1–D10), but also those involving gluon field strengths (D11–D14). Furthermore, the operators can originate from heavy mediators exchange in the s-channel. For instance, the D1' (D5) operators may be originated by the tree-level s-channel exchange of a very heavy scalar (vector) boson S (V_μ), with lagrangians

$$\mathscr{L}_{\text{D1}'} \supset \frac{1}{2} M^2 S^2 - g_q \bar{q} q S - g_\chi \bar{\chi} \chi S \,, \tag{6.8}$$

$$\mathscr{L}_{\text{D5}} \supset \frac{1}{2} M^2 V^\mu V_\mu - g_q \bar{q} \gamma^\mu q V_\mu - g_\chi \bar{\chi} \gamma^\mu \chi V_\mu \,. \tag{6.9}$$

Notice the presence of the "primed" operators $D1'-D4'$, very similar to the ones often considered $D1-D4$, respectively, but with a different normalization, independent of the quark masses. In fact, they may arise from integrating out heavy scalars which do not take a vacuum expectation value and therefore do not give rise to quark masses.[2]

We also consider the following effective operator describing the interactions between Dirac dark matter χ and left-handed quarks q

$$\text{DT1} = \frac{1}{\Lambda^2} \left(\bar{\chi} P_L q \right) \left(\bar{q} P_R \chi \right) \,. \tag{6.11}$$

Only the coupling between dark matter and the first generation of quarks is considered. Including couplings to the other generations of quarks requires fixing the relationships between the couplings and mediator masses for each generation, making such an analysis less general. In principle the dark matter can also couple to the right-handed quark singlet, switching P_R and P_L in the above operator. The inclusion of both of these operators does not modify our results, even if the two terms have different coupling strengths.

---

[2] A normalization proportional to the quark mass is common in many models motivated by flavour physics, but in general the coefficient Λ^3 at the denominator can have a different form. For example, if the effective operators come from a Naturalness-motivated new physics theory like Supersymmetry or Composite Higgs Models, assuming a $U(2)^3$ flavour symmetry [26, 27] the normalization would be

$$\lambda_{t,b} \frac{1}{\Lambda^2} \frac{m_q}{m_{t,b}} \tag{6.10}$$

where Λ is an energy scale of the order some TeV related to the Electroweak Symmetry Breaking and $m_{t,b}$, $\lambda_{t,b}$ are the mass and the Yukawa coupling with the Higgs of the top/bottom quark, depending on whether the quark q is up-like or down-like. In the present work, we will be agnostic about this point, and we'll keep both the primed and unprimed operators into account on the same footing as all others.

Table 6.1 Operators used throughout this work. The nomenclature is mostly taken from [4]

Name	Operator	Coefficient
D1	$\bar{\chi}\chi\,\bar{q}q$	m_q/Λ^3
D1'	$\bar{\chi}\chi\,\bar{q}q$	$1/\Lambda^2$
D2	$\bar{\chi}\gamma^5\chi\,\bar{q}q$	im_q/Λ^3
D2'	$\bar{\chi}\gamma^5\chi\,\bar{q}q$	i/Λ^2
D3	$\bar{\chi}\chi\,\bar{q}\gamma^5 q$	im_q/Λ^3
D3'	$\bar{\chi}\chi\,\bar{q}\gamma^5 q$	i/Λ^2
D4	$\bar{\chi}\gamma^5\chi\,\bar{q}\gamma^5 q$	m_q/Λ^3
D4'	$\bar{\chi}\gamma^5\chi\,\bar{q}\gamma^5 q$	$1/\Lambda^2$
D5	$\bar{\chi}\gamma_\mu\chi\,\bar{q}\gamma^\mu q$	$1/\Lambda^2$
D6	$\bar{\chi}\gamma_\mu\gamma^5\chi\,\bar{q}\gamma^\mu q$	$1/\Lambda^2$
D7	$\bar{\chi}\gamma_\mu\chi\,\bar{q}\gamma^\mu\gamma^5 q$	$1/\Lambda^2$
D8	$\bar{\chi}\gamma_\mu\gamma^5\chi\,\bar{q}\gamma^\mu\gamma^5 q$	$1/\Lambda^2$
D9	$\bar{\chi}\sigma_{\mu\nu}\chi\,\bar{q}\sigma^{\mu\nu}q$	$1/\Lambda^2$
D10	$\bar{\chi}\sigma_{\mu\nu}\gamma^5\chi\,\bar{q}\sigma^{\mu\nu}q$	i/Λ^2
D11	$\bar{\chi}\chi\,G^{\mu\nu}G_{\mu\nu}$	$\alpha_s/4\Lambda^3$
D12	$\bar{\chi}\gamma^5\chi\,G^{\mu\nu}G_{\mu\nu}$	$i\alpha_s/4\Lambda^3$
D13	$\bar{\chi}\chi\,G^{\mu\nu}\tilde{G}_{\mu\nu}$	$i\alpha_s/4\Lambda^3$
D14	$\bar{\chi}\gamma^5\chi\,G^{\mu\nu}\tilde{G}_{\mu\nu}$	$\alpha_s/4\Lambda^3$
DT1	$(\bar{\chi}P_L q)\,(\bar{q}P_R\chi)$	$1/\Lambda^2$

The operator in Eq. (6.11) can be viewed as the low-energy limit of a simplified model describing a quark doublet Q_L coupling to DM, via t-channel exchange of a scalar mediator S_Q,

$$\mathcal{L}_{\text{int}} = g\,\bar{\chi}Q_L S_Q^* + h.c. \qquad (6.12)$$

and integrating out the mediator itself. Since we consider only coupling to the first generation of quarks, $Q_L = (u_L, d_L)$. This model is popular as an example of a simple DM model with t-channel couplings, which exist also in well-motivated models such as supersymmetry where the mediator particle is identified as a squark, and the DM is a Majorana particle. In [28] a version of this model with Majorana DM in place of Dirac DM is used to test the prospects of Z-bosons as a potential search channel. This has been followed up by a dedicated ATLAS search in this channel [29]. Refs. [30–35] have also constrained this model, using both the standard monojet search channel as well as searching for multiple jets arising from direct mediator production. Refs. [31, 35] found that collider constraints on this model were competitive if not stronger than direct detection constraints across most of the parameter space.

The t-channel operator in Eq. (6.11) can be expressed as a sum of s-channel operators using Fierz transformations. For arbitrary Dirac spinors such as $\bar{q}_1$, q_2, $\bar{\chi}_1$, χ_2, and adopting in part the notation of [36], the Fierz transformation can be expressed as

$$(\bar{q}_1 X \chi_2)\, (\bar{\chi}_1 Y q_2) = \frac{1}{4} \sum_B \left(\bar{q}_1 X \Gamma^B Y q_2 \right) \left(\bar{\chi}_1 \Gamma_B \chi_2 \right), \qquad (6.13)$$

where X, Y are some combination of Dirac-matrices, and $\Gamma^B = \{\mathbb{1}, i\gamma_5, \gamma^\mu,$ $\gamma_5\gamma^\mu, \sigma^{\mu\nu}\}$ and $\Gamma_B = \{\mathbb{1}, -i\gamma_5, \gamma_\mu, -\gamma_5\gamma_\mu, \frac{1}{2}\sigma_{\mu\nu}\}$ form a basis spanning 4×4 matrices over the complex number field [36]. Due to the chiral coupling between the quarks and DM, most of the terms in the sum cancel, and we are left with

$$
\begin{aligned}
\mathcal{O} &= \frac{1}{\Lambda^2}\, (\bar{\chi} P_L q)\, (\bar{q} P_R \chi) \\
&= \frac{1}{8\Lambda^2}\, (\bar{\chi}\gamma^\mu \chi)\, (\bar{q}\gamma_\mu q) && (D5) \\
&+ \frac{1}{8\Lambda^2}\, (\bar{\chi}\gamma^\mu\gamma_5 \chi)\, (\bar{q}\gamma_\mu q) && (D6) \\
&- \frac{1}{8\Lambda^2}\, (\bar{\chi}\gamma^\mu \chi)\, (\bar{q}\gamma_\mu\gamma_5 q) && (D7) \\
&- \frac{1}{8\Lambda^2}\, (\bar{\chi}\gamma^\mu\gamma_5 \chi)\, (\bar{q}\gamma_\mu\gamma_5 q) && (D8) \\
&= \frac{1}{2\Lambda^2}\, (\bar{\chi}\gamma^\mu P_R \chi)\, (\bar{q}\gamma_\mu P_L q)\,. && (6.14)
\end{aligned}
$$

This is equivalent to a rescaled sum of the D5, D6, D7 and D8 operators [4]. Thus, it is interesting to see whether the EFT limit of the t-channel model under investigation has similar phenomenology to these s-channel operators.

6.2.2   Analytic Cross Sections

We have computed the tree-level differential cross sections in the transverse momentum p_T and rapidity η of the final jet for the hard scattering process with gluon radiation from the initial state $f(p_1) + \bar{f}(p_2) \to \chi(p_3) + \chi(p_4) + g(k)$, where f is either a quark (for operators D1–D10), or a gluon (for operators D11–D14). For the D1–D14 operators, the results are conveniently written in terms of the momentum transfer in the s-channel

$$Q_{\mathrm{tr}}^2 = (p_1 + p_2 - k)^2 = x_1 x_2 s - \sqrt{s}\, p_T \left(x_1 e^{-\eta} + x_2 e^{\eta} \right), \qquad (6.15)$$

where x_1, x_2 are the fractions of momentum carried by initial partons and η, p_T are the pseudo-rapidity and the transverse momentum of the final state gluon, respectively. The expressions are of course valid for all admitted values of the parameters. It's only when integrated numerically over the PDFs and over η, p_T that the dependence on these values comes in.

For the effective operator DT1, viewed as the low energy limit of the simplified model of Eq. (6.12) in which the mediator is exchanged in the t-channel, the definition of the momentum transfer Q_{tr} is different for each of the diagrams in Fig. 6.2. In the case of gluon emission, the definition of Q_{tr} is given by

$$Q_{tr,g1}^2 = (p_1 - k - p_3)^2, \qquad Q_{tr,g2}^2 = (p_1 - p_3)^2, \tag{6.16}$$

in the first and second diagram of Fig. 6.2, respectively.

For the s-channel operators we obtain

$$\frac{d^2\hat{\sigma}}{dp_T d\eta}\bigg|_{D1'} = \frac{\alpha_s}{36\pi^2} \frac{1}{p_T} \frac{1}{\Lambda^4} \frac{[Q_{tr}^2 - 4m_\chi^2]^{3/2}\left[1 + \frac{Q_{tr}^4}{(x_1 x_2 s)^2}\right]}{Q_{tr}}, \tag{6.17}$$

$$\frac{d^2\hat{\sigma}}{dp_T d\eta}\bigg|_{D4'} = \frac{\alpha_s}{36\pi^2} \frac{1}{p_T} \frac{1}{\Lambda^4} Q_{tr} \left[Q_{tr}^2 - 4m_\chi^2\right]^{1/2}\left[1 + \frac{Q_{tr}^4}{(x_1 x_2 s)^2}\right], \tag{6.18}$$

$$\frac{d^2\hat{\sigma}}{dp_T d\eta}\bigg|_{D5} = \frac{\alpha_s}{27\pi^2} \frac{1}{p_T} \frac{1}{\Lambda^4} \frac{\left[Q_{tr}^2 - 4m_\chi^2\right]^{1/2}\left[Q_{tr}^2 + 2m_\chi^2\right]\left[1 + \frac{Q_{tr}^4}{(x_1 x_2 s)^2} - 2\frac{p_T^2}{x_1 x_2 s}\right]}{Q_{tr}}, \tag{6.19}$$

$$\frac{d^2\hat{\sigma}}{dp_T d\eta}\bigg|_{D8} = \frac{\alpha_s}{27\pi^2} \frac{1}{p_T} \frac{1}{\Lambda^4} \frac{\left[Q_{tr}^2 - 4m_\chi^2\right]^{3/2}\left[1 + \frac{Q_{tr}^4}{(x_1 x_2 s)^2} - 2\frac{p_T^2}{x_1 x_2 s}\right]}{Q_{tr}}, \tag{6.20}$$

$$\frac{d^2\hat{\sigma}}{dp_T d\eta}\bigg|_{D9} = \frac{2\alpha_s}{27\pi^2} \frac{1}{p_T} \frac{1}{\Lambda^4} \frac{\sqrt{Q_{tr} - 4m_\chi^2}\left[Q_{tr}^2 + 2m_\chi^2\right]\left[1 + \frac{Q_{tr}^4}{(x_1 x_2 s)^2} + 4p_T^2\left(\frac{1}{Q_{tr}^2} - \frac{1}{x_1 x_2 s}\right)\right]}{Q_{tr}}, \tag{6.21}$$

$$\frac{d^2\hat{\sigma}}{dp_T d\eta}\bigg|_{D11} = \frac{3\alpha_s^3}{256\pi^2 \Lambda^6} \frac{(x_1 x_2 s)^3}{(Q_{tr}^2 - x_1 x_2 s)^2} \frac{(Q_{tr}^2 - 4m_\chi^2)^{3/2}}{p_T Q_{tr}}\left[1 - 4\frac{Q_{tr}^2 - p_T^2}{x_1 x_2 s} + \frac{8Q_{tr}^4 + 21p_T^4}{(x_1 x_2 s)^2}\right.$$
$$\left. -2Q_{tr}^2 \frac{5Q_{tr}^4 + 4Q_{tr}^2 p_T^2 + 5p_T^4}{(x_1 x_2 s)^3} + Q_{tr}^4 \frac{8Q_{tr}^4 + 8Q_{tr}^2 p_T^2 + 5p_T^4}{(x_1 x_2 s)^4} - 4Q_{tr}^8 \frac{Q_{tr}^2 + p_T^2}{(x_1 x_2 s)^5}\right.$$
$$\left. + \frac{Q_{tr}^{12}}{(x_1 x_2 s)^6}\right], \tag{6.22}$$

$$\frac{d^2\hat{\sigma}}{dp_T d\eta}\bigg|_{D12} = \frac{3\alpha_s^3}{256\pi^2 \Lambda^6} \frac{(x_1 x_2 s)^3}{(Q_{tr}^2 - x_1 x_2 s)^2} \frac{Q_{tr}\sqrt{Q_{tr}^2 - 4m_\chi^2}}{p_T}\left[1 - 4\frac{Q_{tr}^2 - p_T^2}{x_1 x_2 s} + \frac{8Q_{tr}^4 + 21p_T^4}{(x_1 x_2 s)^2}\right.$$
$$\left. -2Q_{tr}^2 \frac{5Q_{tr}^4 + 4Q_{tr}^2 p_T^2 + 5p_T^4}{(x_1 x_2 s)^3} + Q_{tr}^4 \frac{8Q_{tr}^4 + 8Q_{tr}^2 p_T^2 + 5p_T^4}{(x_1 x_2 s)^4} - 4Q_{tr}^8 \frac{Q_{tr}^2 + p_T^2}{(x_1 x_2 s)^5}\right.$$
$$\left. + \frac{Q_{tr}^{12}}{(x_1 x_2 s)^6}\right], \tag{6.23}$$

$$
\left.\frac{\mathrm{d}^2\hat{\sigma}}{\mathrm{d}p_T\mathrm{d}\eta}\right|_{D13} = \frac{3\alpha_s^3}{256\pi^2\Lambda^6}\frac{(x_1x_2s)^3}{(Q_{\mathrm{tr}}^2-x_1x_2s)^2}\frac{(Q_{\mathrm{tr}}^2-4m_\chi^2)^{3/2}}{p_TQ_{\mathrm{tr}}}\left[1-4\frac{Q_{\mathrm{tr}}^2}{x_1x_2s}+\frac{8Q_{\mathrm{tr}}^4+8Q_{\mathrm{tr}}^2p_T^2+5p_T^4}{(x_1x_2s)^2}\right.
$$
$$
-2Q_{\mathrm{tr}}^2\frac{5Q_{\mathrm{tr}}^4+6Q_{\mathrm{tr}}^2p_T^2-3p_T^4}{(x_1x_2s)^3}+Q_{\mathrm{tr}}^4\frac{8Q_{\mathrm{tr}}^4+8Q_{\mathrm{tr}}^2p_T^2+5p_T^4}{(x_1x_2s)^4}-4Q_{\mathrm{tr}}^8\frac{Q_{\mathrm{tr}}^2+p_T^2}{(x_1x_2s)^5}
$$
$$
\left.+\frac{Q_{\mathrm{tr}}^{12}}{(x_1x_2s)^6}\right], \tag{6.24}
$$

$$
\left.\frac{\mathrm{d}^2\hat{\sigma}}{\mathrm{d}p_T\mathrm{d}\eta}\right|_{D14} = \frac{3\alpha_s^3}{256\pi^2\Lambda^6}\frac{(x_1x_2s)^3}{(Q_{\mathrm{tr}}^2-x_1x_2s)^2}\frac{Q_{\mathrm{tr}}\sqrt{Q_{\mathrm{tr}}^2-4m_\chi^2}}{p_T}\left[1-4\frac{Q_{\mathrm{tr}}^2}{x_1x_2s}+\frac{8Q_{\mathrm{tr}}^4+8Q_{\mathrm{tr}}^2p_T^2+5p_T^4}{(x_1x_2s)^2}\right.
$$
$$
-2Q_{\mathrm{tr}}^2\frac{5Q_{\mathrm{tr}}^4+6Q_{\mathrm{tr}}^2p_T^2-3p_T^4}{(x_1x_2s)^3}+Q_{\mathrm{tr}}^4\frac{8Q_{\mathrm{tr}}^4+8Q_{\mathrm{tr}}^2p_T^2+5p_T^4}{(x_1x_2s)^4}-4Q_{\mathrm{tr}}^8\frac{Q_{\mathrm{tr}}^2+p_T^2}{(x_1x_2s)^5}
$$
$$
\left.+\frac{Q_{\mathrm{tr}}^{12}}{(x_1x_2s)^6}\right]. \tag{6.25}
$$

The reader can find the details of the derivation of Eqs. (6.17)–(6.25) in Appendix A. As for the other operators, we get

$$
\left.\frac{\mathrm{d}^2\hat{\sigma}}{\mathrm{d}p_T\mathrm{d}\eta}\right|_{D2'}=\left.\frac{\mathrm{d}^2\hat{\sigma}}{\mathrm{d}p_T\mathrm{d}\eta}\right|_{D4'}\qquad\left.\frac{\mathrm{d}^2\hat{\sigma}}{\mathrm{d}p_T\mathrm{d}\eta}\right|_{D3'}=\left.\frac{\mathrm{d}^2\hat{\sigma}}{\mathrm{d}p_T\mathrm{d}\eta}\right|_{D1'}\qquad\left.\frac{\mathrm{d}^2\hat{\sigma}}{\mathrm{d}p_T\mathrm{d}\eta}\right|_{D6}=\left.\frac{\mathrm{d}^2\hat{\sigma}}{\mathrm{d}p_T\mathrm{d}\eta}\right|_{D8}\tag{6.26}
$$

$$
\left.\frac{\mathrm{d}^2\hat{\sigma}}{\mathrm{d}p_T\mathrm{d}\eta}\right|_{D7}=\left.\frac{\mathrm{d}^2\hat{\sigma}}{\mathrm{d}p_T\mathrm{d}\eta}\right|_{D5}\qquad\qquad\left.\frac{\mathrm{d}^2\hat{\sigma}}{\mathrm{d}p_T\mathrm{d}\eta}\right|_{D9}=\left.\frac{\mathrm{d}^2\hat{\sigma}}{\mathrm{d}p_T\mathrm{d}\eta}\right|_{D10},\tag{6.27}
$$

in the limit of massless light quarks. The operators $D1$–$D4$ are simply related to $D1'$–$D4'$ by a straightforward rescaling

$$
\left.\frac{\mathrm{d}^2\hat{\sigma}}{\mathrm{d}p_T\mathrm{d}\eta}\right|_{D1,D2,D3,D4}=\left(\frac{m_q}{\Lambda}\right)^2\left.\frac{\mathrm{d}^2\hat{\sigma}}{\mathrm{d}p_T\mathrm{d}\eta}\right|_{D1',D2',D3',D4'}.\tag{6.28}
$$

We checked that the differences between the cross sections for $D1'$–$D4'$ computed for $m_q\neq0$ and those reported above assuming $m_q=0$ are at the per-mille level, so the approximation $m_q=0$ which we used in all our analytical calculations is justified. The cross sections for the UV completions of dim-6 operators, with s-channel exchange of a mediator of mass M_{med}, are simply obtained by the replacement $1/\Lambda^4\to g_q^2g_\chi^2/[Q_{\mathrm{tr}}^2-M_{\mathrm{med}}^2]^2$.

In order to get the cross sections initiated by the colliding protons one needs to average over the PDFs. For example, for processes with initial state quarks

$$
\left.\frac{\mathrm{d}^2\sigma}{\mathrm{d}p_T\mathrm{d}\eta}\right|_{Di}=\sum_q\int\mathrm{d}x_1\mathrm{d}x_2[f_q(x_1)f_{\bar{q}}(x_2)+f_q(x_2)f_{\bar{q}}(x_1)]\left.\frac{\mathrm{d}^2\hat{\sigma}}{\mathrm{d}p_T\mathrm{d}\eta}\right|_{Di}.\tag{6.29}
$$

We have performed the analytical calculation only for the emission of an initial state gluon (identified with the final jet observed experimentally). The extension to include also the smaller contribution coming from initial radiation of quarks ($qg \to \chi\chi + q$) is done numerically in Sect. 6.6. We have used the MSTW PDFs from Refs. [37, 38], and checked that the our results are not sensitive to the choice of leading or next-to-leading-order MSTW PDFs.

6.3 An Estimate of the Momentum Transfer

To assess the validity of the EFT, we first adopt a procedure which, albeit not rigorous, gives an idea of the error one might make in adopting the EFT. The advantage of this procedure is that it is model-independent in the sense that it does not depend on the particular UV completion of the EFT theory. A simple inspection of the expansion (6.5) tells us that the EFT is trustable only if $Q_{\mathrm{tr}}^2 \ll M^2$ and we take for the typical value of Q_{tr} the square root of the averaged squared momentum transfer in the s-channel, where the average is computed properly weighting with PDFs [37]

$$\langle Q_{\mathrm{tr}}^2 \rangle = \frac{\sum_q \int \mathrm{d}x_1 \mathrm{d}x_2 \left[f_q(x_1) f_{\bar{q}}(x_2) + f_q(x_2) f_{\bar{q}}(x_1) \right] \theta(Q_{\mathrm{tr}} - 2m_\chi) Q_{\mathrm{tr}}^2}{\sum_q \int \mathrm{d}x_1 \mathrm{d}x_2 \left[f_q(x_1) f_{\bar{q}}(x_2) + f_q(x_2) f_{\bar{q}}(x_1) \right] \theta(Q_{\mathrm{tr}} - 2m_\chi)} . \quad (6.30)$$

The integration in x_1, x_2 is performed over the kinematically allowed region $Q_{\mathrm{tr}} \geq 2m_\chi$ and we have set the renormalization and factorization scales to $p_{\mathrm{T}} + 2[m_\chi^2 + p_{\mathrm{T}}^2/4]^{1/2}$, as often done by the LHC collaborations (see e.g. Ref. [25]). The results are plotted in Fig. 6.3 as a function of the DM mass m_χ and for different choices of p_{T} and η of the radiated jet. From Fig. 6.3 we see that the lower the jet p_{T}, the lower the momentum transfer is, and therefore the better the EFT will work. The same is true for smaller DM masses. These behaviors, which are due to the fact we have restricted the average of the momentum transfer to the kinematically allowed domain, will be confirmed by a more rigorous approach in the next section. Notice that $\langle Q_{\mathrm{tr}}^2 \rangle^{1/2}$ is always larger than about 500 GeV, which poses a strong bound on the

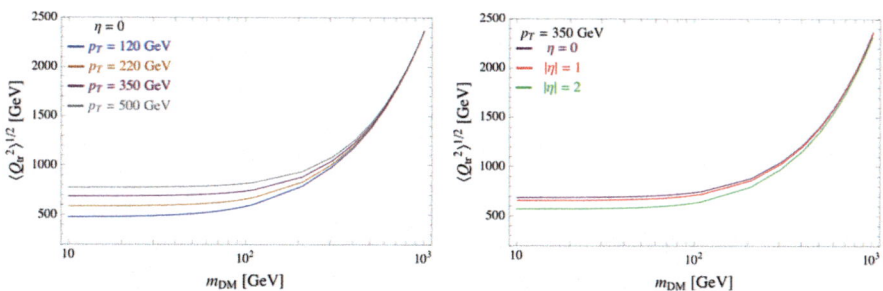

Fig. 6.3 The momentum transfer in the s-channel in Eq. (6.15), weighted with PDFs, as a function of m_χ, for different choices of p_{T}, η of the radiated jet. We considered $\sqrt{s} = 8\,\mathrm{TeV}$

cutoff scale Λ: when the coupling constants g_q and g_χ are close to their perturbative regime, from the condition (6.7) we get $\Lambda \gtrsim 50$ GeV, but when the couplings are of order unity, one gets a much stronger bound $\Lambda \gtrsim 500$ GeV.

6.4 Comparing the Effective Operator with a UV Completion

Let us now turn to quantify the validity of the EFT by comparing cross sections for the production of DM plus mono-jet or mono-photon in the simple example of a theory containing a DM particle χ and a heavy mediator S with the Lagrangian described in Eq. (6.8) with its effective counterpart given by the D1' operator in Table 6.1. The matching condition implies $\Lambda = M/\sqrt{g_q g_\chi}$. Let us study the ratio of the cross sections obtained with the UV theory and with the effective operator

$$
r_{\text{UV/eff}} \equiv \frac{\left.\dfrac{d^2 \sigma_{\text{UV}}}{dp_T d\eta}\right|_{Q_{\text{tr}} < M}}{\left.\dfrac{d^2 \sigma_{\text{eff}}}{dp_T d\eta}\right|_{Q_{\text{tr}} < \Lambda}} . \tag{6.31}
$$

This ratio quantifies the error of using the EFT, truncated at the lowest-dimensional operator, with respect to its UV completion, for given p_T, η of the radiated object. Values of $r_{\text{UV/eff}}$ close to unity indicate the effective operator is accurately describing the high-energy theory, whereas large values of $r_{\text{UV/eff}}$ imply a poor effective description.

For numerical integrations over the PDFs we have regularized the propagator introducing a small width $\Gamma = (g_q^2 + g_\chi^2) M/(8\pi)$ for the scalar mediator. The function $r_{\text{UV/eff}}$ is plotted in Fig. 6.4, for different choices of p_T, η, Λ, m_χ. Once again, one can see that the smaller p_T and m_χ are, the better the EFT works.

Also, we can integrate over p_T, η using cuts commonly used in the experimental analysis (see e.g. [25]): $p_T \geq p_T^{\text{min}}$, $|\eta| \leq 2$

$$
r_{\text{UV/eff}}^{\text{tot}} \equiv \frac{\sigma_{\text{UV}}|_{Q_{\text{tr}} < M}}{\sigma_{\text{eff}}|_{Q_{\text{tr}} < \Lambda}} = \frac{\int_{p_T^{\text{min}}}^{1\text{TeV}} dp_T \int_{-2}^{2} d\eta \left.\dfrac{d^2 \sigma_{\text{UV}}}{dp_T d\eta}\right|_{Q_{\text{tr}} < M}}{\int_{p_T^{\text{min}}}^{1\text{TeV}} dp_T \int_{-2}^{2} d\eta \left.\dfrac{d^2 \sigma_{\text{eff}}}{dp_T d\eta}\right|_{Q_{\text{tr}} < \Lambda}} . \tag{6.32}
$$

The DM pair production is kinematically allowed for $Q_{\text{tr}} > 2m_\chi$; furthermore, when dealing with the processes with mediator exchange, one also has to require $Q_{\text{tr}} < M$ to avoid on-shell mediator. Therefore we have worked with the condition $2m_\chi < Q_{\text{tr}} < M$, which can only be satisfied if $m_\chi < M/2$. The results for $r_{\text{UV/eff}}^{\text{tot}}$ are plotted in Fig. 6.5. We see that one needs a cutoff scale Λ at least larger than about

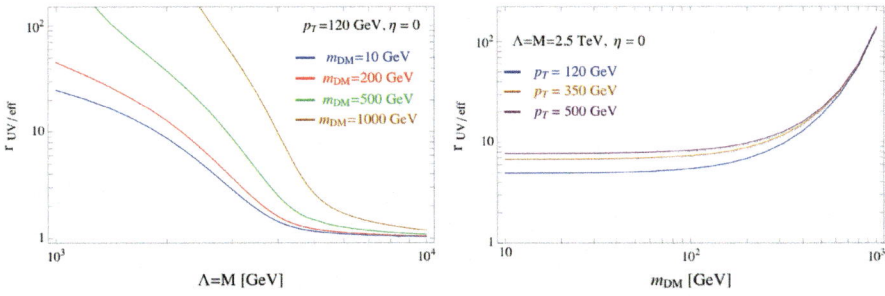

Fig. 6.4 The ratio $r_{\mathrm{UV/eff}}$ defined in Eq. (6.31), for $\sqrt{s} = 8$ TeV, $\eta = 0$ and $M = \Lambda$ (corresponding to $g_q = g_\chi = 1$). *Left panel*: $r_{\mathrm{UV/eff}}$ as a function of Λ for various choices of m_χ and $p_T = 120$ GeV. *Right panel*: $r_{\mathrm{UV/eff}}$ as a function of m_χ for various choices of p_T, and $\Lambda = 2.5$ TeV

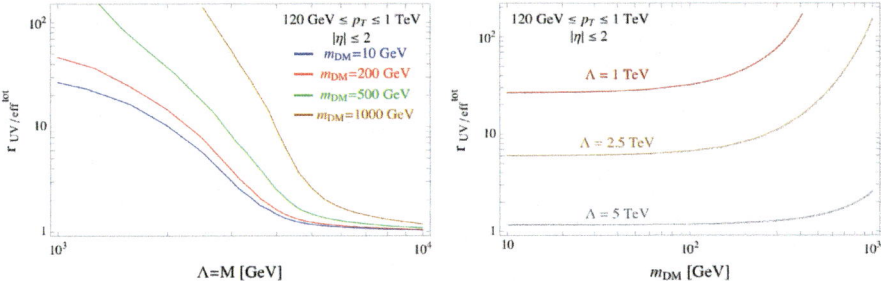

Fig. 6.5 The ratio $r_{\mathrm{UV/eff}}^{\mathrm{tot}}$ defined in Eq. (6.32), as a function of Λ (*left panel*) and m_χ (*right panel*). We have set $p_T^{\mathrm{min}} = 120$ GeV, $|\eta| \le 2$, $M = \Lambda$, $g_q = g_\chi = 1$ and $\sqrt{s} = 8$ TeV

a few TeV in order for the ratios $r_{\mathrm{UV/eff}}$ and $r_{\mathrm{UV/eff}}^{\mathrm{tot}}$ to be of order unity, the best case being attained for the lowest DM masses. Notice that the ratios involving differential and total cross sections ($r_{\mathrm{UV/eff}}$ and $r_{\mathrm{UV/eff}}^{\mathrm{tot}}$) are very similar, as a consequence of the fact that the integrands are very peaked at low p_T and at $\eta = 0$.

6.5 The Effect of the EFT Cutoff

In what regions of the parameter space (Λ, m_χ) is the effective description accurate and reliable? The truncation to the lowest-dimensional operator of the EFT expansion is accurate only if the momentum transfer is smaller than an energy scale of the order of Λ, see Eq. (6.1). Therefore we want to compute the fraction of events with momentum transfer lower than the EFT cutoff scale. To this end we define the ratio of the cross section obtained in the EFT with the requirement $Q_{\mathrm{tr}} < \Lambda$ on the PDF integration domain, over the total cross section obtained in the EFT.

$$R_\Lambda \equiv \frac{\left.\dfrac{\mathrm{d}^2\sigma_{\mathrm{eff}}}{\mathrm{d}p_\mathrm{T}\mathrm{d}\eta}\right|_{Q_{\mathrm{tr}}<\Lambda}}{\dfrac{\mathrm{d}^2\sigma_{\mathrm{eff}}}{\mathrm{d}p_\mathrm{T}\mathrm{d}\eta}} . \tag{6.33}$$

This ratio quantifies the fraction of the differential cross section for $q\bar{q} \to \chi\chi$+gluon, for given p_T, η of the radiated object, mediated by the effective operator (6.3), where the momentum transfer is below the scale Λ of the operator. Values of R_Λ close to unity indicate that the effective cross section is describing processes with sufficiently low momentum transfers, so the effective approach is accurate. On the other hand, a very small R_Λ signals that a significant error is made by extrapolating the effective description to a regime where it cannot be fully trusted, and where the neglected higher-dimensional operators can give important contributions.

This ratio is plotted in Fig. 6.6 as a function of Λ and m_χ, for various choices of p_T and η, for the D1' operator. Our results indicate that if one would measure the cross section for the mono-jet emission process within the EFT, but without taking into account that Q_{tr} should be bounded from above, one makes an error which may even be very large, depending on the values of the DM mass, the scale Λ of the operator and the p_T, η of the emitted object. Of course, the precise definition of the cutoff scale of an EFT is somewhat arbitrary, with no knowledge of the underlying UV theory; therefore one should consider the values of R_Λ with a grain of salt.

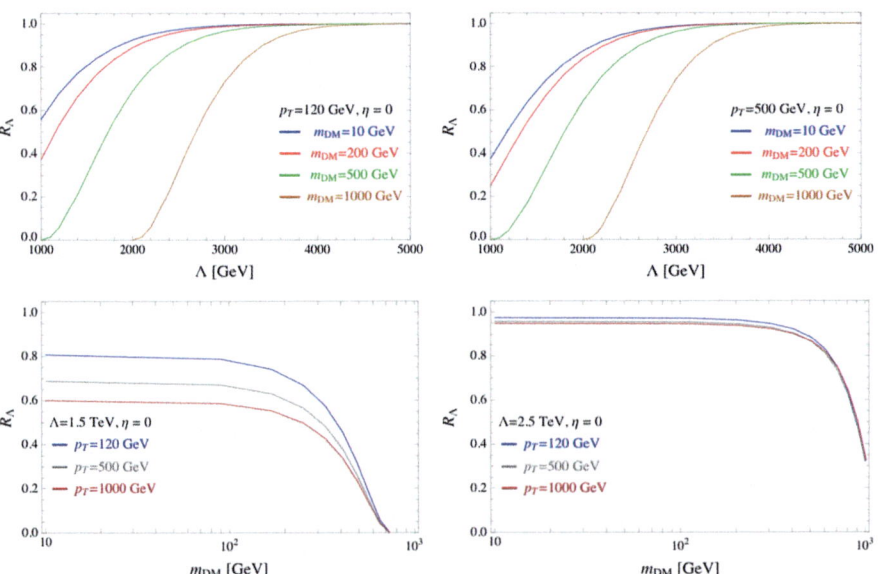

Fig. 6.6 The ratio R_Λ defined in Eq. (6.33) for $\sqrt{s} = 8$ TeV, $\eta = 0$. *Top row*: R_Λ as a function of Λ, for various choices of m_χ, for $p_\mathrm{T} = 120$ GeV (*left panel*), $p_\mathrm{T} = 500$ GeV (*right panel*). *Bottom row*: R_Λ as a function of m_χ, for various choices of p_T, for $\Lambda = 1.5$ TeV (*left panel*), $\Lambda = 2.5$ TeV (*right panel*)

To sum over the possible p_T, η of the jets, we integrate the differential cross sections over values typically considered in the experimental searches.

$$R_\Lambda^{\text{tot}} \equiv \frac{\sigma|_{Q_{\text{tr}}<\Lambda}}{\sigma} = \frac{\int_{p_T^{\min}}^{p_T^{\max}} dp_T \int_{-2}^{2} d\eta \left.\frac{d^2\sigma}{dp_Td\eta}\right|_{Q_{\text{tr}}<\Lambda}}{\int_{p_T^{\min}}^{p_T^{\max}} dp_T \int_{-2}^{2} d\eta \frac{d^2\sigma}{dp_Td\eta}}. \tag{6.34}$$

We consider $p_T^{\min} = 500\,\text{GeV}$ (as used in the signal region SR4 of [25]), $|\eta| < 2$ and the two cases with centre-of-mass energies $\sqrt{s} = 8$ and $14\,\text{TeV}$. For $p_T^{\max}$ we used 1, $2\,\text{TeV}$ for $\sqrt{s} = 8$, $14\,\text{TeV}$, respectively. The sum over quark flavours is performed only considering u, d, c, s quarks.

We first study the behavior of the ratio R_Λ^{tot}, as a function of Λ and m_χ. The results are shown in Fig. 6.7 for representative s-channel operators $D1'$, $D5$, $D9$ and in Fig. 6.8 for the $DT1$ operator. In the t-channel case, where there are two values of Q_{tr}, mixing between diagrams makes it impossible to disentangle a single transferred momentum for any individual event, and so we require that for each event *both* values of Q_{tr} for that process satisfy the requirement that $Q_{\text{tr}}^2 < \Lambda^2$.

The ratio R_Λ^{tot} gets closer to unity for large values of Λ, as in this case the effect of the cutoff becomes negligible. The ratio drops for large m_χ because the momentum transfer increases in this regime. This confirms what noticed in Sect. 6.4, that the

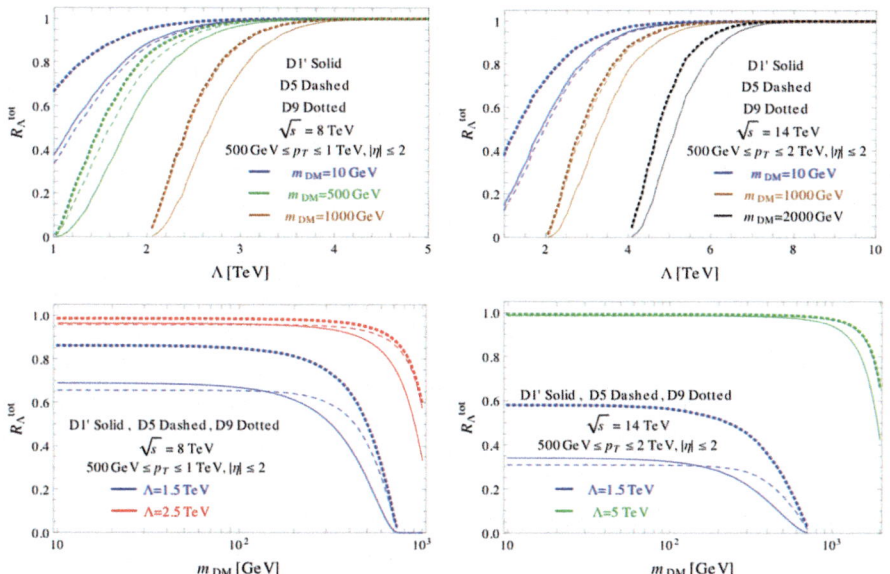

Fig. 6.7 The ratio R_Λ^{tot} defined in Eq. (6.34) for operators $D1'$ (*solid lines*), $D5$ (*dashed lines*) and $D9$ (*dotted lines*) as a function of Λ and m_χ, for $\sqrt{s} = 8\,\text{TeV}$ (*left panel*) and $14\,\text{TeV}$ (*right panel*)

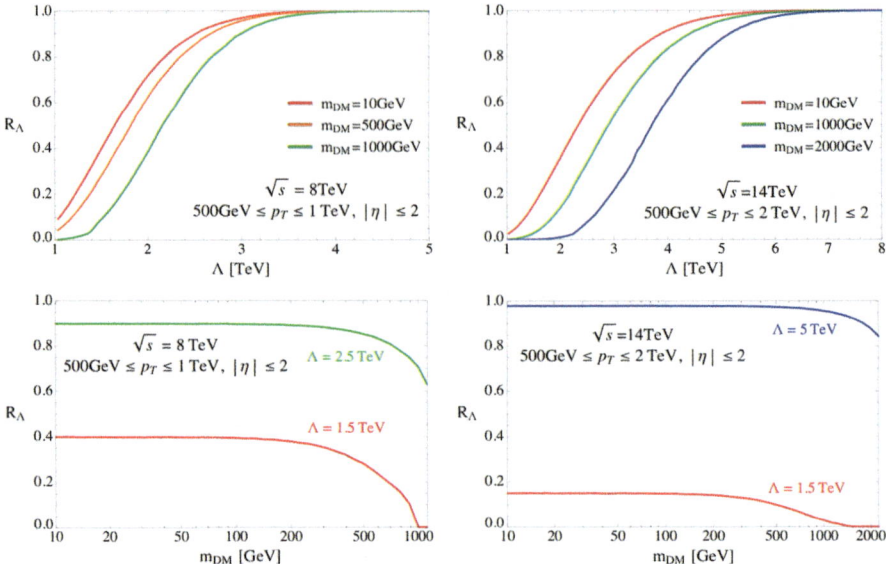

Fig. 6.8 Same as Fig. 6.7, for the DT1 operator

EFT works better for large Λ and small m_χ. Notice also that, going from $\sqrt{s} = 8$ TeV to $\sqrt{s} = 14$ TeV, the results scale almost linearly with the energy, so for the same value of the ratio m_χ/Λ one obtains nearly the same R_Λ^{tot}.

Next, we turn to study the contours of constant values of the quantity R_Λ^{tot}, in the plane (m_χ, Λ). These contour curves for the different operators are shown for the s-channel operators in Fig. 6.9 for $\sqrt{s} = 8$ TeV and in Fig. 6.10 for $\sqrt{s} = 14$ TeV, and in Fig. 6.11 for the t-channel operator $DT1$. The requirement that at least 50% of the events occur with momentum transfer below the cutoff scale Λ requires such a cutoff scale to be above ~ 1 TeV for $\sqrt{s} = 8$ TeV, or above ~ 2 TeV for $\sqrt{s} = 14$ TeV. Note also that the contours for $D1$–$D4$ differ by the corresponding contours for $D1'$–$D4'$ by $\mathcal{O}(1)$ factors, due to the different weighting of the quarks' PDFs. On the other hand, the experimental bounds on the scale of the operators $D1$–$D4$ are much lower (of the order of tens of GeV), as such operators experience an additional suppression of m_q/Λ. This means that the bounds on $D1$–$D4$ are not reliable from the point of view of EFT validity. Contrasted with the s-channel case [19, 21], the ratio R_Λ^{tot} has less DM mass dependence in the t-channel case, being even smaller than in the s-channel case at low DM masses and larger at large DM masses, without becoming large enough to save EFTs.

In Fig. 6.11 we also show the curves corresponding to the correct DM relic density, assuming that interactions between the DM particle and the SM plasma were mediated by the operator (6.11). For given m_χ, larger Λ leads to a smaller self-annihilation cross section and therefore to larger relic abundance. It is evident that the large-Λ region where the EFT is valid typically leads to an unacceptably large DM density.

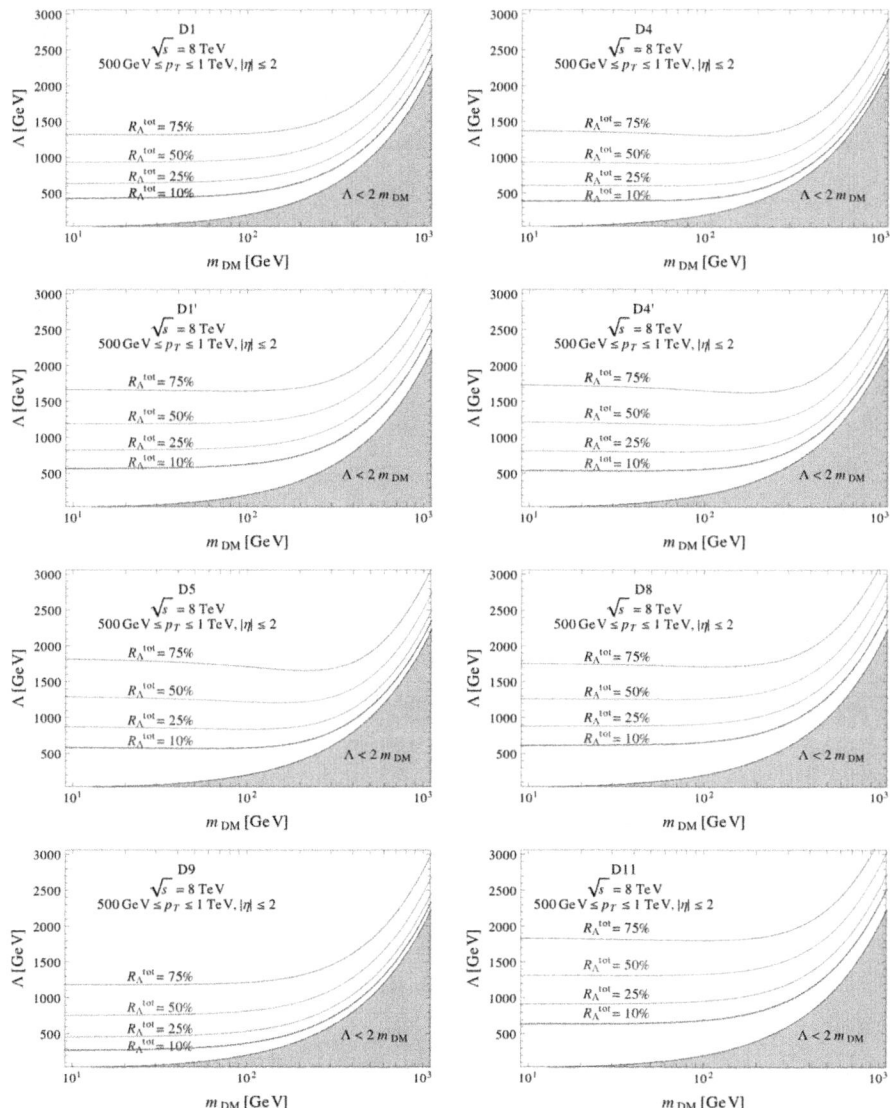

Fig. 6.9 Contours for the ratio R_Λ^{tot}, defined in Eq. (6.34), on the plane (m_χ, Λ), for the different operators. We set $\sqrt{s} = 8$ TeV, $|\eta| \leq 2$ and 500 GeV $< p_T < 1$ TeV

However, it may certainly be that additional annihilation channels and interactions, beyond those described by the operator (6.11) can enhance the cross section and decrease the relic abundance to fit the observations.

We stress once again that the precise definition of a cutoff scale for an EFT is only possible when the details of the UV completion are known. The most conservative

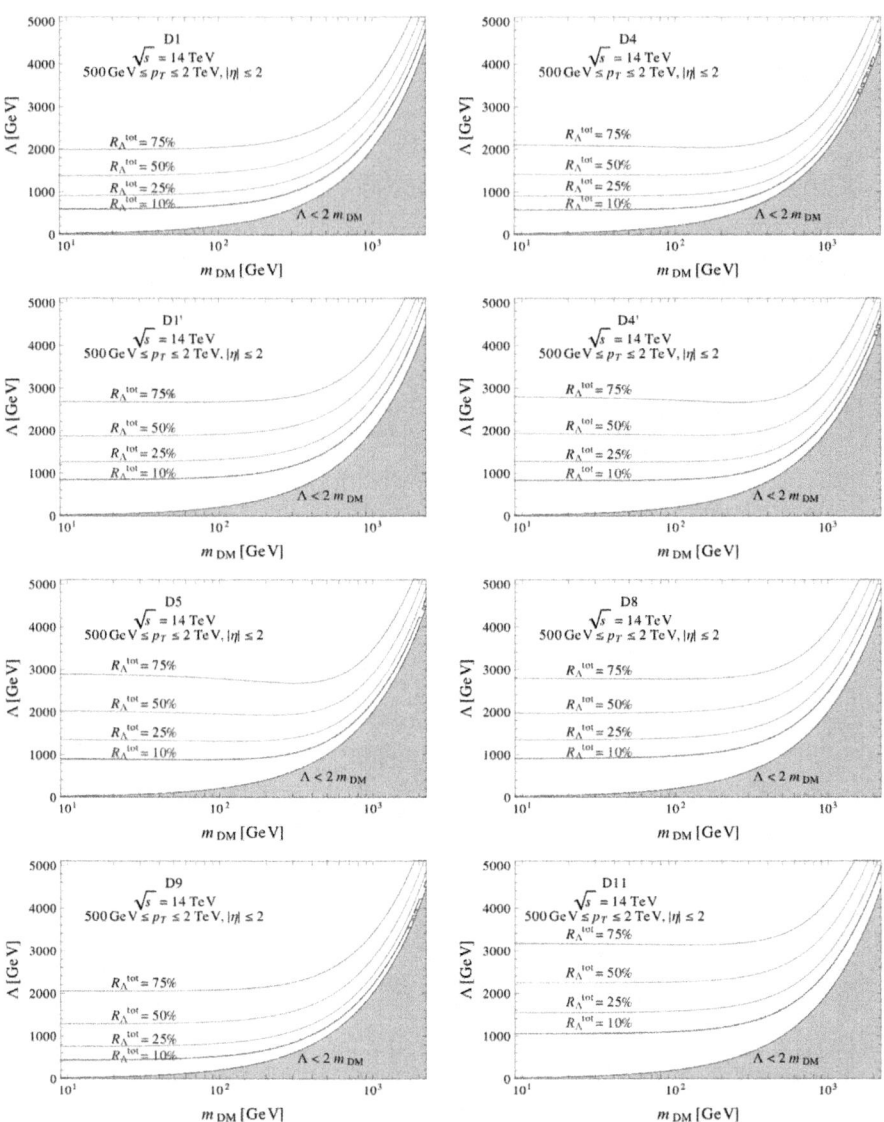

Fig. 6.10 Contours for the ratio R_Λ^{tot}, defined in Eq. (6.34), on the plane (m_χ, Λ), for the different operators. We set $\sqrt{s} = 14\,\mathrm{TeV}$, $|\eta| \leq 2$ and $500\,\mathrm{GeV} < p_{\mathrm{T}} < 2\,\mathrm{TeV}$

regime is when the couplings of the UV theory reach their maximal values allowed by perturbativity. In such a situation, the requirement on the momentum transfer becomes $Q_{\mathrm{tr}} < 4\pi\Lambda$. We show the effect of varying the cutoff scale in Figs. 6.12 and 6.13, for the representative contour $R_\Lambda^{\mathrm{tot}} = 50\%$ of D5 and DT1. As it should be clear, the variation of the cutoff scale is equivalent to a change of the unknown

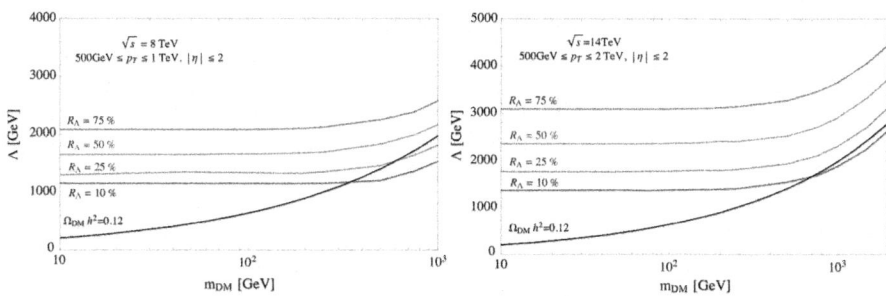

Fig. 6.11 Contours for the ratio R_Λ^{tot}, defined in Eq. (6.34), for the DT1 operator, on the plane (m_χ, Λ). We set $\sqrt{s} = 8\,\text{TeV}$, $|\eta| \leq 2$ and $500\,\text{GeV} < p_T < 1\,\text{TeV}$ in the *left panel*, and $\sqrt{s} = 14\,\text{TeV}$, $|\eta| \leq 2$ and $500\,\text{GeV} < p_T < 2\,\text{TeV}$ in the *right panel*. The *black solid curves* indicates the correct relic abundance

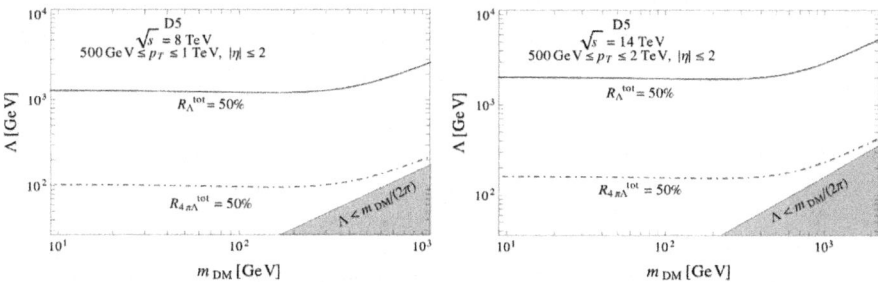

Fig. 6.12 50% contours for the ratio R_Λ^{tot} for the operator $D5$, varying the cutoff $Q_{tr} < \Lambda$ (*solid line*) and $Q_{tr} < 4\pi\Lambda$ (*dot-dashed line*). We have also shown the region corresponding to $\Lambda < m_\chi/(2\pi)$ (*gray shaded area*), often used as a benchmark for the validity of the EFT. We set $\sqrt{s} = 8\,\text{TeV}$ (*left panel*) and $\sqrt{s} = 14\,\text{TeV}$ (*right panel*)

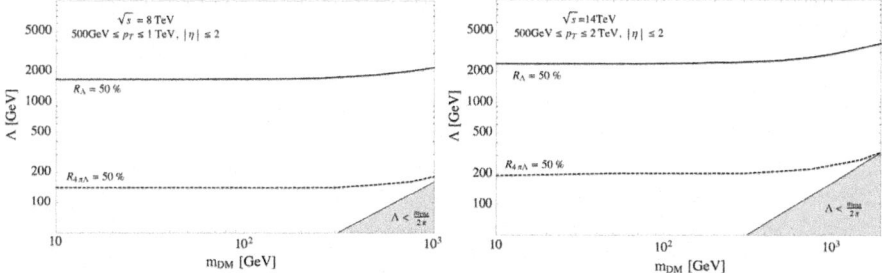

Fig. 6.13 Same as Fig. 6.12 for the DT1 operator

couplings of the UV theory. All the operators have very similar results, as the contours scale linearly with the cutoff. As a comparison, we show as a shaded area the region $\Lambda > m_\chi/(2\pi)$ often used as a benchmark for the validity of the EFT (see Eq. (6.6)). Interestingly, in the t-channel the kinematic constraints on Q_{tr} no longer apply, and the bound on Λ can cross the shaded area. This indicates that at very large DM masses the EFT approximation can become safer than naively assumed - although in practice the ratio is still too low for EFTs to be of any practical use. The 50% contour is above such a region, meaning that the parameter space regions of validity of the effective operator approach is smaller than commonly considered.

6.6 Comparison with Monte Carlo Simulations

In order to perform an alternative check of our analytical results and to be able to compare to the experimental limits as close as possible, we present in this section the results of numerical event simulations. For our purposes, it is not necessary to simulate the event response to the events and their reconstruction.

6.6.1 Simulation and Analysis Description

We made use of `MadGraph` 5 [39] to simulate pp collisions at $\sqrt{s} = 8$ TeV and $\sqrt{s} = 14$ TeV. Both PDF sets CTEQ6L1 and MSTW2008LO (discussed in Ref. [37]) are employed. The PDF choice affects the cross section, but only minimally the acceptance. Hence, the change in contours of the ratio R_Λ^{tot} is negligible. Since MSTW2008LO is used for the analytical calculations, this set is also used where direct comparisons between simulation and calculation are shown. For the comparison to the experimental results, CTEQ6L1 is used instead. As for the analytic calculation, only u, d, c, s quarks were considered in the s-channel, both in the initial and in the final state. In the t- channel analysis only first generation quarks were included in the initial state.

According to the event kinematics we have evaluated whether or not the conditions of validity discussed in Sect. 6.5 are fulfilled. Specifically, we have checked if Eqs. (6.7) and (6.6) are fulfilled, that is, if the following condition is satisfied

$$\Lambda > \frac{Q_{\text{tr}}}{\sqrt{g_q g_\chi}} > 2\frac{m_\chi}{\sqrt{g_q g_\chi}}. \tag{6.35}$$

Samples of 20000 events were simulated for each s-channel operator, scanning DM mass values of 10, 50, 80, 100, 400, 600, 800 and 1000 GeV and cutoff scales of 250, 500, 1000, 1500, 2000, 2500 and 3000 GeV in the case of $\sqrt{s} = 8$ TeV collisions. When increasing the collision energy to $\sqrt{s} = 14$ TeV, the DM mass of 2000 GeV and cutoff scales of 4000 and 5000 GeV were added. In the t-channel analysis, the DM mass scanned the values 10, 50, 100, 200, 300, 500, 1000 and 2000.

From the simulated samples the fraction of events fulfilling $\Lambda > Q_{\text{tr}}/\sqrt{g_q g_\chi}$ for each pair of DM mass and cutoff scale can be evaluated, if one assumes a certain value for the couplings $\sqrt{g_\chi g_q}$ connecting the cutoff scale Λ and the mediator mass M via $\Lambda = M/\sqrt{g_q g_\chi}$. As above, $g_q g_\chi$ was assumed to be 1.

6.6.2 Results

In order to confirm that analytical and numerical results are in agreement, Figs. 6.14 and 6.15 show a comparison for the operators $D1'$, $D4'$, $D5$, $D8$, $D9$ and $DT1$. The results were obtained for the scenario of one radiated gluon jet above 500 GeV within $|\eta| < 2$. The contours of $R_\Lambda^{tot} = 50\%$ from analytical and numerical evaluation agree within less than 7%. The remaining differences could be due to the upper jet p_T cut not imposed during event simulation but needed for the analytical calculation, and the details of the fitting procedures.

Next, we vary the kinematical constraints step by step from the scenario considered in the analytical calculations, namely one radiated gluon jet above 500 GeV within $|\eta| < 2$, to a scenario closest to the analysis cuts applied in the ATLAS monojet analysis [25]. More specifically, the leading jet is allowed to come from either a gluon or a quark being radiated, the leading jet p_T cut is changed from 500 to 350 GeV, a second jet is allowed and its range in η is enlarged to $|\eta| < 4.5$. No further cuts are applied at simulation level.

Fig. 6.14 Comparison of the contour $R_\Lambda^{tot} = 50\%$ for the analytical calculation (*dashed line*) and the simulation (*solid line*) for the different operators $D1'$, $D4'$, $D5$, $D8$ and $D9$. The results agree within less than 7%

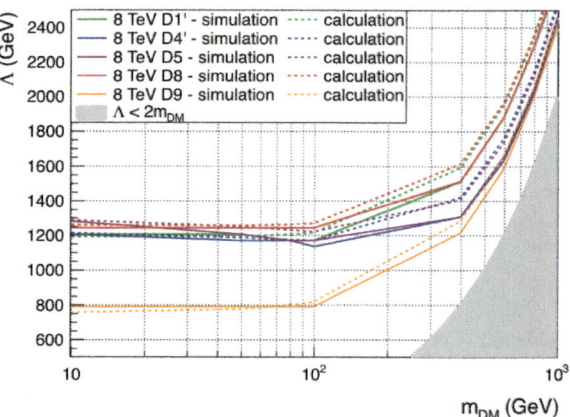

Fig. 6.15 Comparison of the contour $R_\Lambda^{tot} = 50\%$ for the analytical calculation (*solid line*) and the simulation (*dashed line*). The *dotted curve* indicates the correct relic abundance

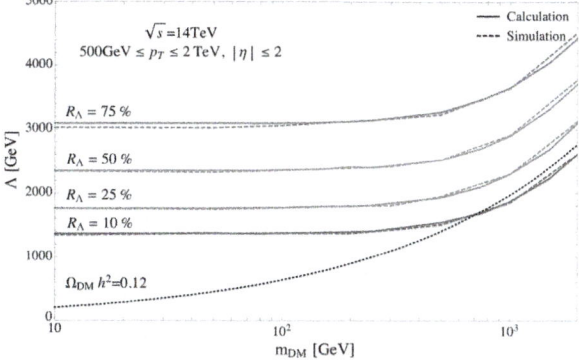

The effect of the variation of the cuts can be seen in Fig. 6.16. Allowing not only for a gluon jet but also taking into account the possibility of a quark jet changes the R_Λ^{tot} contours appreciably. The change from lowering the p_T of the leading jet has a smaller effect. Allowing for a second jet and enhancing its rapidity range barely changes the R_Λ^{tot} contour, especially at large m_χ values.

If the collision energy is augmented to $\sqrt{s} = 14$ TeV, all the R_Λ^{tot} contours increase. As seen for $\sqrt{s} = 8$ TeV, moving to the scenario closer to the experimental analysis leads to contours that are at most $\sim 30\%$ lower in Λ.

After having extracted R_Λ^{tot} for each WIMP and mediator mass, a curve can be fitted through the points obtained in the plane of R_Λ^{tot} and Λ. The following functional form is used for this purpose

Fig. 6.16 The changes of the contour of $R_\Lambda^{tot} = 50\%$ are shown for several variations from the analytically calculated scenario to a scenario close to the cuts used in the ATLAS monojet analysis exemplarily for the operator D5 at $\sqrt{s} = 8$ TeV. In the legend, "g" means only gluon radiation, "j" stands for either quark- or gluon-initiated jets, "j(j)" means a second jet is allowed

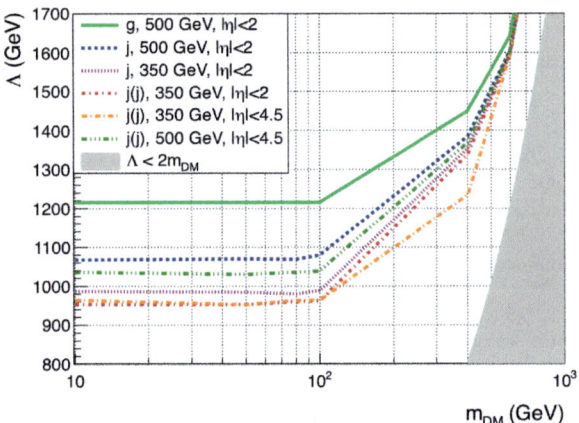

Table 6.2 Coefficient for the fitting functions for R_Λ^{tot} in Eq. (6.36), in the cases $\sqrt{s} = 8$ and 14 TeV. The fitting functions describe processes where quarks and/or gluons are radiated, the final state contains 1 or 2 jets, where the leading jet has minimum p_T of 350 GeV while the second jet is allowed to be within $|\eta| < 4.5$. See text for further details

$\sqrt{s} = 8$ TeV					$\sqrt{s} = 14$ TeV					
Operator	a	b	c	d	e	Operator	a	b	c	d	e
D1	1.32	787.13	1.39	1.08	1.53	D1	0.89	1017.37	1.45	1.28	1.24
D1'	1.30	1008.25	1.49	0.77	1.83	D1'	0.43	909.66	1.59	0.53	1.37
D4	1.65	702.93	1.14	0.65	1.75	D4	1.23	996.82	1.25	0.80	1.48
D4'	1.51	859.83	1.22	0.48	1.92	D4'	0.76	982.75	1.33	0.37	1.63
D5	1.54	816.83	1.18	0.50	1.85	D5	0.78	894.86	1.25	0.39	1.54
D8	1.23	964.62	1.50	0.91	1.59	D8	0.48	945.09	1.55	0.74	1.24
D9	1.43	681.92	1.15	1.02	1.35	D9	0.91	891.65	1.21	1.23	1.04
D11	1.23	1002.33	1.49	0.82	1.69	D11	0.68	1250.49	1.58	0.81	1.35

$$R_\Lambda^{\text{tot}} = \left[1 - e^{-a\left(\frac{\Lambda - 2m_\chi}{b}\right)^c}\right]\left[1 - e^{-d\left(\frac{\Lambda + 2m_\chi}{b}\right)^e}\right]. \qquad (6.36)$$

Further, the parameters are fitted for each DM mass separately. From these fits, the points denoting a cutoff scale where R_Λ^{tot} equals e.g. 50% can be extracted for each DM mass, and the lines of constant R_Λ^{tot} can be plotted in the usual limit-setting plane Λ vs. m_χ. Table 6.2 collects the values of the fitting parameters for all operators except D12-D14 and DT1, for which no experimental analysis exists.

6.7 Implications of the Limited Validity of EFT in DM Searches at LHC

Figure 6.17 shows the experimental limits obtained from the ATLAS monojet analysis [25] in the plane (Λ, m_χ), for the operators D5, D8 and D11. The contours of R_Λ^{tot} for 25, 50 and 75% are superimposed. The experimental limits are placed in a region where about 30% of the events can be expected to fulfill the EFT conditions - the exact number depends on the operator considered. Especially the limit on the gluon operator $D11$ seems questionable. For comparison, dashed lines show the contours of R_Λ^{tot} for the extreme case of couplings $\sqrt{g_q g_\chi} = 4\pi$, presenting the limiting case for which the theory is still considered perturbative.

Unfortuntately, there is no possibility to measure Q_{tr} in data, on an event-by-event basis. So the information on what is the fraction of the events to cut out comes from analytical computations or a numerical simulation. To assess the impact of the limited validity of the EFT on the current collider bounds, we adopt the procedure that relies on the assumption that the p_T (or MET) distributions with the Q_{tr} cut are simply a rescaling of those without the cut. A more refined study should account for possible kinematic shape changes with the jet transverse momentum and/or missing energy and DM mass.

Very naively, neglecting the statistical and systematical uncertainties, the number of signal events in a given EFT model has to be less than the experimental observation, $N_{\text{signal}}(\Lambda, m_\chi) < N_{\text{expt}}$. The cross section due to an operator of mass dimension d scale like $\Lambda^{-2(d-4)}$, so $N_{\text{signal}}(\Lambda, m_\chi) = \Lambda^{-2(d-4)}\tilde{N}_{\text{signal}}(m_\chi)$, and the experimental lower bound in the scale of the operator becomes

$$\Lambda > \left[\tilde{N}_{\text{signal}}(m_\chi)/N_{\text{exp}}\right]^{1/[2(d-4)]} \equiv \Lambda_{\text{expt.}} \ . \qquad (6.37)$$

Now, if we do not consider any information about the shapes of the p_T or MET distributions, the experimental bound only comes from the total number of events passing given cuts. The fact that a fraction of the events involve a transfer momentum exceeding the cutoff scale of the EFT means that the number of signal events for placing a limit gets reduced by a factor R_Λ^{tot}. Therefore, actually $N_{\text{signal}}(\Lambda, m_\chi) \to R_\Lambda^{\text{tot}}(m_\chi)N_{\text{signal}}(\Lambda, m_\chi)$, so the new limit is found by solving the implicit equation

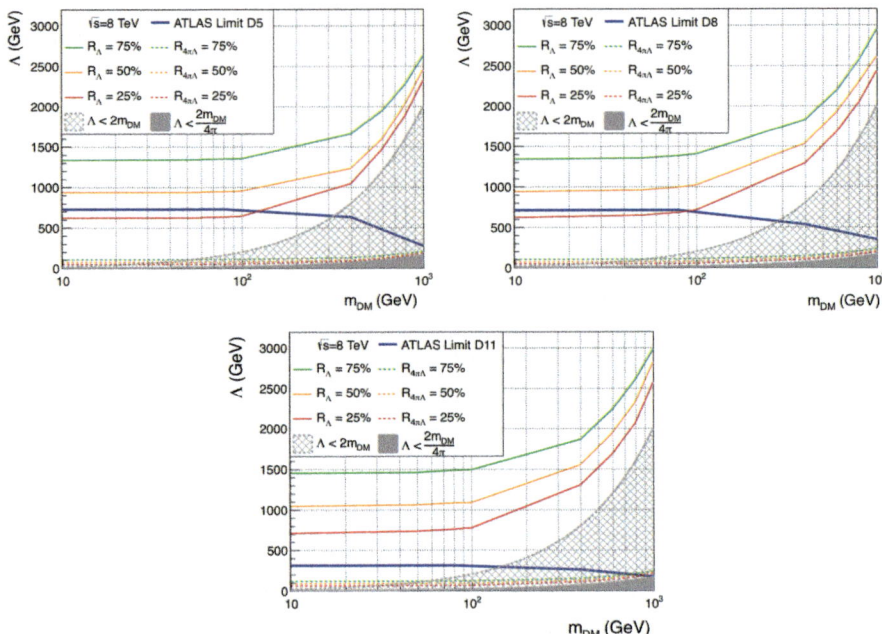

Fig. 6.17 25, 50 and 75% contours for the ratio R_Λ^{tot}, compared to the experimental limits from ATLAS [25] (*blue line*). Also indicated are the contours of R_Λ^{tot} in the extreme case when setting the couplings $\sqrt{g_q g_\chi} = 4\pi$ (*dashed lines*). Results are shown for different operators: D5 (*upper left panel*), D8 (*upper right panel*) and D11 (*lower panel*)

$$\Lambda > [R_\Lambda^{\text{tot}}(m_\chi)]^{1/2[(d-4)]}[N_{\text{signal}}(m_\chi)/N_{\text{exp}}]^{1/[2(d-4)]} = [R_\Lambda^{\text{tot}}(m_\chi)]^{1/[2(d-4)]}\Lambda_{\text{expt}}$$

(6.38)

and it turns out to be weaker than Λ_{expt}. In Fig. 6.18 we show the new limits for the dim-6 operators D5, D8 and the dim-7 operator D11, for the conditions $Q_{\text{tr}} < \Lambda, 2\Lambda, 4\pi\Lambda$, corresponding different choices of the UV couplings: $\sqrt{g_q g_\chi} = 1, 2, 4\pi$, respectively. The curves are obtained solving Eq. 6.38 with R_Λ^{tot}, $R_{2\Lambda}^{\text{tot}}$, $R_{4\pi\Lambda}^{\text{tot}}$ respectively. The ATLAS bound reported is the 90% CL observed limit. The functions R_Λ^{tot} used are taken from the fitting functions described in Table 6.2, which include both quark and gluon jets, and the same cuts as the "Signal Region 3" used by ATLAS. As expected, the weaker is the condition on Q_{tr}, the more the new limits approach the ATLAS bound. In the case of extreme couplings $\sqrt{g_q g_\chi} = 4\pi$, the condition on the momentum transfer is very conservative $Q_{\text{tr}} < 4\pi\Lambda$. For D5 and D8, the new limit is indistinguishable from the ATLAS one, meaning that the experimental results are safe from the EFT point of view, in this limiting situation. For D11, even for extreme values of the couplings, the bound at large DM masses must be corrected. In general, for couplings of order one, the limits which are safe from the EFT point of view are appreciably weaker than those reported.

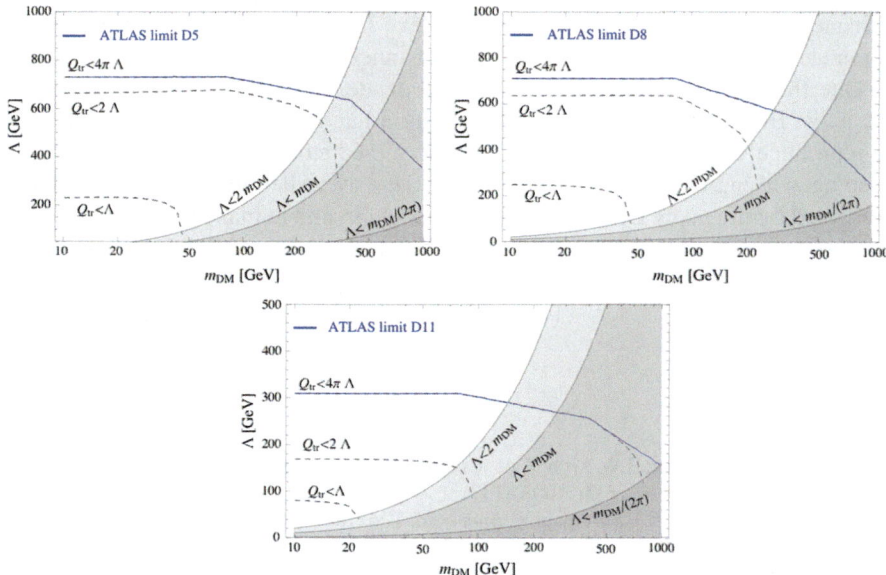

Fig. 6.18 The experimental limits by ATLAS [25] on the suppression scale Λ are shown as *solid blue lines*. The updated limits taking into account EFT validity are shown as *dashed black lines*, for $Q_{tr} < \Lambda, 2\Lambda, 4\pi\Lambda$, corresponding to different choices of the UV couplings: $\sqrt{g_q g_\chi} = 1, 2, 4\pi$, respectively. The corresponding kinematical constraints (Eq. (6.35)) are denoted by *gray bands*. The different plots refer to different operators: D5 (*upper left panel*), D8 (*upper right panel*) and D11 (*lower panel*)

This truncation procedure was first adopted by the ATLAS collaboration in [40], and then inserted in the final recommendations of the *ATLAS-CMS Dark Matter Forum* [41] as a way to present LHC limits on effective operators in DM searches.

One drawback of the procedure presented here is the fact that it assumes the knowledge of the relevant momentum Q_{tr} that has to be compared with the effective scale Λ. This is equivalent to assume a particular UV completion of the effective model. A generalization of our analysis was proposed in [24]. In this paper, the authors impose the condition that the centre of mass energy E_{cm} of the parton-level event is smaller than the cut-off scale of the EFT. In this way they obtain bounds that are fully model independent, at the price of having less stringent limits.

6.8 Conclusions

In this chapter we described the limitations of the EFT approach and the procedures needed to get consistent results out of it, following the line of [19, 21, 23]. The key quantity in this analysis was the ratio R_Λ^{tot}, which estimates the fraction of events in a collider set-up which respect the EFT validity condition $Q_{tr} < \Lambda$. Our results indicate that the range of validity of the naïve EFT is significantly limited in the parameter space (Λ, m_χ), both for s and t-channel operators.

One the one hand, our results clearly indicate the need of an overcoming of the naïve EFT approach, through identifying a handful of classes of models able to reproduce the EFT operators in the heavy mediator limit. On the other hand, keeping working with the EFT allows to avoid the overwhelming model-dependence generated by the many DM models proposed so far, and to make the bounds clearer and easy to present. Nonetheless, as we have shown in Sect. 6.7, the price to pay is a deterioration of the limits.

References

1. M. Beltran, D. Hooper, E.W. Kolb, Z.A. Krusberg, T.M. Tait, Maverick dark matter at colliders. JHEP **1009**, 037 (2010), arXiv:1002.4137
2. J. Goodman, M. Ibe, A. Rajaraman, W. Shepherd, T.M. Tait et al., Constraints on light Majorana dark matter from colliders. Phys. Lett. B **695**, 185–188 (2011), arXiv:1005.1286
3. Y. Bai, P.J. Fox, R. Harnik, The tevatron at the frontier of dark matter direct detection. JHEP **1012**, 048 (2010), arXiv:1005.3797
4. J. Goodman, M. Ibe, A. Rajaraman, W. Shepherd, T.M. Tait et al., Constraints on dark matter from colliders. Phys. Rev. D **82**, 116010 (2010), arXiv:1008.1783
5. P.J. Fox, R. Harnik, J. Kopp, Y. Tsai, LEP shines light on dark matter. Phys. Rev. D **84**, 014028 (2011), arXiv:1103.0240
6. A. Rajaraman, W. Shepherd, T.M. Tait, A.M. Wijangco, LHC bounds on interactions of dark matter. Phys. Rev. D **84**, 095013 (2011), arXiv:1108.1196
7. P.J. Fox, R. Harnik, J. Kopp, Y. Tsai, Missing energy signatures of dark matter at the LHC. Phys. Rev. D **85**, 056011 (2012), arXiv:1109.4398
8. I.M. Shoemaker, L. Vecchi, Unitarity and Monojet bounds on models for DAMA, CoGeNT, and CRESST-II. Phys. Rev. D **86**, 015023 (2012), arXiv:1112.5457
9. H. An, X. Ji, L.-T. Wang, Light dark matter and Z' dark force at colliders. JHEP **07**, 182 (2012), arXiv:1202.2894
10. R. Cotta, J. Hewett, M. Le, T. Rizzo, Bounds on dark matter interactions with electroweak gauge bosons. Phys. Rev. D **88**, 116009 (2013), arXiv:1210.0525
11. H. Dreiner, M. Huck, M. Krämer, D. Schmeier, J. Tattersall, Illuminating dark matter at the ILC. Phys. Rev. D **87**(7), 075015 (2013), arXiv:1211.2254
12. Y.J. Chae, M. Perelstein, Dark matter search at a linear collider: effective operator approach. JHEP **1305**, 138 (2013), arXiv:1211.4008
13. P.J. Fox, C. Williams, Next-to-leading order predictions for dark matter production at Hadron colliders. Phys. Rev. D **87**, 054030 (2013), arXiv:1211.6390
14. A. De Simone, A. Monin, A. Thamm, A. Urbano, On the effective operators for dark matter annihilations. JCAP **1302**, 039 (2013), arXiv:1301.1486
15. H. Dreiner, D. Schmeier, J. Tattersall, Contact interactions probe effective dark matter models at the LHC. Europhys. Lett. **102**, 51001 (2013), arXiv:1303.3348
16. J.-Y. Chen, E.W. Kolb, L.-T. Wang, Dark matter coupling to electroweak gauge and Higgs bosons: an effective field theory approach, Phys. Dark Univ. **2**, 200–218, (2013), arXiv:1305.0021
17. Q.-H. Cao, C.-R. Chen, C.S. Li, H. Zhang, Effective dark matter model: relic density, CDMS II. Fermi LAT and LHC. JHEP **08**, 018 (2011), arXiv:0912.4511
18. J. Fan, M. Reece, L.-T. Wang, Non-relativistic effective theory of dark matter direct detection. JCAP **1011**, 042 (2010), arXiv:1008.1591

19. G. Busoni, A. De Simone, E. Morgante, A. Riotto, On the validity of the effective field theory for dark matter searches at the LHC. Phys. Lett. B **728**, 412–421 (2014), arXiv:1307.2253
20. O. Buchmueller, M.J. Dolan, C. McCabe, Beyond effective field theory for dark matter searches at the LHC. JHEP **1401**, 025 (2014), arXiv:1308.6799
21. G. Busoni, A. De Simone, J. Gramling, E. Morgante, A. Riotto, On the validity of the effective field theory for dark matter searches at the LHC, Part II: complete analysis for the s-channel, JCAP **1406**, 060 (2014), arXiv:1402.1275
22. A. Berlin, T. Lin, L.-T. Wang, Mono-Higgs detection of dark matter at the LHC. JHEP **06**, 078 (2014), arXiv:1402.7074
23. G. Busoni, A. De Simone, T. Jacques, E. Morgante, A. Riotto, On the validity of the effective field theory for dark matter searches at the LHC part III: analysis for the t-channel, JCAP **1409**, 022 (2014), arXiv:1405.3101
24. D. Racco, A. Wulzer, F. Zwirner, Robust collider limits on heavy-mediator dark matter. JHEP **05**, 009 (2015), arXiv:1502.04701
25. ATLAS Collaboration, *Search for New Phenomena in Monojet plus Missing Transverse Momentum Final States using 10fb-1 of pp Collisions at $\sqrt{s} = 8\,TeV$ with the ATLAS detector at the LHC*, ATLAS-CONF-2012-147
26. R. Barbieri, G. Isidori, J. Jones-Perez, P. Lodone, D.M. Straub, U(2) and minimal flavour violation in supersymmetry. Eur. Phys. J. C **71**, 1725 (2011), arXiv:1105.2296
27. R. Barbieri, D. Buttazzo, F. Sala, D.M. Straub, Flavour physics from an approximate $U(2)^3$ symmetry. JHEP **07**, 181 (2012), arXiv:1203.4218
28. N.F. Bell, J.B. Dent, A.J. Galea, T.D. Jacques, L.M. Krauss et al., Searching for dark matter at the LHC with a Mono-Z. Phys. Rev. D **86**, 096011 (2012), arXiv:1209.0231
29. ATLAS Collaboration, G. Aad et al., *Search for dark matter in events with a Z boson and missing transverse momentum in pp collisions at $\sqrt{s} = 8\,TeV$ with the ATLAS detector*, Phys. Rev. D **90**(1), 012004 (2014), arXiv:1404.0051
30. S. Chang, R. Edezhath, J. Hutchinson, M. Luty, Effective WIMPs. Phys. Rev. D **89**, 015011 (2014), arXiv:1307.8120
31. H. An, L.-T. Wang, H. Zhang, *Dark matter with t-channel mediator: a simple step beyond contact interaction*, Phys. Rev. D **89**(11), 115014 (2014), arXiv:1308.0592
32. Y. Bai, J. Berger, Fermion portal dark matter. JHEP **1311**, 171 (2013), arXiv:1308.0612
33. A. DiFranzo, K.I. Nagao, A. Rajaraman, T.M.P. Tait, Simplified models for dark matter interacting with quarks. JHEP **1311**, 014 (2013), arXiv:1308.2679
34. M. Papucci, A. Vichi, K.M. Zurek, *Monojet versus rest of the world I: t-channel Models*, JHEP **1411**, 024 (2014), arXiv:1402.2285
35. M. Garny, A. Ibarra, S. Rydbeck, S. Vogl, *Majorana Dark Matter with a Coloured Mediator: Collider versus Direct and Indirect Searches*, JHEP **1406**, 169 (2014), arXiv:1403.4634
36. N.F. Bell, J.B. Dent, T.D. Jacques, T.J. Weiler, W/Z bremsstrahlung as the dominant annihilation channel for dark matter. Phys. Rev. D **83**, 013001 (2011), arXiv:1009.2584
37. A. Martin, W. Stirling, R. Thorne, G. Watt, Parton distributions for the LHC. Eur. Phys. J. C **63**, 189–285 (2009), arXiv:0901.0002
38. http://mstwpdf.hepforge.org/
39. J. Alwall, M. Herquet, F. Maltoni, O. Mattelaer, T. Stelzer, MadGraph 5: going beyond. JHEP **1106**, 128 (2011), arXiv:1106.0522
40. ATLAS Collaboration, G. Aad et al., Search for new phenomena in final states with an energetic jet and large missing transverse momentum in pp collisions at $\sqrt{s}$ =8 TeV with the ATLAS detector. Eur. Phys. J. C**75**(7) 299 (2015), arXiv:1502.01518. [Erratum: Eur. Phys. J. C75, no.9, 408 (2015)]
41. D. Abercrombie et al., *Dark Matter Benchmark Models for Early LHC Run-2 Searches: Report of the ATLAS/CMS Dark Matter Forum*, arXiv:1507.00966

Chapter 7
Simplified Models

7.1 Introduction

In this chapter we will discuss the construction and the use of simplified models for DM searches at the LHC. First we are going to describe the philosophy of simplified models in relation to the EFT approach and to complete new physics models. Then we are going to list the set of benchmarks models which have emerged as the most common ones in the recent literature. Finally we are going to summarize results obtained with these models, without entering into details about LHC searches, implementation of the models for experimental analyses, or recasting of existing bounds.

The first part of this chapter will be based on a series of white papers appeared in the last three years [1–3], in which the machinery of simplified models was set up, and on the reports of the ATLAS/CMS DM Forum [4] and the subsequent LHC Dark Matter Working Group [5]. The idea of simplified DM models and a description of its state of the art is provided in [6].

As in the case of the EFT, the idea beyond simplified models is to provide a good representation of possibly all realistic WIMP scenarios within the energy reach of the LHC, restricting to the smallest possible set of benchmark models, each with the least possible number of free parameters. Simplified models should then be complete enough to give an accurate description of the physics at the scale probed by colliders, but at the same time they must have a limited number of new states and parameters. More over, they should satisfy all constraints posed by low-p_T analysis, such as those coming from flavour physics. A recipe to build a simplified model is:

- Besides the SM, they must contain a stable DM candidate and a mediator that couples the two sectors. All additional states should be decoupled.
- The Lagrangian should contain all renormalizable terms consistent with Lorentz invariance, gauge symmetry and DM stability.
- *Ad hoc* simplifications may be achieved by setting some parameters to zero or taking some of them to be equal, but this should be implemented in such a way that the phenomenology is not totally altered, in order not to prejudice the credibility of the constraints on the model itself.

© Springer International Publishing AG 2017
E. Morgante, *Aspects of WIMP Dark Matter Searches at Colliders and Other Probes*, Springer Theses, https://doi.org/10.1007/978-3-319-67606-7_7

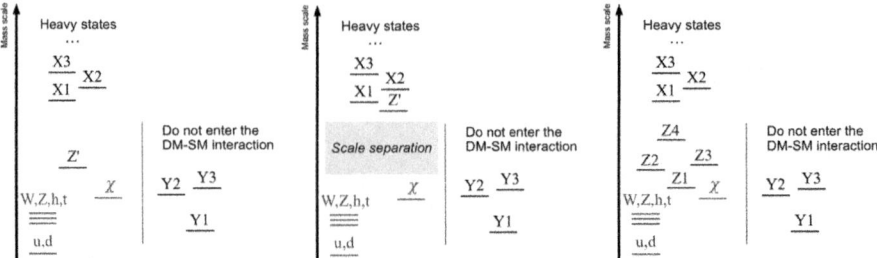

Fig. 7.1 *Left* a simplified model viewed as a sector of a more general new physics scenario. The SM is complemented by the DM particle χ and a mediator Z'. Other heavy states $X_1, X_2, X_3, \ldots$ may be integrated out because they are very heavy, while lighter states $Y_1, Y_2, Y_3, \ldots$ do not play any role and may be ignored. *Centre* in the case in which the mediator Z' itself has a very large mass, it may be integrated out as well and the interaction is mediated by effective operators. *Right* when the DM-SM interaction is mediated by a number of additional operators, possibly interfering with each other, the simplified model approach is no more valid, and a complete description of the model is necessary

- Interactions that violate the accidental global symmetries of the SM model (both exact and approximate) must be handled with great care. Indeed, constraints on processes that violate these symmetries are typically very strong, and may overcome those coming from DM searches or even rule out all of the interesting parameter space of the simplified model. For this reason, lepton and baryon number conservation is typically assumed, together with Minimal Flavour Violation (MFV).[1] Even with this assumption, there are cases in which constraints from flavour physics may be stronger than those coming from mono-X searches [8] (see also [9] for a discussion of a non-minimally flavour violating dark sector).

Most simplified models of interest may be understood as the limit of a more general new-physics scenario, where all new states but a few are integrated out because they have a mass larger than the energy scale reachable at the LHC or because they have no role in DM interactions with the SM. Similarly, in the limit where the mass of the mediator is very large, the EFT framework may be recovered by integrating out the mediator. On the contrary, there are new physics models which can not be recast in terms of simplified models, typically because more than just one operators are active at the same time, and possibly interfere with each other. The situation is summarized in Fig. 7.1.

---

[1]Constraints on BSM models from CP and flavour violating observables are very strong, and the energy scale at which new physics may show up must be larger than tens of TeV in the best case, if the flavour structure of the model is generic. Minimal Flavour Violation is a way to reconcile these constraints with possible new physics at the TeV scale [7]. The basic idea is that the structure of flavour changing interactions must reproduce that of the SM. The SM is invariant under the flavour group $\mathcal{G}_F = SU(3)_q \times SU(3)_u \times SU(3)_d$, except from a small breaking associated to the Yukawa matrices Y_u and Y_d. The invariance is restored if these matrices are regarded as "spurions" with transformation law $Y_u \sim (3, \bar{3}, 1)$ and $Y_d \sim (3, 1, \bar{3})$. Imposing MFV amounts to requiring that new physics is invariant under $\mathcal{G}_F$.

It should be noticed that the correspondence between simplified models and EFT is not one to one. Different simplified models may give rise to the same effective operator. To illustrate this point, let us consider the two following examples [10]. The first model is characterized by a Dirac DM particle and a vector mediator Z' which interacts only with the axial current of DM and quarks:

$$\mathcal{L}_A = \mathcal{L}_{SM} - \frac{1}{4} Z'_{\mu\nu} Z'^{\mu\nu} + \frac{1}{2} m^2_{Z'} Z'_\mu Z'^\nu + \frac{1}{2} \bar{\chi} (\slashed{\partial} - m_\chi) \chi + Z'_\mu \left(g_q \sum_q \bar{q} \gamma^\mu \gamma^5 q + g_\chi \bar{\chi} \gamma^\mu \gamma^5 \chi \right) \tag{7.1}$$

In the low momentum limit $p^2 \ll m^2_{Z'}$, the solution of the classical equation of motion for the Z' reads

$$Z'_\mu = -\frac{1}{m^2_{Z'}} J_\mu, \quad \text{where} \quad J_\mu = \left(g_q \sum_q \bar{q} \gamma_\mu \gamma^5 q + g_\chi \bar{\chi} \gamma_\mu \gamma^5 \chi \right), \tag{7.2}$$

which, substituted in Eq. (7.1), yields an effective Lagrangia for the DM-quarks interaction

$$\mathcal{L}_A^{EFT} = -\frac{g_q g_\chi}{m^2_{Z'}} \bar{\chi} \gamma_\mu \gamma^5 \chi \sum_q (\bar{q} \gamma^\mu \gamma^5 q) \tag{7.3}$$

The second model we want to consider is inspired by Supersymmetry. We add to the SM a Majorana fermion to play the role of DM (similar to the supersymmetric neutralino) and a set of "squarks" that mediate the interaction with the SM. In particular, we consider three families of scalar particles degenerate in mass, which we denote by $(\tilde{u}_{iL}, \tilde{d}_{iL}, \tilde{u}_{iR}, \tilde{d}_{iR})$, where $i = 1, 2, 3$ are family indices. The quantum numbers are the same of the quarks $(u_{iL}, d_{iL}, u_{iR}, d_{iR})$, respectively. The Lagrangian in this model reads

$$\mathcal{L}_B = \mathcal{L}_{SM} + \mathcal{L}_\chi + \mathcal{L}_{\tilde{q}} + \mathcal{L}^B_{int}, \tag{7.4}$$

$$\mathcal{L}_{\tilde{q}} = \sum_{i=1}^3 \left[(\partial^\mu \tilde{u}_{iL})^\dagger (\partial_\mu \tilde{u}_{iL}) + (\partial^\mu \tilde{d}_{iL})^\dagger (\partial_\mu \tilde{d}_{iL}) + (\partial^\mu \tilde{u}_{iR})^\dagger (\partial_\mu \tilde{u}_{iR}) + (\partial^\mu \tilde{d}_{iR})^\dagger (\partial_\mu \tilde{d}_{iR}) \right.$$
$$\left. - \tilde{m}^2 \left(\tilde{u}^\dagger_{iL} \tilde{u}_{iL} + \tilde{d}^\dagger_{iL} \tilde{d}_{iL} + \tilde{u}^\dagger_{iR} \tilde{u}_{iR} + \tilde{d}^\dagger_{iR} \tilde{d}_{iR} \right) \right] + \dots, \tag{7.5}$$

$$\mathcal{L}^B_{int} = -g_\chi \left[\sum_{i=1}^3 \left(\tilde{u}_{iL} \bar{u}_{iL} + \tilde{d}_{iL} \bar{d}_{iL} + \tilde{u}_{iR} \bar{u}_{iR} + \tilde{d}_{iR} \bar{d}_{iR} \right) \chi + \text{h.c.} \right], \tag{7.6}$$

where the dots in the second line of Eq. (7.5) stands for the gauge interactions of the squarks. Solving the classical equations of motion for the fieldsone gets, in the low momentum limit,

$$\tilde{u}_{iL} = -\frac{g_{DM}}{\tilde{m}^2}\, \bar{\chi} u_{iL}\,, \quad \tilde{u}_{iR} = -\frac{g_{DM}}{\tilde{m}^2}\, \bar{\chi} u_{iR}\,, \quad \tilde{d}_{iL} = -\frac{g_{DM}}{\tilde{m}^2}\, \bar{\chi} d_{iL}\,, \quad \tilde{d}_{iR} = -\frac{g_{DM}}{\tilde{m}^2}\, \bar{\chi} d_{iR}\,. \tag{7.7}$$

which, substituting into $\mathcal{L}_B$, yields

$$\begin{aligned}
\mathcal{L}_B^{\mathrm{EFT}} &= \frac{g_{DM}^2}{\tilde{m}^2} \sum_{i=1}^{3} \left[(\bar{\chi} u_{iL})(\bar{u}_{iL}\chi) + (\bar{\chi} u_{iR})(\bar{u}_{iR}\chi) + (\bar{\chi} d_{iL})(\bar{d}_{iL}\chi) + (\bar{\chi} d_{iR})(\bar{d}_{iR}\chi) \right] \\
&= -\frac{g_{DM}^2}{4\tilde{m}^2} \left(\bar{\chi}\gamma^\mu\gamma^5\chi \right) \left[\sum_{i=1}^{3} \left(\bar{u}_i\gamma_\mu\gamma^5 u_i + \bar{d}_i\gamma_\mu\gamma^5 d_i \right) \right],
\end{aligned} \tag{7.8}$$

where the second line follows from the use of Fierz identities and the fact that for a Majorana fermion the vector bilinear $\bar{\chi}\gamma^\mu\chi$ vanishes identically. The effective Lagrangians of Eqs. (7.3), (7.8) contain the same operator and may be identified with an opportune choice of coefficients.

7.2 Classification of Simplified Models

7.2.1 Mediator Exchange in the s-Channel

The first set of simplified models we want to consider is the one where the DM interacts with quarks through the exchange of a mediator in the s-channel. Assuming the DM particle χ to be a fermion (either Dirac or Majorana), and assuming CP-conservation, the Lagrangian of our models are

$$\begin{aligned}
\mathcal{L}_S &\supset -\frac{1}{2}M_{\mathrm{med}}^2 S^2 - y_\chi S\bar{\chi}\chi - y_q^{ij} S\bar{q}_i q_j + \mathrm{h.c.}\,, \\[4pt]
\mathcal{L}_{S'} &\supset -\frac{1}{2}M_{\mathrm{med}}^2 S'^2 - y_\chi' S'\bar{\chi}\gamma_5\chi - y_q'^{ij} S'\bar{q}_i\gamma_5 q_j + \mathrm{h.c.}\,, \\[4pt]
\mathcal{L}_V &\supset \frac{1}{2}M_{\mathrm{med}}^2 V_\mu V^\mu - g_\chi V_\mu\bar{\chi}\gamma^\mu\chi - g_q^{ij} V_\mu\bar{q}_i\gamma^\mu q_j\,, \\[4pt]
\mathcal{L}_{V'} &\supset \frac{1}{2}M_{\mathrm{med}}^2 V_\mu' V'^\mu - g_\chi' V_\mu'\bar{\chi}\gamma^\mu\gamma_5\chi - g_q'^{ij} V_\mu'\bar{q}_i\gamma^\mu\gamma_5 q_j\,. \tag{7.9}
\end{aligned}$$

where S, S', V, V' stand for a scalar, a pseudo-scalar, a vector or an axial-vector mediator respectively, $q = u, d$ and $i, j = 1, 2, 3$ are flavor indices.

As concerns the mediator couplings to quarks, the existence of off-diagonal coupling is tightly constrained by various FCNC processes [11]. For this reason, a good choice is to force the couplings to be diagonal: $g_q^{ij} = g_q^i \delta^{ij}$. As a further simplification, one could fix the couplings to be flavour blind

$$g_d^i = g_u^i \equiv g_q \quad \text{for} \quad i = 1, 2, 3, \tag{7.10}$$

or take the coupling to the third generation to be stronger than the others (as described at the end of this section) or absent. We further assume that the only available decay channels of the mediator are into quarks and DM particles.

It should be noticed that the scalar and pseudo-scalar models of Eq. (7.9) are not gauge invariant. This may lead to spurious results in processes where a W/Z boson is emitted, but results of jets + MET searches are expected to be only mildly affected by this issue [5]. Moreover, in the axial vector model perturbative unitarity is violated in a large portion of parameter space [12], and the indication of where the violation happens should be clearly shown when presenting constraints on this model. We will return to these issues at the end of this chapter.

In the low momentum limit, the simplified models of Eq. (7.9) give rise to the s-channel effective operators D1', D4', D5 and D8 of Table 6.1. Notice that, even if as was shown in the previous section the reversed correspondence do not strictly hold, the simplified models of Eq. (7.9) are the most obvious UV completion of the D1', D4', D5 and D8 operators, and so we can refer to them as the "corresponding" simplified models as is often done in the literature.

In the limit of low quark mass, the operators (D2', D3', D6, D7) have the same partonic cross section for $\bar{q}q \to \bar{\chi}\chi g$ as the operators (D4', D1', D8, D5). The same is true for the corresponding simplified models, and therefore we do not consider models in which the interaction vertices of quarks and DM with the mediator have a different Dirac structure. Moreover, the (D1-D4) operators of Table 6.1 would be obtained by a simplified model in which the Yukawa couplings scale as $y_q \simeq m_q/M_{\text{med}}$.

The differential cross sections at the parton level (with respect to the pseudo-rapidity (η) and transverse momentum (p_T) of the final jet) for the s-channel process $f(p_1) + \bar{f}(p_2) \to \chi(p_3) + \chi(p_4) + g(k)$ are given in the Appendix A. It should be noticed that this calculation is not suited for the case in which the Yukawa couplings of the scalar mediator scale as the particle mass (corresponding to the operators D1-D4). In this case the production will happen predominantly via gluon fusion through a loop of top quarks, where the final jet is originated by a gluon emitted from the quarks in the loop or by an initial state quark (see Fig. 7.2).

As discussed earlier, the EFT approach is appropriate in processes with low energy transfer:

$$M_{\text{med}} \gtrsim Q_{\text{tr}} \geq 2m_\chi. \tag{7.11}$$

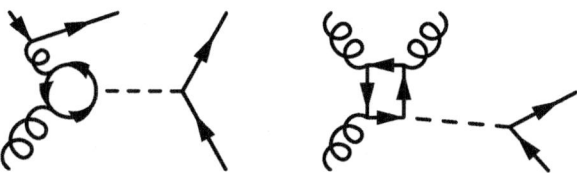

Fig. 7.2 In a simplified model with a scalar mediator with Higgs-like couplings, the dominant production mechanism is gluon fusion, with a loop of top quarks

At the partonic level the differences between the cross sections of the effective theory and the full theory are

$$\left(\frac{\mathrm{d}^2\hat{\sigma}}{\mathrm{d}\eta\,\mathrm{d}p_\mathrm{T}}\right)_\mathrm{full} \Big/ \left(\frac{\mathrm{d}^2\hat{\sigma}}{\mathrm{d}\eta\,\mathrm{d}p_\mathrm{T}}\right)_\mathrm{EFT} = \frac{M^4_\mathrm{med}}{\left(Q^2_\mathrm{tr} - M^2_\mathrm{med}\right)^2 + \Gamma^2 M^2_\mathrm{med}}, \qquad (7.12)$$

where $\Lambda = M_\mathrm{med}/\sqrt{g_q g_\chi}$ was used, independently of the spin of the mediator. The ratio between the EFT resulting cross section and the full theory at $8\,\mathrm{TeV}$ centre of mass energy was studied in Sect. 6.4. This ratio is smaller than 50% for both scalar and vector interactions if $\Lambda \gtrsim 2 - 3\,\mathrm{TeV}$ and $m_\chi \lesssim 1\,\mathrm{TeV}$. A more refined analysis based on MC simulations is presented in [1]. As it is intuitive, the EFT and the simplified model approach coincide in the limit in which the mass of the mediator is larger than the DM mass and the p_T of the jet (which gives a rough measure of Q_tr). On the other hand, if the mediator is light, the two differ for both the total normalization of the cross section and for the kinematical distribution.

To find the most convenient and enlightening set of simplified models, one needs to study the sensitivity of the observables to the helicity structure of the mediator couplings. As a rough estimate, the calculation of Appendix A shows that, in the massless quark limit and including only gluon emission, at the parton level

$$\left(\frac{\mathrm{d}^2\hat{\sigma}}{\mathrm{d}\eta\,\mathrm{d}p_\mathrm{T}}\right)_s \Big/ \left(\frac{\mathrm{d}^2\hat{\sigma}}{\mathrm{d}\eta\,\mathrm{d}p_\mathrm{T}}\right)_{s'} = \left(1 - \frac{4m^2_\mathrm{DM}}{Q^2_\mathrm{tr}}\right), \qquad (7.13)$$

and

$$\left(\frac{\mathrm{d}^2\hat{\sigma}}{\mathrm{d}\eta\,\mathrm{d}p_\mathrm{T}}\right)_v \Big/ \left(\frac{\mathrm{d}^2\hat{\sigma}}{\mathrm{d}\eta\,\mathrm{d}p_\mathrm{T}}\right)_{v'} = \left(\frac{Q^2_\mathrm{tr} + 2m^2_\mathrm{DM}}{Q^2_\mathrm{tr} - 4m^2_\mathrm{DM}}\right). \qquad (7.14)$$

In the limit $Q_\mathrm{tr} \gg 2m_\mathrm{DM}$, the two differential cross sections share the same η and p_T distribution, and therefore one among scalar and pseudo-scalar mediator can be neglected in favour of the other for the sake of simplicity (and the same for vector and axial-vector).[2] However, these kinematical regions are suppressed by the parton distribution functions (PDFs), and therefore it is not clear a priori that this estimate gives the correct answer. An analysis based on Monte Carlo simulations was performed in [1], and it shows that the ratio is indeed independent of the jet p_T. Nevertheless, the two cross sections differ in normalization, with non trivial dependence on M_med for heavy DM masses, as a result of the PDFs. Since the scalar and axial vector interactions result in smaller cross-sections it is sufficient, as a

---

[2]At the energy scales involved at particle colliders, the valence quarks are practically massless, thus their helicities coincide with their chiralities. Vector and axial vector Lorentz bilinears can be written respectively as the sum and the difference of bilinears made up with L-handed or R-handed quarks. Since the helicity is a physical observable, the total cross section can be decomposed in terms of cross sections with given initial helicity states, and their interference term vanishes. Since the latter is the only difference that could arise between the cross sections for the vector or axial vector quark current, the two are practically identical.

first step, to explore the scalar and axial vector mediation resolving the s-channel DM pair-production at the LHC. If a signal is discovered, further analysis of the jet angular distribution could differentiate between the different particles mediating the DM production.

Constraints on s-channel simplified models have been obtained by numerous groups, in particular for exchange of a vector mediator (scalar mediators are more problematic, see the discussion in Sect. 7.3). In the mono-jet channel, the analysis in [1] shows that, for $M_{med} \lesssim 2m_\chi$, the LHC at 14 TeV with 300 fb^{-1} is sensitive to $\mathcal{O}(1)$ couplings only for $m_\chi \lesssim \mathcal{O}(100\,\text{GeV})$, while for $m_\chi \sim 1$ TeV it is sensitive only to couplings of order $g_\chi \cdot g_q \gtrsim 10$. Mono-jet and mono-X constraints are discussed in details in [13, 14], in which a rescaling procedure is proposed in order to set limits on the model's couplings in the plane $m_{DM} - M_{med}$ without the need of a three-dimensional scanning. In [15, 16] mono-jet searches are compared to dijet searches (see also [17]), direct detection limits, dark matter overproduction in the early universe and constraints from perturbative unitarity.

A very interesting phenomenology arises in the case where the couplings to third generation quarks is larger than the couplings to the first two. This may happen for example in models in which a scalar mediator is exchanged and MFV (or the $U(2)^3$ symmetry proposed in [18, 19]) is assumed. The Yukawa couplings of the mediator are then proportional to the fermions' mass, resulting in an enhanced coupling to b and t quarks. Because of the peculiar signature of events with these quarks in the final state, very strong constraints on these models come from searches for one or two b-tagged jet + MET and $t\bar{t}$ + MET (see [20] for an early proposal within the EFT framework and [4, 21] for a discussion in terms of simplified models).

An interesting possibility is that the scalar mediator of the DM-SM interaction is the Higgs boson itself, as it happens in the "Higgs portal" models (see e.g. [22]). Many manifestations of Higgs portal models would lead to a reduction or suppression of the Higgs boson couplings to SM particles, in favor of its interactions with new particles [23]. Precision measurements of the Higgs couplings that can be undertaken in future LHC phases and future accelerators can further constrain Higgs portal models [24]. Alternatively, the Higgs' coupling to DM can be constrained by measurements of the Higgs partial width to invisible particles. Current ATLAS and CMS limits on invisible Higgs decay at the 95% C.L. are around 70%; they are expected to decrease to 20–30% by the end of the upcoming 300 fb^{-1} LHC run [25].

7.2.2 Mediator Exchange in the t-Channel

In this section we consider a coloured fermionic mediator with an interaction vertex between quarks and the WIMP resulting in a t-channel exchange. A concrete model is that of a squark exchange in supersymmetric models, that was already introduced in Sect. 7.1

$$\mathcal{L} = \mathcal{L}_{SM} + g_M \sum_i \left(\bar{Q}_L^i \tilde{Q}_L^i + \bar{u}_R^i \tilde{u}_R^i + \bar{d}_R^i \tilde{d}_R^i \right) \chi + \text{mass terms} + c.c. \qquad (7.15)$$

where Q_L^i, u_R^i, d_R^i are the usual SM quarks, $\tilde{Q}_L^i$, $\tilde{u}_R^i$, $\tilde{d}_R^i$ correspond to the respective squarks (from here on the "mediators"), and i represents a flavour index. Unlike the usual case in Superysmmetry, here the WIMP χ can be taken to be either Dirac or Majorana fermion. This model is extensively analysed in [26], and a comparison with its effective operator limit DT1 (in the notation of Table 6.1) is performed.

Two extreme cases may be highlighted: (1) all mediator flavors are present or (2) only $\tilde{d}_R^i$ are present. Simply due to multiplicity these two cases maximize and minimize the mediator production cross-section, respectively.

An important difference with the s-channel models discussed in Sect. 7.2.1 is the fact that, being the mediators coloured, gluons may be emitted not only as initial state radiation but also from the mediator itself. This process is suppressed in the EFT limit by two powers of M_{med}, and this make a large qualitative difference in the kinematic distribution within the simplified model and the DT1 operator.

When the mediator is light enough, its pair production becomes kinematically accessible, and an event like

$$p\,p \to \tilde{q}\,\tilde{q} \to q\chi\, q\chi \qquad (7.16)$$

leads to a di-jet + MET signature (or in general jets + MET, when additional jet radiation is taken into account). It was shown in [26] that, on a large portion of parameter space, this kind of signature with two high-p_T jets leads to constraints stronger then the mono-jet one, even when the effect of additional sub-dominant jets is taken into account. More over, the effect of off-shell $\tilde{q}$ production is important, and the effect of a finite width should be taken into account.

As for the s-channel models presented in Sect. 7.2.1, strong constraints come from searches for b, t quarks in the final state in the case in which the coupling to third generation quarks are enhanced (see [1] and references therein).

The parameter space of this model consists of only three parameters: the masses m_χ, M_{med} and the coupling g_M. A convenient way to present results is in the form of a colour density plot for g_M in the m_χ, M_{med} plane. The most interesting range is $100\,\text{GeV} \leq M_{\text{med}} \leq 2\,\text{TeV}$ and $100\,\text{GeV} \leq m_\chi \leq 1\,\text{TeV}$. Outside this region the sensitivity decreases until the interpretation of the model as the exchange of a scalar resonance is lost, either because $g_M \sim 4\pi$ or because the minimal decay width $\Gamma_{\text{min}} \sim M_{\text{med}}$ (the latter typically happens before the former) [26].

7.2.3 Simplified Models for DM - Gluons Interaction

In the previous sections we have described a list of simplified models which, in the low momentum limit, give rise to the effective operators D1–D8 and D1'–D4' of

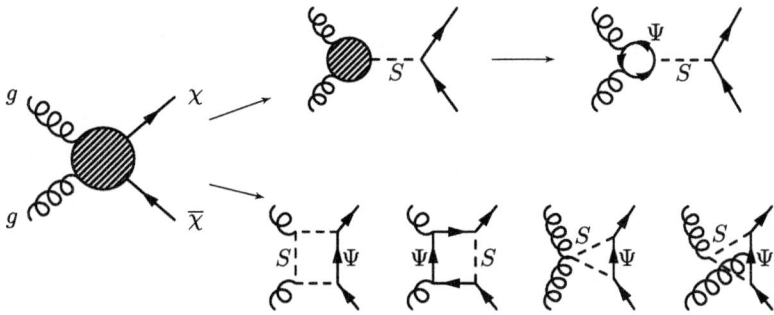

Fig. 7.3 Two different simplified models leading to the operators of Eq. (7.17). *Top* the dimension-7 operator may originate from the s-channel exchange of a heavy scalar S. The coupling of the scalar to the gluon field strength may be in turn be due to a loop of heavy coloured fermions Ψ. *Bottom* if also the heavy scalar is coloured, the dimension-7 operator can be originated by a loop including both S and Ψ. Diagrams with external legs exchanged are not shown

Table 6.1. In this section we want to discuss the operators D11–D13, i.e. the ones in which the DM couples to the gluon field strength. In particular we are interested in the two CP conserving operators

$$\frac{\alpha_s}{4\Lambda^3}\mathrm{tr}\left(G_{\mu\nu}G^{\mu\nu}\right)\bar{\chi}\chi, \quad\text{and}\quad \frac{i\alpha_s}{4\Lambda^3}\epsilon^{\mu\nu\alpha\beta}\mathrm{tr}\left(G_{\mu\nu}G_{\alpha\beta}\right)\bar{\chi}\gamma_5\chi, \tag{7.17}$$

where α_s is the strong coupling and Λ denotes a high-energy scale. As it was shown in Chap. 6, the EFT approximation is badly broken at the LHC for these operators, and a reinterpretation in terms of simplified models is necessary.

The resolution of these operators in terms of simplified models is less straightforward than for the previous ones. The first possibility is illustrated in the top row of Fig. 7.3. A (pseudo-)scalar particle may be introduced, that couples to the DM fermion with Yukawa-like couplings and to the gluon field strength with a dimension-5 effective operator:

$$\mathcal{L} \supset y_\chi S\bar{\chi}\chi + \frac{\alpha_s}{\Lambda_s}SG_{\alpha\beta}G^{\alpha\beta} \tag{7.18}$$

for the scalar and

$$\mathcal{L} \supset iy'_\chi S'\bar{\chi}\gamma_5\chi + \frac{\alpha_s}{\Lambda_{s'}}S'G_{\mu\nu}G_{\alpha\beta}\epsilon^{\mu\nu\alpha\beta} \tag{7.19}$$

for the pseudo-scalar. Here Λ_s is some mass scale associated with the dimension-5 operator and the trace over the colour indices has been left implicit. The scale that appears in the dimension-7 operators (D11–14) is given by

$$\frac{\alpha_s}{4\Lambda^3} \sim \frac{1}{m_S^2}\frac{y_\chi\alpha_s}{\Lambda_s} \tag{7.20}$$

where m_S is the mass of the scalar S. A similar expression holds for the case of a pseudo-scalar or for multiple scalars.

From the point of view of EFT, the dimension-5 operator that is left is safer than the dimension-7 one. Indeed, for a sufficiently light scalar S, the effective scale Λ_s can be high enough that an UV completion is not needed. On the other hand, resolving the dimension-5 operator $S G_{\alpha\beta} G^{\alpha\beta}$ can be done if the scalar S is coupled through Yukawa coupling to some new heavy coloured states (this is completely analogous to the Higgs coupling to gluons via the top quark loop). In the limit of heavy mediators' mass the dimension-5 coupling is related to the heavy coloured states' mass and coupling through

$$\frac{\alpha_s}{\Lambda_s} \propto \frac{\alpha_s}{8\pi} \sum_{\Psi} \left(\frac{y_\Psi}{M_\Psi} \right) \tag{7.21}$$

where the sum runs over all heavy coloured fermions, y_Ψ is the Yukawa coupling of these fermions to the scalar S, M_Ψ is the mass of the heavy fermion Ψ. A similar expression holds for the case of a pseudo-scalar. So, this model can be resolved into a fully renormalizable model by introducing new heavy (vector-like) quarks that couple to the scalar mediator. The relations of Eqs. (7.21) and (7.20) require a mediator mass m_s which is not too heavy or the coloured states are far too light and would have already been observed in searches for new coloured states.

It is also possible to resolve the dimension-7 operators directly into renormalizable interactions with coloured mediators as was done for example in Refs. [27, 28] and is shown in the bottom panel of Fig. 7.3. A simple example of such a model is one with new coloured scalars and fermions that couple to the WIMP through a Yukawa-type interaction. The coupling of the dimension-7 operator is then related to the mass and coupling of these new states through,

$$\frac{\alpha_s}{4\Lambda^3} \propto \frac{\alpha_s \lambda_\chi^2}{M_{\mathrm{med}}^3} \tag{7.22}$$

where M_{med} is the mass of the mediators and λ_χ is their coupling to the WIMP. Evidently, one needs fairly light mediators to generate the scale bounded by searches at the LHC, $\Lambda \sim 350\,\mathrm{GeV}$ as in Ref. [29]. Such new coloured states are much easier to search for in other channels by producing them directly.

To conclude this section we reiterate that resolving dimension-7 operators of the type discussed above (D11–D14 of Ref. [30]) in terms of simplified models is not as straightforward as it is for operators associated with quarks (e.g. D1–D10). Because of their high dimensionality using these EFT operators at the LHC is particularly problematic as was shown in Chap. 6. Perhaps the simplest way of making sense of such operators is through a new higgs-like scalar (or pseudoscalar) that couples directly to the WIMP through a Yukawa coupling and to gluons through a dimension five operator as in Eqs. (7.18) and (7.19).

7.3 Simplified Models - a Critical Look

The list of simplified models presented in the previous section was constructed keeping in mind the EFT approach and its limitations. In this sense, simplified models can be viewed as an improvement of effective operators, where the effective scale Λ^4 is replaced by a propagator's denominator $(p^2 - M^2)^2 + \Gamma^2 M^2$ in order to avoid energy limitations and exploit resonance enhancement in the production cross section. In a bottom-up approach, this is a small step above. As soon as we start looking at them more carefully, anyway, the situation turns out not to be so simple. Since the arguments we are going to discuss have to do with perturbative unitarity, it is not useless to review this issue first.

7.3.1 Perturbative Unitarity

Let us consider a $2 \rightarrow 2$ scattering process in the centre of mass frame between an initial state i and a final state f. The matrix element can then be decomposed in partial waves as [31][3]

$$\mathcal{M}_{if}^J(s) = \frac{1}{32\pi} \beta_{if} \int_{-1}^{1} d\cos\theta \, d_{\mu_i \mu_j}^J(\theta) \, \mathcal{M}_{if}(s, \cos\theta), \qquad (7.23)$$

where $d_{\mu\mu'}^J$ is the Jth Wigner d-function and $\mu_i = \lambda_{i_1} - \lambda_{i_2}$ and $\mu_f = \lambda_{f_1} - \lambda_{f_2}$ are defined in terms of the helicity of the initial and final state particles. The factor β_{if} is a kinematical factor given by

$$\beta_{if} = \frac{\beta^{1/4}(s, m_{f_1}^2, m_{f_2}^2)\beta^{1/4}(s, m_{i_1}^2, m_{i_2}^2)}{32\pi s} \qquad (7.24)$$

where $\beta(x, y, z) = x^2 + y^2 + z^2 - 2xy - 2yz - 2zx$. In the high energy limit $\sqrt{s} \rightarrow +\infty$ the factor β_{if} tends to 1, so that we can neglect it from now on. Unitarity of the S matrix implies

$$\text{Im}(\mathcal{M}_{ii}^J) = \sum_f |\mathcal{M}_{if}^J|^2 = |\mathcal{M}_{ii}^J|^2 + \sum_{f \neq i} |\mathcal{M}_{if}^J|^2 \geq |\mathcal{M}_{ii}^J|^2 \qquad (7.25)$$

for all J and all s. The sum over f in the first line runs over all possible final states, including an integral over the phase space and the sum over final state particle's spin.

---

[3]The expansion in terms of Wigner d-functions has the advantage, with respect to the standard one in terms of Legendre polynomials, that the matrix element is expanded in the basis of helicity states of the particles, instead that considering total spin states.

Fig. 7.4 The matrix element $\mathcal{M}_{ii}^J$ is forced by unitarity to fall in the circle of radius $1/2$, centred in $+i/2$. As far as internal particles are off-shell, the matrix element at tree level is real, and loop corrections puts it back into the circle. If $(\mathcal{M}_{ii}^J)_{\text{tree}} \gtrsim 1/2$, the necessary correction is larger than roughly 40%, signalling the breakdown of perturbativity

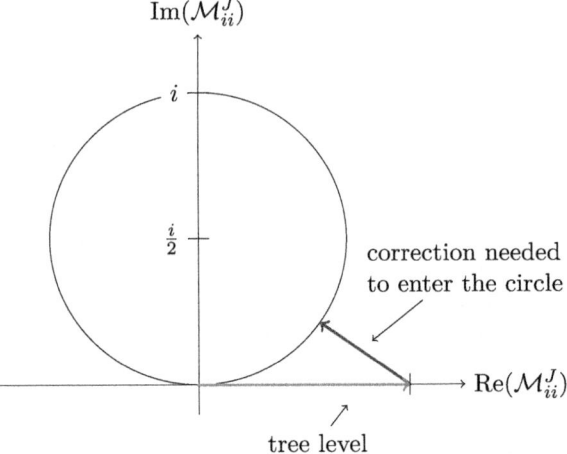

This is equivalent to saying that the matrix element $\mathcal{M}_{ii}^J$ must fall, in the Argand plane, inside a circle of radius $1/2$ centred in $i/2$, as depicted in Fig. 7.4. In general, as far as internal particles are off shell and the $+i\epsilon$ prescription does not play a role, tree level matrix elements are real, and loop corrections move the matrix element back into the circle. A good criterion to keep perturbativity under control is suggested by these considerations:

$$(\mathcal{M}_{ii}^J)_{\text{tree}} \gtrsim 1/2 \,. \tag{7.26}$$

In fact, larger values of $(\mathcal{M}_{ii}^J)_{\text{tree}}$ would imply that higher order corrections are larger than 40%, signalling the loss of perturbativity. This condition is often referred to as "perturbative unitarity".

For practical purposes, we can restrict ourselves to the case $J = 0$, since it is the one that gives the stronger constraint, and for simplicity take equal helicities $(\mu_i = \mu_f = 0)$, so that the Wigner functions simplify to Legendre polynomials $d_{00}^J = P_J$.

7.3.2 Unitarity Issues with Simplified DM Models

The first problem lies in the fact that the simplified models of Eq. (7.9) are not invariant under the full SM gauge group $SU(3)_c \times SU(2)_L \times U(1)_Y$ but only under the unbroken subgroup $SU(3)_c \times U(1)_{\text{e.m.}}$. This is true for models with a scalar mediator (in which Yukawa couplings break $SU(2)_L$), and for models with a vector mediator in which the couplings to up and down quarks are different (so that the mediator does not couple to the left handed quark doublet but to its two components separately). Violation of the electroweak gauge symmetry can lead to spuriously enhanced cross section for DM production with the initial state radiation of a W boson [32, 33]. This problem does not only affect the mono-W searches: the W can indeed decay hadronically, enhancing the signal in the mono-jet search. For example, in the case of a vector mediator with opposite sign couplings to up and down quarks, this process dominates the mono-jet cross section for $\not{E}_T > 400\,\text{GeV}$ [34]. This means that, even when restricting to a particular MET search, constraints descending from the internal consistency of the model can not be neglected.

There are of course ways to cure the models by adding new particles or new interactions. Again referring to the case of a vector mediator, different couplings of the up and down quarks can be made compatible with perturbative unitarity if an appropriate vertex WWZ' is added (where Z' is the new vector mediator), in similarity to what happens for the Z boson in the SM. The situation is more complex in the case of the scalar mediator. Here gauge invariance can not be simply restored with a choice of the couplings. The reason is that a singlet S can couple to DM but not to quarks, while if S is a doublet it can couple only to quarks, but not to DM. A possible solution is to add a mixing of S with the Higgs boson, via a quadrilinear term $H^\dagger H S^2$. In this case, constraints on the Higgs width to invisible particles and direct detection force the mixing angle ε (and therefore the coupling $g_q \sim \sin \varepsilon$) to be very small, making LHC constraints weak. In turn, this may be overcome by adding to the model a second Higgs doublet and letting S mixing with it, but then again the phenomenology is altered.

A second issue is again related to perturbative unitarity, in the very same way as it would appear in the SM without the Higgs [35, 36]. Following [12], let us consider a model with a spin-1 mediator with both vectorial and axial couplings to DM and to quarks:

$$\mathcal{L} = - \sum_{f=q,l,\nu} Z'^\mu \bar{f} \left[g_f^V \gamma_\mu + g_f^A \gamma_\mu \gamma^5 \right] f - Z'^\mu \bar{\psi} \left[g_{\text{DM}}^V \gamma_\mu + g_{\text{DM}}^A \gamma_\mu \gamma^5 \right] \psi \ . \quad (7.27)$$

Applying the criterion described in the previous section to the elastic scattering of fermions (both SM fermions or DM) within this model one gets the perturbative unitarity bound

$$m_f \lesssim \sqrt{\frac{1}{2} \frac{m_{Z'}}{g_f^A}}, \quad (7.28)$$

where f may stand for both a SM fermion or the DM particle. In a similar way, perturbative unitarity is violated in the process of 2 fermions annihilation into $Z'Z'$,[4] which is important for the calculation of the relic density, but has only a minor impact on LHC results, since the unitarity violating effects are proportional to the mass of

---

[4]The unitarity criterion proposed above deals with the diagonal matrix elements $\mathcal{M}_{ii}^J$, in which the initial and final states are equal. It can be adapted as well to the case of off-diagonal matrix elements, by restricting to the 2×2 subspace spanned by the states $\psi\bar{\psi}$ and $Z'_L Z'_L$, and noticing that for $s \to +\infty$ only the off-diagonal elements survive, and hence the eigenvalues of the matrix become equal to the off-diagonal element.

the initial fermions squared, which are light in a collider setup. In order to restore unitarity new physics has to be invoked. In particular, what violates unitarity is the longitudinal mode of the Z' boson, therefore the addition to the model of a scalar particle that give rise to its mass via Higgs mechanism would serve the purpose. In this case, the condition on the mass of the Z' would read

$$\sqrt{\pi} \frac{m_{Z'}}{g_{\mathrm{DM}}^A} \geq \max[m_s, \sqrt{2} m_{\mathrm{DM}}], \tag{7.29}$$

where m_s is the mass of the new scalar. Notice that, in any case, the problem of unitarity affects only the axial coupling of the Z', while the vector coupling g_f^V is not affected by these constraints (which indeed become trivial in the limit $g_f^A \to 0$).

7.3.3 Complementarity of LHC and Other DM Searches

When comparing LHC results with those of other probes, and in particular with those of indirect searches, it becomes of primary importance to carefully consider the consistency of the simplified models under consideration. As we already discussed in this chapter, the simplest models suffer from issues related to the loss of gauge invariance and perturbative unitarity, which need additional fields and interactions in order to be cured. In the example of a Z' gauge mediator, to which we will return in Chap. 9 [37], not only an additional Higgs boson s is necessary to give mass to the Z' and to cure the unitarity issues, but also the SM Higgs boson must be charged under the new gauge group for the Yukawa couplings to be gauge invariant. As we will show, this leads to the necessary presence of a $Z'Zh$ vertex. While one can, to a certain extent, ignore these complications at the LHC, this is not possible for indirect searches since tree level annihilation into Zh [37] and Zs [38] are easily the dominant channels, together with loop annihilation into EW gauge bosons. Moreover, leptonic annihilation channels $\ell\bar{\ell}$ are at least as important as $q\bar{q}$ ones, and the coupling of the mediator to leptons can not be ignored as it is typically done for LHC searches. The same is true for the calculation of the relic density.

We would like to spend a last comment on the issue of gauge anomalies. If the interaction of DM with SM fermions is due to an extended gauge symmetry, in order for the theory to be consistent at the quantum level the charge assignment under the new gauge group has to be decided in such a way that gauge anomalies are avoided. For example, in the case of an extra U(1) group the only consistent assignment to avoid anomalies involving the SM generators is a linear combination of $B - L$ and hypercharge [39]. Alternatively, additional heavy fermions may be added to the model. If they are heavy enough, their impact on LHC searches is negligible, and the details of this part of the dark sector can be ignored. Unfortunately this is not the case when we want to compare LHC results with those from indirect searches: in the latter, indeed, loop annihilation channels are relevant, but these can not computed including

SM fermions alone because divergences do not cancel, and the full knowledge of the dark sector is needed.

To conclude this chapter, let us ask a somewhat more philosophical question on the approach of simplified DM models. In the way they were originally introduced, they were meant to be a small step beyond the EFT approach, in such a way to avoid their intrinsic energy limitation and to exploit resonant production of the mediator in order to improve the constraining power of LHC searches. It was successively realised that, in order not to deal with unphysical results, the vanilla picture had to be supplied with additional constraints, couplings and states, in a kind of second order improvement. On the one hand, the typical consequence is that the strong LHC constraints on the dark sector do not come from DM production processes (as in mono-X searches) but from other observables (di-jet and di-lepton resonances, mixing with Z boson and electroweak precision tests, Higgs width to invisibles, perturbative unitarity etc. [16]). This comes with no surprise, since the high energy reach of the LHC consents to explore a large variety of phenomena above the weak scale, without restricting to the lightest stable state of this new physics sector. This is quite the opposite with respect to what happens with direct and indirect searches, which are intrisically limited to constrain the properties of the DM particle. On the other hand, simplified models can not (or only partially) be viewed as an exhaustive toolbox to constraint all possible WIMP scenarios at once. For this reason, it is of extreme importance that the LHC collaborations publish their results on simple, search-specific, models in such a way that they are recastable for any other model (as it is for cut-and-count analyses). In turn, theoreticians should keep working in close contact with experimentalists in order to maximise the utility of the simplified models toolkit. Finally, the use of (truncated) EFT should not be disregarded, since this is the most model independent approach and it is economical from the point of view of the computational effort because of the reduced dimensionality of its parameter space (and therefore of parameter scannings).

References

1. J. Abdallah et al., Simplified models for dark matter and missing energy searches at the LHC, arXiv:1409.2893
2. S. Malik, C. McCabe, H. Araujo, A. Belyaev, C. Boehm, et al., Interplay and characterization of dark matter searches at colliders and in direct detection experiments, Phys. Dark. Univ. 9–10, 51–58 (2015), arXiv:1409.4075
3. J. Abdallah et al., Simplified models for dark matter searches at the LHC. Phys. Dark Univ. **9–10**, 8–23 (2015), arXiv:1506.03116
4. D. Abercrombie et al., Dark matter benchmark models for early LHC run-2 searches: report of the ATLAS/CMS dark matter forum, arXiv:1507.00966
5. G. Busoni et al., Recommendations on presenting LHC searches for missing transverse energy signals using simplified s-channel models of dark matter, arXiv:1603.04156
6. A. De Simone, T. Jacques, Simplified models vs. effective field theory approaches in dark matter searches, Eur. Phys. J. C 76(7), 367 (2016), arXiv:1603.08002
7. G. D'Ambrosio, G.F. Giudice, G. Isidori, A. Strumia, Minimal flavor violation: an effective field theory approach. Nucl. Phys. B **645**, 155–187 (2002), arXiv:hep-ph/0207036

8. M.J. Dolan, F. Kahlhoefer, C. McCabe, K. Schmidt-Hoberg, A taste of dark matter: flavour constraints on pseudoscalar mediators. JHEP **03**, 171 (2015), arXiv:1412.5174. [Erratum: JHEP **07**, 103 (2015)]

9. P. Agrawal, M. Blanke, K. Gemmler, Flavored dark matter beyond minimal flavor violation. JHEP **10**, 72 (2014), arXiv:1405.6709

10. D. Racco, A. Wulzer, F. Zwirner, Robust collider limits on heavy-mediator dark matter. JHEP **05**, 009 (2015), arXiv:1502.04701

11. G. Isidori, Y. Nir, G. Perez, Flavor physics constraints for physics beyond the standard model. Ann. Rev. Nucl. Part. Sci. **60**, 355 (2010), arXiv:1002.0900

12. F. Kahlhoefer, K. Schmidt-Hoberg, T. Schwetz, S. Vogl, Implications of unitarity and gauge invariance for simplified dark matter models. JHEP **02**, 016 (2016), arXiv:1510.02110. [JHEP **02**, 016 (2016)]

13. T. Jacques, K. Nordström, Mapping monojet constraints onto simplified dark matter models. JHEP **06**, 142 (2015), arXiv:1502.05721

14. A.J. Brennan, M.F. McDonald, J. Gramling, T.D. Jacques, Collide and conquer: constraints on simplified dark matter models using mono-X collider searches, JHEP **1605**, 112 (2016), arXiv:1603.01366

15. M. Chala, F. Kahlhoefer, M. McCullough, G. Nardini, K. Schmidt-Hoberg, Constraining dark sectors with monojets and dijets. JHEP **07**, 089 (2015), arXiv:1503.05916

16. M. Duerr, F. Kahlhoefer, K. Schmidt-Hoberg, T. Schwetz, S. Vogl, How to save the WIMP: global analysis of a dark matter model with two s-channel mediators, JHEP **1609**, 042 (2016), arXiv:1606.07609

17. M. Fairbairn, J. Heal, F. Kahlhoefer, P. Tunney, Constraints on Z' models from LHC dijet searches, JHEP **1609**, 018 (2016), arXiv:1605.07940

18. R. Barbieri, G. Isidori, J. Jones-Perez, P. Lodone, D.M. Straub, $U(2)$ and minimal flavour violation in supersymmetry. Eur. Phys. J. C **71**, 1725 (2011), arXiv:1105.2296

19. R. Barbieri, D. Buttazzo, F. Sala, D.M. Straub, Flavour physics from an approximate $U(2)^3$ symmetry. JHEP **07**, 181 (2012), arXiv:1203.4218

20. T. Lin, E.W. Kolb, L.-T. Wang, Probing dark matter couplings to top and bottom quarks at the LHC. Phys. Rev. D **88**(6), 063510 (2013), arXiv:1303.6638

21. M.R. Buckley, D. Feld, D. Goncalves, Scalar simplified models for dark matter, Phys. Rev. D **91**, 015017 (2015), arXiv:1410.6497

22. M. Duerr, P. Fileviez Pérez, J. Smirnov, Scalar dark matter: direct vs. indirect detection. JHEP **06**, 152 (2016), arXiv:1509.04282

23. C. Englert, T. Plehn, D. Zerwas, P.M. Zerwas, Exploring the Higgs portal. Phys. Lett. B **703**, 298–305 (2011), arXiv:1106.3097

24. C. Englert, A. Freitas, M.M. Mühlleitner, T. Plehn, M. Rauch, M. Spira, K. Walz, Precision measurements of Higgs couplings: implications for new physics scales. J. Phys. G **41**, 113001 (2014), arXiv:1403.7191

25. ATLAS Collaboration, Sensitivity to New Phenomena via Higgs Couplings with the ATLAS Detector at a High-Luminosity LHC. Technical report ATL-PHYS-PUB-2013-015, CERN, Geneva, October 2013

26. M. Papucci, A. Vichi, K.M. Zurek, Monojet versus rest of the world I: t-channel models, JHEP **1411**, 024 (2014), arXiv:1402.2285

27. N. Weiner, I. Yavin, UV completions of magnetic inelastic and Rayleigh dark matter for the Fermi Line(s). Phys. Rev. D **87**(2), 023523 (2013), arXiv:1209.1093

28. M.T. Frandsen, U. Haisch, F. Kahlhoefer, P. Mertsch, K. Schmidt-Hoberg, Loop-induced dark matter direct detection signals from gamma-ray lines. JCAP **1210**, 033 (2012), arXiv:1207.3971

29. ATLAS Collaboration, G. Aad et al., Search for dark matter candidates and large extra dimensions in events with a jet and missing transverse momentum with the ATLAS detector. JHEP **04**, 075 (2013), arXiv:1210.4491

30. J. Goodman, M. Ibe, A. Rajaraman, W. Shepherd, T.M. Tait et al., Constraints on dark matter from colliders. Phys. Rev. D **82**, 116010 (2010), arXiv:1008.1783

31. M. Jacob, G.C. Wick, On the general theory of collisions for particles with spin. Ann. Phys. **7**, 404–428 (1959). [Ann. Phys. **281**, 774 (2000)]
32. N.F. Bell, Y. Cai, J.B. Dent, R.K. Leane, T.J. Weiler, Dark matter at the LHC: effective field theories and gauge invariance. Phys. Rev. D **92**(5), 053008 (2015), arXiv:1503.07874
33. N.F. Bell, Y. Cai, R.K. Leane, Mono-W dark matter signals at the LHC: simplified model analysis. JCAP **1601**(01), 051 (2016), arXiv:1512.00476
34. U. Haisch, F. Kahlhoefer, T.M.P. Tait, On mono-W signatures in spin-1 simplified models, Phys. Lett. B **760**, 207–213, (2016), arXiv:1603.01267
35. B.W. Lee, C. Quigg, H.B. Thacker, The strength of weak interactions at very high-energies and the Higgs Boson mass. Phys. Rev. Lett. **38**, 883–885 (1977)
36. B.W. Lee, C. Quigg, H.B. Thacker, Weak interactions at very high-energies: the role of the Higgs Boson mass. Phys. Rev. D **16**, 1519 (1977)
37. T. Jacques, A. Katz, E. Morgante, D. Racco, M. Rameez, A. Riotto, Complementarity of DM searches in a consistent simplified model: the case of Z, JHEP **1610**, 071 (2016), arXiv:1605.06513
38. N.F. Bell, Y. Cai, R.K. Leane, Dark forces in the sky: signals from Z' and the dark Higgs, JCAP **1608**, 011 (2016), arXiv:1605.09382
39. S. Weinberg, *The Quantum Theory of Fields. Volume 2: Modern Applications* (Cambridge University Press, Cambridge, 2013)

Chapter 8
The Relic Density Constraint

8.1 Introduction

The calculation of the relic abundance of DM particles in the standard freeze-out mechanism was discussed in Sect. 1.2.1. The final result, Eq. (1.9) can be approximated as

$$\Omega_{\mathrm{DM}} h^2 \simeq \frac{2 \times 2.4 \times 10^{-10}\,\mathrm{GeV}^{-2}}{\langle \sigma v \rangle_{\mathrm{ann}}}, \tag{8.1}$$

where $\langle \sigma v \rangle_{\mathrm{ann}}$ is the total thermally-averaged annihilation cross section, and the factor of 2 in the numerator is made explicit to emphasize the fact that, in this chapter, we will assume a non-self-conjugate DM particle.

A fundamental question which one should ask is the following: under the optimistic hypothesis that the LHC gives evidence for a new weakly interacting particle with a lifetime that exceeds about a microsecond, how confident can we be in claiming we have finally revealed the true nature of the DM? On top of this question, another one arises in the context of simplified DM models that were introduced in Chap. 7. Indeed, being the parameter space enlarged with respect to the simple EFT approach, it becomes of primary importance to figure out the best way in which to present experimental constraints and to prioritize promising regions of parameter space with respect to others.

A tentative answer to both questions may come from the relic density. On the one hand, for a new stable particle discovered at the LHC to be assigned the label of thermal relic DM, it must fall in a region of parameter space where the right abundance is attained. If this is not the case, either the particle is not a viable thermal relic DM, or the physics responsible for its equilibrium in the early universe is not the same that is in play at the LHC. On the other hand, identifying the regions of the parameter space of a given model where the DM abundance fits the observed one might be useful to set priorities for the LHC collaborations when comparing the future data with the plethora of models.

© Springer International Publishing AG 2017   143
E. Morgante, *Aspects of WIMP Dark Matter Searches at Colliders and Other Probes*, Springer Theses, https://doi.org/10.1007/978-3-319-67606-7_8

Following the analysis of [1], in this chapter we will consider a simplified model with a vector mediator exchanged in the s-channel and compare the region of the parameter space which gives the correct relic density with the predicted exclusion regions for the LHC Run 2 and the 5σ discovery potential regions. The chapter is structured as follows. In Sect. 8.2 we provide some general considerations and state our assumptions, along with a description of the model we consider. In Sect. 8.3 we compare ATLAS 14 TeV sensitivity with the region of parameter space consistent with thermal relic DM. In Sect. 8.4 we extend this analysis to simplified models. Finally, we collect our concluding remarks in Sect. 8.5.

8.2   Working Assumptions

8.2.1   DM Abundance Considerations

Let us consider an hypothetical DM candidate χ, efficiently pair-produced at the LHC. Assuming that the main production channel is through annihilation of light quark - antiquark pairs, one can define the thermally averaged DM annihilation cross section[1]

$$\langle \sigma v \rangle_* \equiv \langle \sigma v \rangle_{\chi\bar{\chi} \to u\bar{u}} + \langle \sigma v \rangle_{\chi\bar{\chi} \to d\bar{d}}, \tag{8.2}$$

which also sets a reference for DM production at the LHC. In the early universe, besides annihilations into quarks, there can be additional annihilation channels, so that the total DM annihilation cross section which is relevant for the relic abundance is

$$\langle \sigma v \rangle_{\text{ann}} \geq \langle \sigma v \rangle_*. \tag{8.3}$$

So, by requiring that the abundance of χ and $\bar{\chi}$ match the observed relic abundance, we get

$$\Omega_{\text{DM}}^{\text{obs}} h^2 \simeq \frac{2 \times 2.4 \times 10^{-10}\,\text{GeV}^{-2}}{\langle \sigma v \rangle_{\text{ann}}} \leq \frac{2 \times 2.4 \times 10^{-10}\,\text{GeV}^{-2}}{\langle \sigma v \rangle_*}, \tag{8.4}$$

or, plugging in the measured value $\Omega_{\text{DM}} h^2 = 0.1197 \pm 0.0022$ [2],

$$\langle \sigma v \rangle_* \lesssim 4.0 \times 10^{-9}\,\text{GeV}^{-2}. \tag{8.5}$$

Given the assumption that the DM particle is efficiently produced at the LHC, we can make the further hypothesis that DM annihilates dominantly into SM fermions and that the coupling to the first generation of quarks is not less than the coupling to other SM fermions. This assumption is clearly debatable, but can be considered as a

---

[1]Gluons and other quarks can of course contribute to DM production at the LHC, so the $*$ subscript defines a reference channel rather than all possible channels of DM production at the LHC.

working hypothesis. We will show later how weakening this assumption affects our results. In this case, we get

$$\langle \sigma v \rangle_{\text{ann}} \leq \sum_{\text{quark gen.}} \langle \sigma v \rangle_* + \sum_{\text{lepton gen.}} \frac{1}{3} \langle \sigma v \rangle_* = 4 \langle \sigma v \rangle_*, \tag{8.6}$$

and therefore

$$\Omega_{\text{DM}}^{\text{obs}} h^2 \simeq \frac{2 \times 2.4 \times 10^{-10}\,\text{GeV}^{-2}}{\langle \sigma v \rangle_{\text{ann}}} \geq \frac{6.0 \times 10^{-11}\,\text{GeV}^{-2}}{\langle \sigma v \rangle_*}, \tag{8.7}$$

or

$$\langle \sigma v \rangle_* \gtrsim 1.0 \times 10^{-9}\,\text{GeV}^{-2}. \tag{8.8}$$

Let us consider a simple example to illustrate the relevance of these inequalities. Assume that the interactions between DM and SM quarks are described within an Effective Field Theory (EFT), where the basic parameters are the DM mass m_{DM} and the UV scale Λ. Let us also imagine that the annihilation controlling the thermal abundance takes place in the s-wave. One therefore expects roughly that $\langle \sigma v \rangle_* \simeq 10^{-1} m_{\text{DM}}^2 / \Lambda^4$. We then obtain, from Eqs. (8.5) and (8.8),

$$0.7 \left(\frac{m_{\text{DM}}}{10^2\,\text{GeV}} \right)^{1/2} \text{TeV} \lesssim \Lambda \lesssim 1.0 \left(\frac{m_{\text{DM}}}{10^2\,\text{GeV}} \right)^{1/2} \text{TeV}. \tag{8.9}$$

Curves corresponding to the correct relic abundance have been used as a benchmark or comparison for EFT constraints since the early usage of EFTs [3, 4]. However, these relic density constraints on thermal DM are usually considered not to be robust: for a given set of parameters, the relic density can be smaller if the cross section is enhanced by inclusion of other annihilation channels, such as annihilation to leptons; conversely, the true relic density can be larger if there is a larger dark sector including other types of DM. However, these constraints can become more powerful if presented as a region in parameter space in which a given model satisfies a modest set of assumptions, which we list here:

1. the DM candidate χ makes up 100% of the DM of the universe;
2. the DM annihilation rate is related to the observed density today via the standard thermal production mechanism;
3. the dominant annihilation channel is to SM fermions, via one dark mediator;
4. the DM couples to u, d quarks, so that it can be produced at the LHC;
5. the coupling to the first generation of quarks is no less than the coupling to other SM fermions.

In this situation, the relic density constraint gives a range within which the dark sector parameters should lie. It is clear that assumption 5 is by no means a certainty, and so we will show how our results are sensitive to relaxing this assumption. In the event of a signal, this assumption can instead be used to learn about the flavour structure of a thermal relic model that attempts to explain the signal. If the signal falls into the region where DM would be overproduced, then there must be enhanced couplings to other SM particles relative to u, d quarks in order to avoid overproduction, or alternatively, the DM is produced by some mechanism other than thermal production.

Assumption 2 can break down if either the DM was not produced thermally in the early universe, or if some other effect breaks the relationship between the DM density and annihilation rate. For example, unusual cosmologies between freezeout and today can influence the relic density of DM [5].

To summarize, under our generic assumptions 1–5 the DM production cross section must satisfy the bounds

$$1.0 \times 10^{-9}\, \text{GeV}^{-2} \simeq \frac{1}{4} \langle \sigma v \rangle_{\text{ann}} \leq \langle \sigma v \rangle_* \leq \langle \sigma v \rangle_{\text{ann}} \simeq 4.0 \times 10^{-9}\, \text{GeV}^{-2}. \quad (8.10)$$

where the value of the annihilation cross section is dictated by ensuring the correct relic abundance.

The tidy inequalities presented in Eq. (8.10) were derived with a number of simplifications: in particular, in order to get a more accurate result, one should include the effect of the top quark mass, mediator widths, and a more accurate expression for the relic density. This will be implemented later in the text, although the principle behind Eq. 8.10 remains the same.

The two limits on the cross section describe two contours in the parameter space: if $\langle \sigma v \rangle_*$ is too large, then DM will be underproduced, we call this the *underproduction line*; if $\langle \sigma v \rangle_*$ is too small, then DM will be overproduced; this is called the *overproduction line*. This information is summarised in the table below, where $g_{(\text{DM},f)}$ generically indicate the mediator couplings to DM and SM fermions, respectively.

Overproduction	$\langle \sigma v \rangle_{\text{ann}} \simeq 4 \langle \sigma v \rangle_*$	**EFT:** Max Λ, min m_{DM}. **Simp. model:** Max M, min $g_{(\text{DM},f)}$ and m_{DM}.
Underproduction	$\langle \sigma v \rangle_{\text{ann}} = \langle \sigma v \rangle_*$	**EFT:** Min Λ, max m_{DM} **Simp. model:** Min M, max $g_{(\text{DM},f)}$ and m_{DM}.

8.2.2   Models and Cross Sections

To illustrate our point, we focus on a class of simplified models where the DM is a Dirac particle annihilating to SM fermions in the s-channel via a Z'-type mediator. We choose to study a Z'-type model as it has the best prospective LHC Run-II constraints with which to compare.

We consider the general interaction term in the Lagrangian for a vector mediator Z',

$$\mathcal{L} = -\sum_f Z'_\mu [\bar{f}\gamma^\mu (g_f^V - g_f^A \gamma_5) f] - Z'_\mu [\bar{\chi}\gamma^\mu (g_{DM}^V - g_{DM}^A \gamma_5)\chi], \qquad (8.11)$$

where f is a generic SM fermion, the kinetic and gauge terms have been omitted, and the sum is over the quark and lepton flavours of choice (see e.g. Ref. [6]).

The LHC searches are only mildly sensitive to the ratios g_f^V/g_f^A and g_{DM}^V/g_{DM}^A, however the distinction is important for relic density calculations, and so we consider a pure vector coupling ($g_{f,DM}^A = 0$). In the EFT limit, we also consider pure axial ($g_{f,DM}^V = 0$) interactions. In the low-energy limit, the Lagrangian (8.11) leads to the effective operators

$$\mathcal{O}_V = \frac{1}{\Lambda^2}[\bar{\chi}\gamma^\mu \chi][\bar{f}\gamma_\mu f], \qquad \text{(D5)} \qquad (8.12)$$

$$\mathcal{O}_A = \frac{1}{\Lambda^2}[\bar{\chi}\gamma^\mu \gamma^5 \chi][\bar{f}\gamma_\mu \gamma^5 f]. \qquad \text{(D8)} \qquad (8.13)$$

The effective operators $\mathcal{O}_V$ and $\mathcal{O}_A$ correspond to the usual D5 and D8 operators respectively, defined in Ref. [3].

The process relevant for relic density calculations is the annihilation of DM particles of mass m_{DM} into SM fermions of mass m_f

$$\chi\bar{\chi} \to f\bar{f}. \qquad (8.14)$$

In the effective operator limit, the relative cross sections per SM fermion flavour, expanded up to order v^2, are

$$(\sigma v)_*^V \simeq \frac{N_C m_{DM}^2}{2\pi \Lambda^4}\left(\sqrt{1 - \frac{m_f^2}{m_{DM}^2}}\left(\frac{m_f^2}{m_{DM}^2} + 2\right) + v^2\frac{11 m_f^4/m_{DM}^4 + 2m_f^2/m_{DM}^2 - 4}{24\sqrt{1 - m_f^2/m_{DM}^2}}\right), \qquad (8.15)$$

$$(\sigma v)_*^A \simeq \frac{N_C}{2\pi \Lambda^4}\left(m_f^2\sqrt{1 - \frac{m_f^2}{m_{DM}^2}} + v^2\frac{23 m_f^4/m_{DM}^2 - 28 m_f^2 + 8m_{DM}^2}{24\sqrt{1 - m_f^2/m_{DM}^2}}\right). \qquad (8.16)$$

where the colour factor N_C is equal to 3 for quarks and 1 for colourless fermions. The full expressions relative to the process (8.14) with Z' exchange, and the corresponding mediator widths, are reported in Appendix B.

8.3 Results: Effective Operator Limit

In the EFT limit, neglecting the mass of the annihilation products, the DM annihilation cross section is proportional to $g_{DM}^2 g_f^2 m_{DM}^2/M^4 \equiv m_{DM}^2/\Lambda^4$, where M is the mediator mass, and g_f is its coupling with fermion species f. Thus, in general, the underproduction contour is a contour of maximum g_{DM}, g_q, and m_{DM}, and of minimum M, and vice-versa for the overproduction contour.

The LHC constraints in the EFT limit scenario are generally valid in the range $\pi \lesssim \sqrt{g_{DM}g_q} \lesssim 4\pi$ [7]. Since the annihilations relevant to relic density calculations take place when the DM is non-relativistic, the effective operator approximation is valid as long as $M \gg 2m_{DM}$, or $\sqrt{g_{DM}g_q} \gg 2m_{DM}/\Lambda$, while direct detection constraints are valid across the entire parameter space of interest.

We summarize our results in Fig. 8.1, where the *under-* and *over-*production lines defined above are compared with the projected exclusion and discovery reach by ATLAS, which we describe in the following.

8.3.1   ATLAS Reach

The exclusion and discovery reach of ATLAS at 14 TeV are presented in Ref. [7], in the missing energy + jets channel.

The limits from Ref. [7] are given for two values of m_{DM}, namely $m_{DM} = \{50, 400\}$ GeV; however, the variation in the constraint between the two masses is minimal, so we interpolate constraints on Λ between these two points. These limits are determined for the vector operator, but are expected to be the same for the axial-vector operator [8].

The 1 and 5% labels indicate projected limits assuming a 1 or 5% systematic uncertainty in the SM background, respectively. Achieving 1% systematics may be overly optimistic, and can be considered a "best-case scenario". Other labels indicate the results at a given collision energy and integrated luminosity. The red bands in the plots indicate the potential significance of an observed signal, from 3 to 5σ.

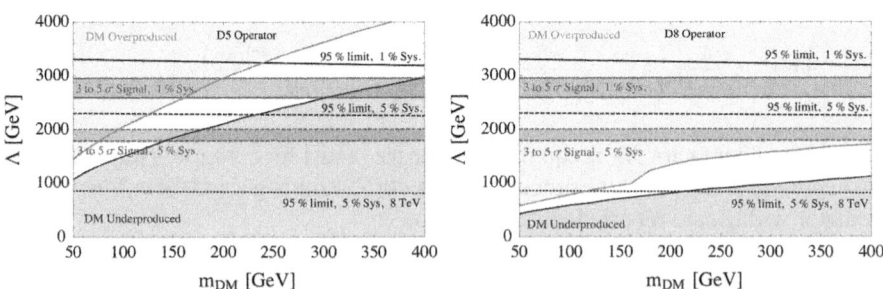

Fig. 8.1   *Blue* and *orange lines* show the over- and under-production lines respectively, defined in the text, for the Vector (D5) (*left*) and Axial-Vector (D8) (*right*) operators. The *black lines* show prospective ATLAS exclusion limits for various energies and systematic uncertainties, and for luminosities of (3000, 300, 20) fb^{-1} from top to bottom. The *red bands* show the 3 to 5 σ discovery potential [7]. EFT approximation is valid for $\pi < \sqrt{g_{DM}g_q} < 4\pi$ for ATLAS prospects, and $\sqrt{g_{DM}g_q} \gg 2m_{DM}/\Lambda$ for the relic density constraints. See text for more details. Direct detection constraints are not shown, but for the vector operator D5 they would rule out the entire visible space (cf. Sect. 8.3.2) (color figure online)

8.3.2 Direct Detection Constraints

Limits on the spin-dependent (SD) and spin-independent (SI) cross sections are easily translated into limits on the effective operator parameter Λ by the relations given in Ref. [3]. We adopted limits from LUX [9] (SI cross section) and Xenon100 [10] (SD cross section). For our simplified models, constraints on Λ correspond to a constraint on $M / \sqrt{g_{\mathrm{DM}} g_q}$.

As discussed in Chap. 2, the vector operator $\mathcal{O}_V$ leads to a spin-independent cross section, which is strongly constrained by DD experiments. Constraints are significantly stronger than prospective LHC bounds on this operator, ruling out the entire region displayed in Fig. 8.1 (left). However, the strength of direct detection constraints falls of quickly below $m_{\mathrm{DM}} \simeq 10\,\mathrm{GeV}$, while LHC constraints are expected to be relatively flat below $m_{\mathrm{DM}} = 50\,\mathrm{GeV}$. If the prospective LHC constraints in Fig. 8.1 (left) can be extrapolated down, they will become stronger than direct detection constraints at around $m_{\mathrm{DM}} = 10\,\mathrm{GeV}$. Conversely, the axial-vector operator $\mathcal{O}_A$ is subject to much weaker constraints on the spin-dependent scattering cross section. In this range they are barely distinguishable from the $\Lambda = 0$ line and thus are not shown.

8.3.3 Relic Density Bounds

In Fig. 8.1, we show the *under-* and *over-*production lines defined in the previous Section, for the vector ($\mathcal{O}_V$, D5) and axial-vector ($\mathcal{O}_A$, D8) operator, under the assumptions 1–5 of Sect. 8.2.1. The range between the orange and blue lines shows the region of parameter space in which any observed χ can also be thermal relic DM. This marks a good starting point for WIMP searches. For example, we can see that pure vector DM will be difficult to observe for larger DM masses, and in any case it is ruled out by direct detection constraints. Conversely, axial-vector DM is unconstrained by direct detection, but it is already heavily constrained by 8 TeV collider bounds, and it is accessible to the 14 TeV searches even for DM masses above 500 GeV. The jump in the orange line is the point where annihilation into top quarks becomes kinematically allowed.

The overproduction lines in Fig. 8.1 rely on the assumption that the DM coupling to the first generation of quarks is not less than the coupling to other SM fermions ($g_f \leq g_{u,d}$), while the underproduction line only depends on the couplings $g_{u,d}$ to the first-generation quarks. Relaxing/strengthening the assumption 5 of Sect. 8.2.1 means allowing the couplings to other SM fermions to span over a wider/smaller range and correspondingly the upper limit is Eq. (8.8) is changed. The effect on the overproduction lines is shown in Fig. 8.2. We see that if the constraint on g_f is relaxed, the orange line of Fig. 8.1 gradually becomes too strong, and correspondingly the region in which to search for DM becomes broader (green curves of Fig. 8.2). In the event that a signal compatible with DM is observed, the region where it falls on the plot can be used to infer something about its nature.

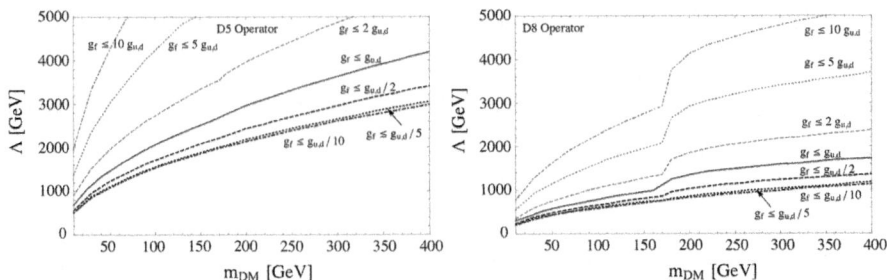

Fig. 8.2 Analogues of the orange overproduction line in Fig. 8.2 (corresponding to $g_f \lesssim g_{u,d}$), changing the relative value of the coupling between u, d quarks and other SM fermions

8.4 Results: Simplified Models

We now focus on the simplified model scenario, restricting our attention to the case of pure vector couplings. Thus we consider $g_{DM}^A = g_f^A = 0$, and we define $g_{DM}^V \equiv g_{DM}$, $g_f^V \equiv g_f$. The annihilation rates and mediator decay widths have been computed and are shown in Appendix B.

In order to compare directly with prospective ATLAS constraints, in Figs. 8.3 and 8.4 we show lines for specific choices of $\sqrt{g_{DM} \cdot g_f}$ =0.5, 1, π and m_{DM} =50, 400 GeV respectively. The ATLAS constraints are again from Ref. [7] and refer to a vector mediator model. These constraints have some degeneracy in M for low values of $\sqrt{g_{DM} \cdot g_f}$, and so we do not show a line corresponding to $\sqrt{g_{DM} \cdot g_f} = 0.5$ In order to compare with their prospective constraints, the relic density constraints assume the same (arbitrary) widths as ATLAS.

Again, the logic in our constraints is that, for the overproduction line, any change in parameters which decreases the cross section will lead to overproduction of DM. Conversely, for the underproduction line, any change in parameters which increases the cross section will lead to underproduction of DM.

While the annihilation rate of DM particles only depends on the product $g_{DM} \cdot g_f$, the mediator decay widths depend on each coupling individually. So we are forced to fix the ratio g_f / g_{DM}, in addition to keeping the product $g_{DM} \cdot g_f$ as a parameter. For fixed values of the mediator width, a bound on the product $\sqrt{g_{DM} \cdot g_f}$ can be recast into a bound on the ratio g_f / g_{DM}.

An important remark should be made about the arbitrary widths used in Figs. 8.3 and 8.4. These can be compared to the physical minimal widths (defined as the decay width into SM fermions + DM, with no other annihilation state) to fix the ratio g_f / g_{DM}. This is shown in Fig. 8.5. In some regions there is no solution, and the width used by ATLAS is in fact not physical. For this reason it is recommendable to avoid the use of arbitrary mediator widths, and to fix the widths to their minimal value given by the decay channels to SM particles and to DM particles.

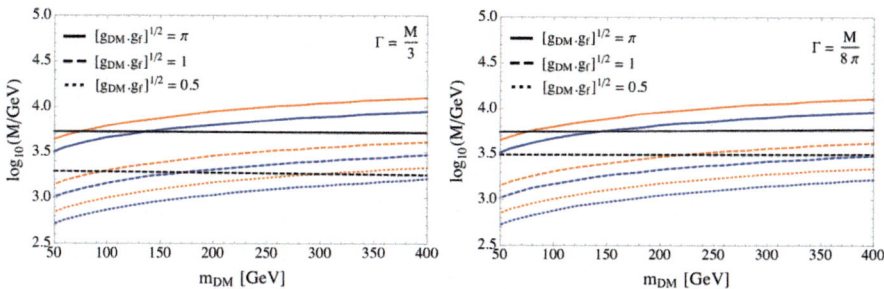

Fig. 8.3 Over- (*orange*) and under- (*blue*) production boundary lines for thermal relic dark matter, for three different choices of the coupling strengths, and a Z'-type mediator with pure vector couplings. *Black lines* are ATLAS projected 95% lower bounds after $25\,\text{fb}^{-1}$ at 14 TeV, assuming 5% systematic uncertainties

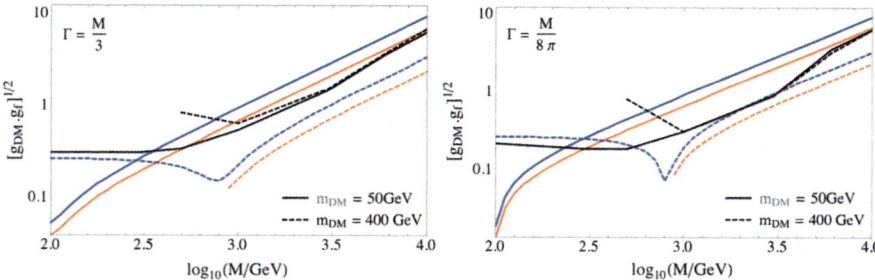

Fig. 8.4 Over- (*orange*) and under- (*blue*) production boundary lines for thermal relic dark matter, compared with projected ATLAS reach (*black*), for two values of the dark matter mass, and a Z'-type mediator with pure vector couplings. *Black lines* are ATLAS projected 95% upper bounds after $25\,\text{fb}^{-1}$ at 14 TeV, assuming 5% systematic uncertainties

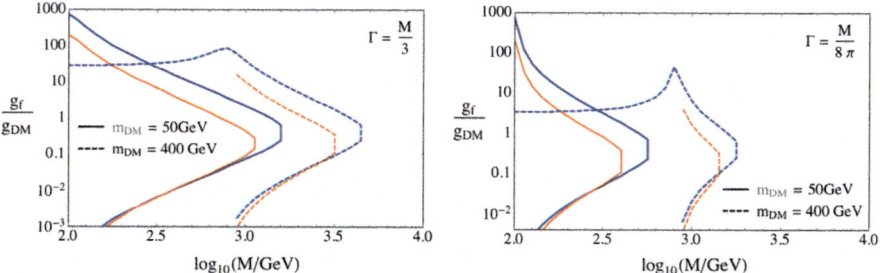

Fig. 8.5 The solution to the ratio g_f/g_{DM} corresponding to the bounds on the product $g_f \cdot g_{\text{DM}}$ combined with fixed mediator widths (as represented in Fig. 8.4). At large mediator masses, no solution exists and the widths are unphysical for the coupling strengths in Fig. 8.4

8.5　Conclusion

In this chapter we have discussed the importance of the relic density as a powerful search tool in the searches for DM at the LHC. In fact, in order to reveal the true nature of DM, any future signal of a new weakly interacting particle possibly produced in the collider must be confronted with the requirement the new particle has a relic abundance compatible with observations before assigning it the label of thermal DM. Of course, the matching of the relic density would not be sufficient on itself, and a positive signal from other probes (in particular direct searches) would be necessary to firmly establish the DM nature of the putative new invisible particle.

We have followed both the approach of effective operators and the approach of simplified models, for a reference case of a vector mediator. We have found that, in both situations, Run II of LHC has the potential to explore a large portion of the parameter space of thermal-relic DM, either in terms of claiming discovery or in terms ruling out models.

Our results are twofold. One the one hand, they can be used by LHC collaborations as a guidance into the parameter space of DM models; in fact, simple relic density considerations help to set priorities and parameter choices when analysing future data in terms of DM.

On the other hand, our results provide clear messages in case of observation of a new stable particle: if the new particle is not compatible with our thermal relic curves, either it is not the DM or one of our working assumptions is not valid. In any case, very interesting lessons about the nature of DM will be learned from LHC data.

References

1. G. Busoni, A. De Simone, T. Jacques, E. Morgante, A. Riotto, Making the most of the relic density for dark matter searches at the LHC 14 TeV run. JCAP **1503**(03), 022 (2015), arXiv:1410.7409
2. Planck Collaboration, P.A.R. Ade et al., Planck 2015 results. XIII. Cosmological parameters, Astron. Astrophys. A **13**, 594 (2016), arXiv:1502.01589
3. J. Goodman, M. Ibe, A. Rajaraman, W. Shepherd, T.M. Tait et al., Constraints on dark matter from colliders. Phys. Rev. D **82**, 116010 (2010), arXiv:1008.1783
4. J. Goodman, M. Ibe, A. Rajaraman, W. Shepherd, T.M. Tait et al., Constraints on light majorana dark matter from colliders. Phys. Lett. B **695**, 185–188 (2011), arXiv:1005.1286
5. G.B. Gelmini, P. Gondolo, Neutralino with the right cold dark matter abundance in (almost) any supersymmetric model. Phys. Rev. D **74**, 023510 (2006), arXiv:hep-ph/0602230
6. O. Buchmueller, M.J. Dolan, S.A. Malik, C. McCabe, Characterising dark matter searches at colliders and direct detection experiments: vector mediators, JHEP **1501**, 037 (2015), arXiv:1407.8257
7. Sensitivity to WIMP Dark Matter in the Final States Containing Jets and Missing Transverse Momentum with the ATLAS Detector at 14 TeV LHC. Technical report, ATL-PHYS-PUB-2014-007, CERN, Geneva, June 2014

8. ATLAS Collaboration, G. Aad et al., Search for dark matter candidates and large extra dimensions in events with a jet and missing transverse momentum with the ATLAS detector. JHEP **1304**, 075 (2013), arXiv:1210.4491

9. LUX Collaboration, D.S. Akerib et al., First results from the LUX dark matter experiment at the Sanford underground research facility. Phys. Rev. Lett. **112**, 091303 (2014), arXiv:1310.8214

10. XENON100 Collaboration, E. Aprile et al., Limits on spin-dependent WIMP-nucleon cross sections from 225 live days of XENON100 data. Phys. Rev. Lett. **111**(2), 021301 (2013), arXiv:1301.6620

Chapter 9
A U(1)' Gauge Mediator

9.1 Introduction

As we saw in the previous chapters, simplified models are a useful tool to interpret LHC and other experiments' data in terms of a minimal theory of Dark Matter, which contains only the relevant degrees of freedom that can be probed in our searches. At the same time, as discussed in Sect. 7.3, some care should be taken in constructing simplified models in order to be theoretically consistent and useful to compare the results of different searches.

In this chapter, based on Ref. [1], we are going to construct an example of a well motivated and theoretically consistent simplified model, and we are going to discuss how different experimental searches can be exploited to constrain its parameters. In particular, we will be careful in defining our model in such a way that it satisfies unitarity conditions and it is free of gauge anomalies.

As we saw in Chap. 2, direct detection experiments are already probing some of the most interesting regions of WIMP parameter space, and the simplest WIMP picture is already suffering some tension. An important caveat is that the scattering is typically assumed to be spin independent, because non relativistic operators depending on the particles' spins are suppressed, and the SI component, if present, is dominant. However, if for some reason the scattering is purely spin dependent, upper limits from direct detection experiments are increased by order of magnitudes: from the level of $\sigma_{\chi p} \sim 10^{-44} - 10^{-45} \, \mathrm{cm}^2$ for SI cross section [2] to that of $\sigma_{\chi p}^{\mathrm{SD}} \lesssim 10^{-38} - 10^{-39} \, \mathrm{cm}^2$ for the SD one [3].

Interestingly, the strongest bounds on spin dependent nucleon-DM scattering do not come from direct detection, but rather from other probes. Strongest limits have been claimed by Fermi-LAT, IceCube and by the LHC collaborations. However, these constraints are more model dependent because they involve the interactions of DM with other SM particles, and their comparison to direct detection operators is not straightforward.

There are two main issues when comparing results from different kind of experiments, namely those from direct detection and those from indirect and collider

© Springer International Publishing AG 2017
E. Morgante, *Aspects of WIMP Dark Matter Searches at Colliders and Other Probes*, Springer Theses, https://doi.org/10.1007/978-3-319-67606-7_9

searches. Let us first focus on neutrino telescopes. Under the assumption of equilibrium between DM capture and annihilation in the solar core, the total annihilation rate is proportional to the scattering cross section, and not on the annihilation one. However, in order to predict the neutrino flux, one should also know the annihilation channels and the branching ratios into each of them, since each channel is going to yield a different flux of neutrinos. As benchmarks, the IceCube collaboration reports its results [4, 5] considering annihilations into WW, $\tau\tau$ (optimistic scenarios with hard neutrinos) and $b\bar{b}$ (pessimistic scenario with soft neutrinos) with 100% branching ratio. However, there is no guarantee that in a full UV complete picture any of these channels is dominant. In fact, if the DM is a Dirac fermion, it will most likely annihilate into a pair of SM fermions, and assuming flavor universality these will most likely be light flavours, leaving no distinctive signature at IceCube. On the other hand, if the DM is Majorana and the fermion current preserves chirality, the light fermion channels are velocity-suppressed,[1] giving way to other channels: WW, ZZ, hh, Zh and $t\bar{t}$. The same problem holds for indirect detection bounds, which depend on the branching fractions into all possible annihilation channel.

The second problem has to do with LHC searches. When they are compared to direct detection bounds, collider exclusions are claimed to be the strongest ones for SD scattering. This claim is very model dependent, because the LHC searches are sensitive to the ratio between the couplings of the mediator to DM and SM particles. Moreover, collider exclusions weaken in the high DM mass region, where other probes are still strong.

9.2 Z'-Mediated Spin-Dependent Interactions

The non relativistic operator $\mathcal{O}_4 = \vec{S}_\chi \cdot \vec{S}_N$, which is responsible for spin dependent scattering, descends from the axial vector - axial vector effective operator

$$\mathcal{L} = \left(\overline{\chi}\gamma_\mu\gamma^5\chi\right)\left(\overline{N}\gamma^\mu\gamma^5 N\right). \tag{9.1}$$

In order to compare direct detection searches for DM with the constraints from the LHC and IceCube an UV completion for this operator is need. There are several options, the simplest and most economical being to mediate the interaction via a massive neutral vector boson, which couples to the SM axial current and to the DM axial current. This might either be a Z-boson of the SM (corresponding to a standard WIMP scenario) or a new Z' boson.

---

[1]The initial configuration $\chi\chi$ is a CP eigenstate with eigenvalue -1:the DM particles are identical fermions and their wavefunction must be antisymmetric. In the limit of zero velocity $L = 0$ requires an antisymmetric opposite spin configuration corresponding to $S = 0$. Thus the CP eigenvalue of the $\chi\chi$ is $(-1)^{2L+S+1} = -1$. In the final state $f\bar{f}$ f and $\bar{f}$ have opposite chiralities, implying opposite helicities if we neglect the fermion mass m_f, and the total spin along that axis is $+1$. Thus the CP eigenvalue for the final state is $(-1)^{2L+S+1} = +1$. CP conservation thus forbids the annihilation of two DM particles into two massless fermions in the limit of zero velocity, if the fermion current preserves chirality (see also [6]).

The first realization is that of a "classical" WIMP, in which a Majorana fermion is charged under the EW gauge group (with zero electric charge). A supersymmetric electroweakino, doublet or triplet of $SU(2)_L$, possibly mixed with a singlet (bino0, is the prototype of this kid of DM.[2] In a SUSY model of this kind, assuming very small mixings between the electroweakinos and suppressed squark interactions,[3] the scattering cross section is spin dependent and therefore the bounds from direct detection experiments are relatively weak. Moreover, being DM a Majorana fermion, the annihilation cross section into light fermions is helicity suppressed. In the neutralino case, the annihilation into the EW gauge bosons proceeds at the leading order and in this case the WW, ZZ and $t\bar{t}$ channels dominate the annihilation branching ratios. The only relevant NLO annihilation channels for the neutralino spin-dependent DM are $\gamma\gamma$ and γZ, which have been fully analyzed in Refs. [13, 14].

An alternative UV completion of Eq. 9.1, which we are going to analyze here, is one in which the interaction is mediated by a heavy vector Z', remnant of a new gauge symmetry at the TeV scale, spontaneously broken by the vev of some scalar field. A generic renormalizable interaction lagrangian of the Z' with the DM and the SM fermions is of the form

$$\mathcal{L} = ig_{Z'}g_\chi \bar{\chi}\gamma^\mu\gamma^5\chi Z'_\mu + ig_{Z'}g_f^A \bar{\psi}\gamma^\mu\gamma^5\psi Z'_\mu + ig_{Z'}g_f^V \bar{\psi}\gamma^\mu\psi Z'_\mu, \qquad (9.2)$$

where χ is a Majorana DM particle and ψ stands for the SM fermions. The vector coupling to the DM vanishes exactly, as a consequence of the Makorana nature of χ. For a Dirac particle this coupling would be present, leading to a spin independent interaction which would be strongly constrained by direct detection experiments.

This Z' scenario has already in a vast literature, see e.g. [15–36], but a comparison of different searches is still unclear. In particular, the signal from neutrino telescopes is poorly studied, and needs a careful treatment because it is very sensitive to the annihilation branching ratios of the DM into the EW bosons. The annihilation cross sections of DM into the SM fermions are helicity suppressed (except for $m_\chi \sim m_f$, with f a SM fermion, where there is no suppression into $f\bar{f}$), and therefore bosonic channels are expected to be the dominant source of neutrinos that can be detected by IceCube from DM annihilation in the Sun.

If the Z' is a gauge boson of a broken symmetry at multi-TeV scale, we should be sure that we are analyzing an anomaly free theory. This is crucial when one wants to compute the annihilation rates of DM into SM gauge bosons, which occur via loop diagrams. The value of these loops can only be calculable in a renormalizable theory, and the requirement of anomaly cancellation is essential.

A flavour-blind, renormalizable, anomaly free $U(1)'$ model is easily parametrized as a linear combination of the SM hypercharge and $U(1)_{B-L}$ (see Sect. 22.4 of [37]). We can parametrize the generators of the new symmetry as

---

[2]The "Minimal Dark Matter" model of [7, 8] can be seen as a generalization of this scenario to arbitrary representations.

[3]This can naturally happen in various scenarios, e.g. split SUSY [9, 10] or recently proposed mini-split, motivated by the 125 GeV higgs [11, 12].

Table 9.1 Charges of the SM matter content under the gauge symmetries of the SM and the gauge $U(1)'$ with the generator (9.3). i stands for the family index

	$SU(3)$	$SU(2)$	$U(1)_Y$	$U(1)_{B-L}$	$U(1)'$
$\begin{pmatrix} \nu_L^{\ell_i} \\ \ell_L^i \end{pmatrix}$	1	2	$-\frac{1}{2}$	-1	$-\frac{1}{2}\cos\theta - \sin\theta$
$\left(\ell_R^i\right)^C$	1	1	1	$+1$	$\cos\theta + \sin\theta$
$\begin{pmatrix} u_L^i \\ d_L^i \end{pmatrix}$	3	2	$\frac{1}{6}$	$+\frac{1}{3}$	$\frac{1}{6}\cos\theta + \frac{1}{3}\sin\theta$
$\left(u_R^i\right)^C$	$\bar{3}$	1	$-\frac{2}{3}$	$-\frac{1}{3}$	$-\frac{2}{3}\cos\theta - \frac{1}{3}\sin\theta$
$\left(d_R^i\right)^C$	$\bar{3}$	1	$\frac{1}{3}$	$-\frac{1}{3}$	$\frac{1}{3}\cos\theta - \frac{1}{3}\sin\theta$
$\Phi = \begin{pmatrix} \phi^+ \\ \phi^0 \end{pmatrix}$	1	2	$\frac{1}{2}$	0	$\frac{1}{2}\cos\theta$

$$\cos\theta \, t_Y + \sin\theta \, t_{B-L} \tag{9.3}$$

where t_Y and t_{B-L} stand for the generators of the hypercharge and $B-L$ symmetries respectively. In order for the SM Yukawas to respect the extended gauge symmetry, also the Higgs must have a charge under $U(1)'$ equal to $\frac{1}{2}\cos\theta$, which is unambiguously determined by those of SM fermions. For completeness we list all the charges under the gauge symmetries, including the $U(1)'$, in Table 9.1.

These charges have a strong impact on the DM phenomenology in this scenario. Because the SM Higgs couples to the Z', tree level couplings between the Z' and the EW gauge bosons are induced after EW symmetry breaking. In this case Z' mixes with the Z. This allows annihilations of the DM to EW gauge bosons at the tree level.

We can now write the full lagrangian of the model as

$$\mathcal{L} = -\frac{1}{4}\text{Tr}\left(G_{\mu\nu}G^{\mu\nu}\right) - \frac{1}{4}\text{Tr}\left(W_{\mu\nu}W^{\mu\nu}\right) - \frac{1}{4}B_{\mu\nu}B^{\mu\nu} - \frac{1}{4}Z'_{\mu\nu}Z'^{\mu\nu} \tag{9.4}$$
$$+ \sum_{\text{SM fermions } \psi} i\bar{\psi}\not{D}\psi + \text{h.c.}$$
$$+ \sum_{\text{SM fermions } \psi} \bar{\psi}^i Y^{ij}\psi^j \Phi + \text{h.c.}$$
$$+ (D_\mu\Phi)^\dagger(D^\mu\Phi) + V(\Phi)$$
$$+ \mathcal{L}_{\text{DM}}.$$

where we have assumed that the kinetic mixing term $Z'_{\mu\nu}B^{\mu\nu}$ is negligible. The covariant derivative of the Higgs field is given by $D_\mu = D_\mu^{\text{SM}} - \frac{i}{2}\cos\theta \, g_{Z'}Z'_\mu$. The couplings of the Z' to the EW gauge bosons, in particular to Zh and W^+W^-, arise at the tree level after EW symmetry breaking, when the Higgs Goldstone modes are

Table 9.2 Coefficients of the vector and axial vector bilinear currents for the SM fermions. With obvious meaning of the notation, the coefficients g^V and g^A are obtained from g^L, g^R of Table 9.1 via $g^L P_L + g^R P_R = \frac{g^L + g^R}{2} + \frac{-g^L + g^R}{2}\gamma_5$, so that $g^V = \frac{g^L + g^R}{2}$, $g^A = \frac{-g^L + g^R}{2}$

SM fermion f	g_f^V: coeff. of $g_{Z'}\overline{f}\slashed{Z}'f$	g_f^A: coeff. of $g_{Z'}\overline{f}\slashed{Z}'\gamma_5 f$
leptons	$-\frac{3}{4}\cos\theta - \sin\theta$	$-\frac{1}{4}\cos\theta$
neutrinos	$-\frac{1}{4}\cos\theta - \frac{1}{2}\sin\theta$	$\frac{1}{4}\cos\theta + \frac{1}{2}\sin\theta$
up quarks	$\frac{5}{12}\cos\theta + \frac{1}{3}\sin\theta$	$\frac{1}{4}\cos\theta$
down quarks	$-\frac{1}{12}\cos\theta + \frac{1}{3}\sin\theta$	$-\frac{1}{4}\cos\theta$

"eaten" by the massive gauge bosons. From the charges in Table 9.1 one can deduce the vector and axial vector couplings of the Z' to SM fermions, which are listed in Table 9.2.

The Lagrangian describing DM is

$$\mathcal{L}_{\text{DM}} = \frac{1}{2}\overline{\chi}\Big(i\slashed{\partial} - m_\chi\Big)\chi + \frac{1}{2}g_{Z'}g_\chi Z'_\mu\Big(\overline{\chi}\gamma^\mu\gamma_5\chi\Big), \tag{9.5}$$

where g_χ is the coupling of χ to the Z'.[4] We do not consider the effects of renormalization group equations that could mix Z and Z' via loop effects. We also do not

---

[4]The simplest model that could provide a massive Majorana particle interacting with the Z' could be the following. We use the 2-component notation. We introduce a Majorana fermion ψ_0 and two Weyl fermions ψ_1, ψ_2. ψ_1 and ψ_2 interact with the Z' with opposite charges, and we write a mass term $(\psi_{2\alpha}\psi_1^\alpha +$ h.c.) that is both gauge invariant and Lorentz invariant. With just one Weyl spinor we would not manage to write a gauge invariant term. (A term like $\psi_1^\dagger\psi_1$ would violate Lorentz symmetry.) The Lagrangian for these fields would be

$$\begin{aligned}\mathcal{L} =&\, \psi_1^\dagger i\sigma_\mu\partial^\mu\psi_1 + \psi_2^\dagger i\sigma_\mu\partial^\mu\psi_2 + m(\psi_2\psi_1 + \psi_2^\dagger\psi_1^\dagger)\\ &+ g_{Z'}g_\psi Z'^\mu(\psi_1^\dagger\partial_\mu\psi_1 + \psi_2^\dagger\partial_\mu\psi_2)\\ &+ \psi_0^\dagger i\sigma_\mu\partial^\mu\psi_0 + \Phi\,\psi_0(y_1\psi_1 + y_2\psi_2)\,,\end{aligned} \tag{9.6}$$

where Φ is the scalar field responsible of the BEH mechanism in the $U(1)'$ sector. Notice that the third line, independently from the value of y_1, y_2, prevents us from writing (9.6) in terms of a Majorana spinor $\binom{\psi_1}{\psi_2}$. Once Φ assumes a vev v, the mass matrix for the triplet (ψ_1, ψ_2, ψ_0) is

$$\frac{1}{2}\begin{pmatrix} 0 & m & y_1 v \\ m & 0 & y_2 v \\ y_1 v & y_2 v & (M) \end{pmatrix}, \tag{9.7}$$

where M is a mass term for ψ_0 that can possibly appear in (9.6). The diagonalisation of the mass matrix brings to a Majorana massive particle interacting with Z', which can be close to ψ_0 if $M, y_i v \ll m$. Notice that the terms proportional to v in (9.7) introduce a splitting in the mass matrix between what could be the L- and R-handed components of a Dirac spinor. Some theories of inelastic dark matter elaborate on the possibility that this mass splitting is small, and the lightest eigenstate could convert into the heavier one via inelastic scattering.

take into account the running of the operator coefficients due to the RG flow because the quantitative effect is expected to be mild (see recent Ref. [38] for the details). The latter is indeed a minor effect, if the model is such that the mixing is zero at a scale close to the electroweak one. We do not specify the dynamics of the spontaneous symmetry breaking sector of $U(1)'$, and for our purposes we just assume that it provides a mass term $\frac{1}{2}m_{Z'}^2 Z'^2$.

As we expand the covariant derivatives in Eq. 9.4 we find that the Z' mixes with Z, with a mixing angle ψ fully determined by the mass $m_{Z'}$ of the physical mass eigenstate, the value of the coupling $g_{Z'}$ and the angle θ. Indeed, the Higgs kinetic term can be rewritten as

$$(D_\mu \Phi)^\dagger (D_\mu \Phi) = \frac{1}{2}\partial_\mu h \partial^\mu h + \frac{1}{4}g^2(v+h)^2 W^- \cdot W^+ + \frac{1}{2}(v+h)^2 \left(-\frac{1}{2}g_Z Z_\mu + \frac{1}{2}\cos\theta\, g_{Z'}\, Z'_\mu \right)^2, \tag{9.8}$$

where $g_Z = \sqrt{g^2 + g'^2} = 2m_Z/v$ with $v = 246$ GeV. Equation 9.8 contains a non diagonal mass term for Z and Z', which induces the mixing. If we denote the mass of the lighter mass eigenstate by $m_Z = 91.2$ GeV, then ψ turns out to be of order $\psi \sim g_{Z'}\cos\theta\left(\frac{m_Z}{m_{Z'}}\right)^2 \ll 1$ in the regime $m_{Z'} \gg m_Z$ that we consider here. For this reason, in the remainder of the discussion we ignore the mixing for simplicity of notation, and denote both the interaction or mass eigenstate equivalently by Z', and the lighter mass eigenstate identifiable with the SM vector boson by Z.

Due to the Z-Z' mixing, the heavy Z' couples at tree level to Zh with a vertex

$$\frac{1}{2}\cos\theta g_{Z'} g_Z v \left(Z' Z h \right), \tag{9.9}$$

and with $W^+ W^-$ with a vertex equal to $\sin\psi$ times the SM vertex for ZW^+W^-,

$$\sin\psi \cdot ig\cos\theta_W \left[\left(W_\mu^+ W_\nu^- - W_\nu^+ W_\mu^- \right)\partial^\mu Z'^\nu + W_{\mu\nu}^+ W^{-\mu} Z'^\nu - W_{\mu\nu}^- W^{+\mu} Z'^\nu \right], \tag{9.10}$$

where $W_{\mu\nu}^\pm = \partial_\mu W_\nu^\pm - \partial_\nu W_\mu^\pm$. Notice that both (9.9) and (9.10) are proportional to $\cos\theta$, thus they vanish in the pure $U(1)_{B-L}$ limit. Interaction (9.10) turns out to be velocity suppressed, thus it gives a small cross section in the low DM velocity regime of DD and IC, and both (9.9) and (9.10) are proportional to $\cos\theta$, thus vanish in the pure $U(1)_{B-L}$ limit. Because of these suppressions, loop channels are also relevant at low kinetic energy, as discussed in Sect. 9.3.6. In particular, the annihilation of $\chi\chi$ to Zh occurs at tree level, except for the case in which the $U(1)'$ extension is a pure $U(1)_{B-L}$ gauge symmetry. At low velocity, annihilation into W^+W^- is instead dominantly driven by diagrams with a fermionic loop (see Appendix C).

(Footnote 4 continued)
We conclude by stressing that the vector-like coupling of Z' to ψ_1 and ψ_2 ensures the cancellation of the gauge anomaly in the diagrams involving these fermions.

9.2.1 Calculation of DM Relic Density

Now that we have all the tree level couplings of the Z', we are ready to calculate the thermal relic abundance of the Majorana DM. We calculate the annihilation rates and perform the thermal average using the procedure of [39]. The result for the thermally averaged self-annihilation cross section as a function of temperature T is

$$\langle \sigma v \rangle = \frac{x}{8\, m_\chi^5} \frac{1}{\left(\mathscr{K}_2(x)\right)^2} \int_{4m_\chi^2}^{\infty} \sigma_{\mathrm{ann}} \sqrt{s}\,\left(s - 4m_\chi^2\right)\,\mathscr{K}_1\left(\frac{x\sqrt{s}}{m_\chi}\right)\, ds\,, \qquad (9.11)$$

where $x = m_\chi/T$, and $\mathscr{K}_i$ is the modified Bessel function of order i.

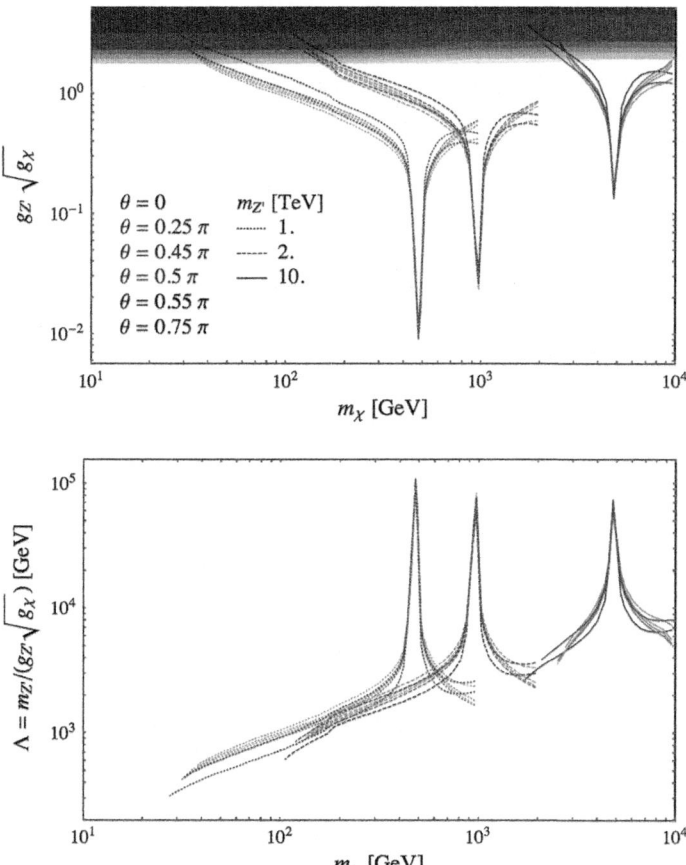

Fig. 9.1 *Top* Value of $g_{Z'}\sqrt{g_\chi}$ that yields to the correct relic density, for three masses of the Z' and different values of θ. The regions coloured with multiple shades of *gray* in the upper part of the plot (one for each θ, $m_{Z'}$) show the regions where $\Gamma_{Z'}$ becomes of the order of $m_{Z'}$, signalling the transition to the non-perturbative regime. *Bottom* Same data as left, in the plane Λ *vs.* m_χ

We do not approximate the thermally averaged cross section with a low DM velocity expansion, since close to the resonance $m_\chi \lesssim m_{Z'}/2$, terms of higher order can also yield important contributions to the relic density [40].

Once we fix the values of the angle θ and $m_{Z'}$, we are left with $g_{Z'}$ as the only free parameter. In Fig. 9.1 we show the value of $g_{Z'}$ that yields the correct relic density as measured by [41]. The areas above (below) the lines correspond to points of the parameter space where the DM is under- (over-) abundant in the thermal scenario.

For the calculation of the relic density we relied only on tree level cross sections, which we confirmed are dominant at typical freeze-out energies. Varying the value of g_χ (which for this computation was set to 1) while keeping the product $g_{Z'}\sqrt{g_\chi}$ fixed only very mildly affects the decay width $\Gamma_{Z'}$, and would have practically no effect on Fig. 9.1.

The lines in Fig. 9.1 stop at $m_\chi = m_{Z'}$: above that threshold, annihilation into $Z'Z'$ opens up, and in principle one would need to specify the details of the spontaneous symmetry breaking sector of $U(1)'$ in order to compute the relic abundance precisely. This is not required for our purpose of understanding the plausible range of values for $g_{Z'}$ that yield a relic abundance close to the observed one.

9.3 Overview of Direct and Indirect Bounds

The constraints on the scenario that we have described above come from three different primary sources: direct detection, neutrino telescopes and the LHC. Since we assume a Majorana DM particle, all interactions that we get between the nuclei and the DM are either spin dependent or halo velocity suppressed, or both. We will comment on these interactions in detail in Sect. 9.3.1. This particular feature renders the direct detection results much less efficient than in the case of spin-independent interactions, while other experiments, notably neutrino telescopes and the LHC, become competitive. In the following section we will carefully go through all of these different experiments and discuss the bounds that they produce.

9.3.1 Direct Detection Experiments

In this type of experiment, in a model with a Z' mediator the effective DM theory is always valid, because the transferred momentum never exceeds hundreds of MeV, well below the mediator scale. The effective contact terms that one gets between the DM and SM quarks are

$$\mathcal{L}_{eff} = \frac{g_q^V}{\Lambda^2}\bar{\chi}\gamma^\mu\gamma^5\chi\,\bar{q}\gamma_\mu q + \frac{g_q^A}{\Lambda^2}\bar{\chi}\gamma^\mu\gamma^5\chi\,\bar{q}\gamma_\mu\gamma^5 q \tag{9.12}$$

with coefficients g given in Table 9.2, and $1/\Lambda^2 = g_{Z'}^2 g_\chi / m_{Z'}^2$.

It is straightforward to translate these interactions to the more intuitive language of the NR effective theory. Using the dictionary of [42, 43] and considering a nucleus N instead of the partons q we get

$$\bar{\chi}\gamma^{\mu}\gamma^5\chi\bar{N}\gamma_{\mu}\gamma^5 N \rightarrow -4\vec{S}_{\chi}\cdot\vec{S}_N = 4\mathcal{O}_4 \tag{9.13}$$

$$\bar{\chi}\gamma^{\mu}\gamma^5\chi\bar{N}\gamma_{\mu}N \rightarrow 2\vec{v}\cdot\vec{S}_{\chi} + 2i\vec{S}_{\chi}\cdot\left(\vec{S}_N\times\frac{\vec{q}}{m_N}\right) = 2\mathcal{O}_8 + 2\mathcal{O}_9. \tag{9.14}$$

For a generic Z' (arbitrary θ angle in (9.3)) we get both axial current - axial current (AA) and axial current - vector current (AV) interactions with the coefficients being roughly of the same order of magnitude. However, the latter induces interactions that are halo velocity suppressed ($\mathcal{O}_8$) and both nuclear spin and halo velocity suppressed ($\mathcal{O}_9$). This usually renders the AV interactions smaller than the AA interections for direct detection and solar capture, although we do include AV interactions when we derive our bounds. In fact, the halo-velocity suppression in $\mathcal{O}_8$ is sometimes comparable to the suppression due to the spin-dependence in $\mathcal{O}_4$. The operator $\mathcal{O}_9$, which is both spin-dependent and halo velocity dependent via the exchange momentum $\vec{q}$ is completely negligible. The one-loop contributions may induce a spin-independent scattering cross section of the DM with the nucleons, but their quantitative impact is negligible in our model [44].

For direct detection, there is a special point in parameter space where the usual spin-dependent operator completely shuts down. This happens when our new gauge symmetry is exactly $U(1)_{B-L}$, or, in our language, $\theta = \frac{\pi}{2}$. In this case the Z' couples only to the vector current of the SM, and therefore the NR interaction operators proceed only via $\mathcal{O}_8$ and subdominantly via $\mathcal{O}_9$.

We derive the exclusion bounds obtained by the experiments LUX, XENON100, CDMS-Ge, COUPP, PICASSO, SuperCDMS with the help of the tables made available by [43], and the bounds by PICO from [3]. The strongest constraints among them are shown in Figs. 9.7 and 9.8, together with the constraints from IceCube and monojet searches.

9.3.2 Direct Constraints on Z' from LHC Searches

Here we review direct constraints on this Z' from the LHC. In addition, we will consider monojet constraints on DM production in Sect. 9.3.3. The easiest way to spot a Z' at a collider is via an analysis of the leptonic modes, unless they are highly suppressed. For these purposes we recast a CMS search for a narrow Z' in the leptonic channel [45], which conveniently phrases the constraints in terms of

$$R_{\sigma} \equiv \frac{\sigma(pp\rightarrow Z')\times BR(Z'\rightarrow l^+l^-)}{\sigma(pp\rightarrow Z)\times BR(Z\rightarrow l^+l^-)} \tag{9.15}$$

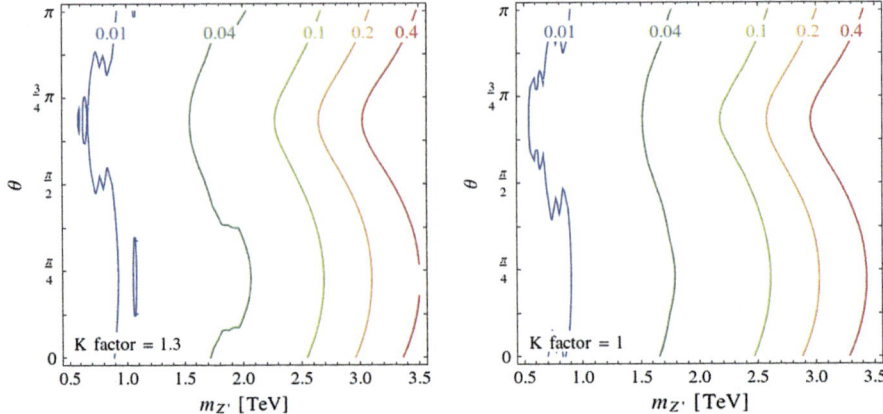

Fig. 9.2 Maximal allowed gauge couplings $g_{Z'}$ of the hidden $U(1)'$ as a function of the angle θ (Eq. (9.3)) and of $m_{Z'}$. On the LH side we assume the nominal LO cross sections, on the RH side we apply a flat $k = 1.3$ factor

and for the reference point we take $\sigma(pp \rightarrow Z) \times BR(Z \rightarrow l^+l^-) = 1.15$ nb at $\sqrt{s} = 8$ TeV [46].

We show the results of our recast in Fig. 9.2 as a function of the mass $m_{Z'}$ of the Z' and the angle θ as defined in Eq. (9.3). Note that, for a given $g_{Z'}$ and θ, all couplings to the SM are fixed, therefore both production cross sections and BRs are unambiguously determined by these values. In Fig. 9.2 we show the exclusions both for nominal cross sections, as we get from `MadGraph 5` [47], as well as for cross sections one gets by applying the flat k-factor $k = 1.3$ (similar to the suggestion in [48]). It is apparent from these lines that, in order to have a Z' with $\mathcal{O}(1)$ couplings, its mass must be $m_{Z'} \gtrsim 3$ TeV. Note that the constraints from LEP, coming from the mixing between the Z and Z', which further affect the T-parameter, are much weaker than the direct LHC constraints.

9.3.3 LHC Monojet Constraints

In this section, we explore the bounds on our $U(1)'$ model coming from LHC searches for DM, analyzing events with one hard jet plus missing transverse energy ($\not{E}_T$).

Currently, the strongest exclusion limits are from the CMS analysis [49], which analyzes 19.7 fb^{-1} at a collision energy of 8 TeV.

Despite the fact that the use of EFT to investigate dark matter signatures through missing energy is severely limited at the LHC energies, we can nevertheless use the EFT interpretation of the exclusion bounds, as it is consistent for $m_{Z'} \gtrsim 2$ TeV [50–53]. The effective operators describing the interaction between DM and quarks, in the high $m_{Z'}$ limit, are

$$\frac{1}{\Lambda^2} \sum_q g_q^V \left(\overline{\chi}\gamma^\mu\gamma_5\chi\right)\left(\overline{q}\gamma_\mu q\right) \qquad (9.16)$$

and

$$\frac{1}{\Lambda^2} \sum_q g_q^A \left(\overline{\chi}\gamma^\mu\gamma_5\chi\right)\left(\overline{q}\gamma_\mu\gamma_5 q\right), \qquad (9.17)$$

where $\Lambda \equiv m_{Z'}/(g_{Z'}\sqrt{g_\chi})$ and the coefficients g_q^V, g_q^A are given in Table 9.2.

At LHC energy scales, the occurrence of a vector or an axial vector current in the fermion bilinears in (9.16) and (9.17) does not affect the cross section for the production of DM. This is also apparent from the experimental exclusion limits reported in [49].

The CMS analysis recasts the exclusion bound as a function of the coefficient Λ of Eq. (9.17), assuming that **1)** all the g_q^A coefficients are equal to 1, and **2)** that χ is a Dirac fermion (with canonically normalized kinetic term). These two assumptions are not true for our analysis. The second assumption gives an overall factor of 2 in the cross section for the Majorana case relative to the Dirac case. We take into account both the first assumption and the convolution with the parton distribution functions (PDFs) of quarks, by means of a parton level simulation performed with MadGraph 5 [47]. We simulate the signal both with the EFT defined by CMS and with our model, and we compute for each value of θ (which determines g_q^V, g_q^A) and m_χ a rescaling factor that we use to rescale the nominal limit reported by the CMS analysis.

The final result[5] for the bounds on Λ from monojet searches are reported in Fig. 9.3.

9.3.4 Constraints from Observations of γ-ray Spectrum

We now examine the exclusion bounds that can be obtained from the analysis of the γ-ray continuum spectrum. Limits coming from γ-ray lines are irrelevant for our model because the $\gamma\gamma$, $Z\gamma$ and $h\gamma$ channels are strongly suppressed. The most stringent and robust bounds on the γ-ray continuum spectrum come from the observation of a set of 15 Dwarf Spheroidal Galaxies (dSph) performed by Fermi-LAT [54, 55]. The robustness of these bounds against astrophysical uncertainties comes mostly from the fact that the photon flux is integrated over the whole volume of the dwarf galaxy, and in this way the inherent uncertainty due to the choice of the DM profile is largely diluted (as a reference, results are presented for the Navarro-Frenk-White

---

[5] We remark the following. If the Z' mass $m_{Z'}$ is larger than a few TeV, the bounds shown in Fig. 9.3 and the following fall in a region in which the product $g_{Z'}g_\chi^{1/2}$ is necessarily $\gtrsim 1$. This is in contrast with the fact that our Z' model has a rather large mediator width, and it must be $g_{Z'}g_\chi^{1/2} \lesssim 1$ in order to have $\Gamma_{Z'} \lesssim m_{Z'}$. For this reason, with the present experimental sensitivity the lines of Fig. 9.8 correspond to a realistic physical situation only for $m_{Z'} \sim$ few TeV.

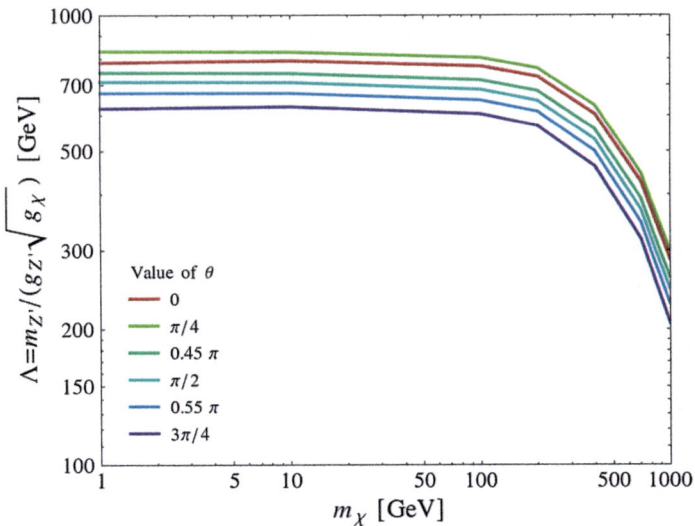

Fig. 9.3 Bound on $\Lambda = m_{Z'}/(g_{Z'}\sqrt{g_\chi})$ as a function of m_χ, for the six values of θ shown in the legend, from the CMS monojet search [49]

profile [56]). However, the bounds are practically independent on the profile choice and the variation of the bounds due to J-factor uncertainties typically does not exceed 30% [55].

Here we should also briefly comment on HESS searches for DM using diffuse γ-rays from the Galactic Centre. These bounds, claimed by the HESS collaboration [57], are nominally much stronger than Fermi dSph bounds for heavy DM, $m_\chi \gtrsim 1$ TeV. However one should also consider the uncertainties on these bounds. Unlike Fermi-LAT searches for emission from the Galactic Centre, which mask a large region around the Galactic Centre,[6] HESS merely masks a tiny region of 0.3° around the Galactic Centre, mainly to avoid the cosmic ray photon background, which is of course not present for the space-based Fermi-LAT. This makes the search much more vulnerable both to the astrophysical uncertainties and to the choice of the DM profile. HESS assumes cuspy DM profiles in its search (NFW and Einasto), and if the profile is cored, the bounds can be attenuated by orders of magnitude. This point was nicely illustrated in the context of a different (wino) DM candidate in Refs. [60, 61]. Therefore we decided not to show HESS' bounds.

In order to properly recast the results of [55] for our model two ingredients are necessary. The first one is a knowledge of the spectrum of γ-rays from DM annihilations, which can be computed using the tables provided in Ref. [62]. Results of this

---

[6]For instance, Ref. [58] practically exclude an area of 10° from consideration. Similarly, theoretical studies (see e.g. [59]) mask out 1° to 2° around the Galactic Centre. We do not show the Fermi-LAT Galactic Centre bounds on our plots because they are inferior to the Fermi-LAT dSph bounds. It is also worth mentioning that measurements of the dSphs are essentially foregrounds-free, which renders them extremely robust.

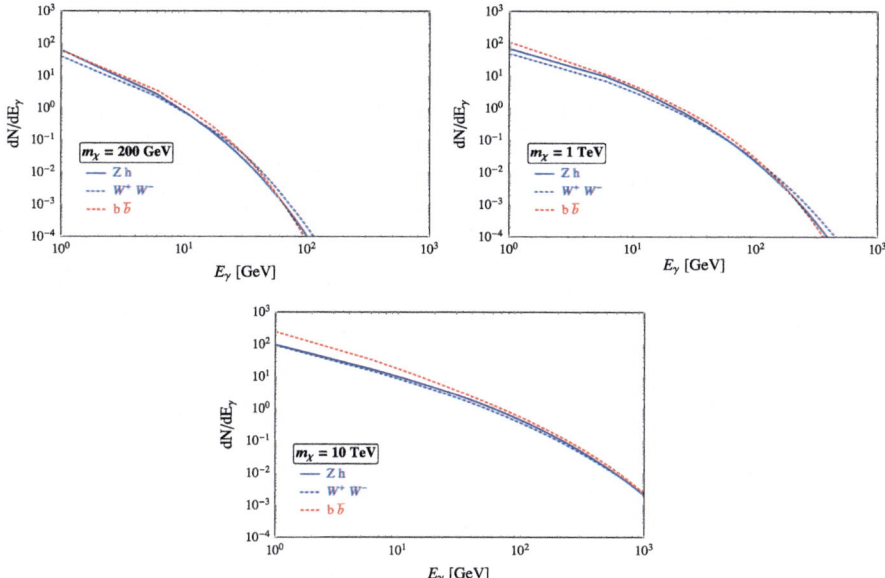

Fig. 9.4 Secondary photon spectra for different primary annihilation channels, for three reference DM masses

calculation are shown in Fig. 9.4, for three reference values of the DM mass. The second ingredient would be the exclusion limits on the flux of γ rays, information which is not provided by the Fermi-LAT collaboration. For this reason, we adopted a simplified recasting procedure. Firstly, we identified in each interval in m_χ the leading annihilation channel providing secondary photons, and approximated the total annihilation cross section with the one into that particular channel (or, in the case of multiple relevant primary channels with a similar γ-ray spectrum, we considered their sum). Secondly, we used the results of [62] to compare the photon flux from our dominant primary channel to the benchmark fluxes, namely e^+e^-, $\mu^+\mu^-$, $\tau^+\tau^-$, $u\bar{u}$, $b\bar{b}$ and W^+W^-, which are shown in the left panel of Fig. 9.5. We used the limit on $\langle \sigma v \rangle$ from the channel with the most similar photon flux as the limit on our channel. Finally, the limit on $\langle \sigma v \rangle$ is converted into a limit on Λ. Though rough, we expect our procedure to provide bounds with at least an order of magnitude accuracy on $\sigma_{\chi p}^{SD}$ (which translates into a factor of $\lesssim 2$ on Λ).

As we will discuss in Fig. 9.6, there are two regions of interest, the first for $m_\chi < m_W$ and the second for $m_\chi > m_W$. Leaving aside for the moment the peculiar case $\theta = \pi/2$, for $m_\chi < m_W$ the dominant channels are $b\bar{b}$ and $c\bar{c}$, which give a similar γ spectrum, so the Fermi-LAT limit on $b\bar{b}$ can be assumed. On the other hand, for $m_\chi > m_W$ the dominant annihilation channel is Zh, complemented by W^+W^-, ZZ and $t\bar{t}$, all of which give a similar photon flux. Since the flux of photons in the Zh channel is similar (up to a factor $\lesssim 2$) to that in the $b\bar{b}$ channel, we again picked the Fermi-LAT limit on the $b\bar{b}$ channel. In the peculiar case $\theta = \pi/2$, the dominant

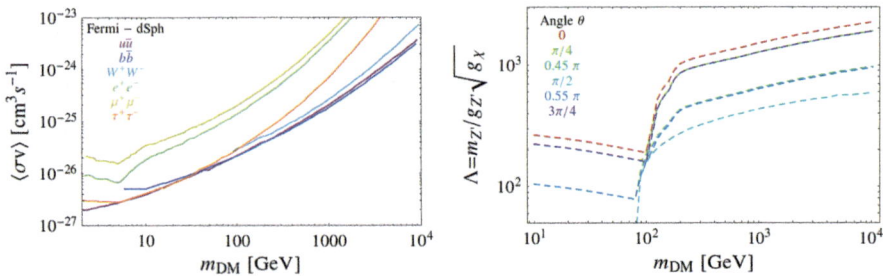

Fig. 9.5 *Left* bounds reported by the Fermi-LAT [55] collaboration assuming that DM annihilates in the specified channel. *Right* recast of these limits in our model, for the different values of θ

channels are leptonic for $m_\chi < m_W$ and W^+W^-, ZZ for $m_\chi > m_W$. Therefore, for $m_\chi < m_W$ we picked the $\tau^+\tau^-$ channel (which, among leptons, gives the strongest bounds), while for $m_\chi > m_W$ we summed the W^+W^- and ZZ contributions and compared with the limits on the W^+W^- channel. Figure 9.5 shows the result of this recast in the right panel.

9.3.5   Neutrino Telescopes – IceCube

The IceCube experiment, located at the South Pole, is a neutrino telescope observing high energy neutrinos by detecting Cherenkov photons radiated by charged particles produced in their interactions [63]. Muons from ν_μ and $\overline{\nu}_\mu$ charged current interactions leave long visible tracks within the detector, which can be easily reconstructed to estimate the direction of the incoming neutrino. IceCube has an angular resolution of a few degrees for ~ 100 GeV ν_μ (and $< 2°$ for $\sim$ for a 1 TeV ν_μ), allowing it to search for an excess of GeV-TeV neutrinos from the direction of the Sun [4]. The physics related to DM annihilations in the Sun was already analysed in Chap. 3.

Results for the benchmark channels $b\bar{b}$, WW and $\tau\tau$ can be recast for the scenario of known annihilation channels by the following method. The search utilizes the Unbinned Maximum Likelihood ratio method [64, 65], for which the sensitivity improves as Signal/$\sqrt{\text{Background}}$. For an unbinned maximum likelihood search of variable resolution, the background level varies as Ψ^2 where Ψ is the median angular resolution [66]. For a given differential (anti)-muon neutrino flux $\mathcal{F}(E)$, the total number of signal events expected within a sample can be calculated as

$$n_s(\mathcal{F}) = \int_{E_{\text{threshold}}}^{m_{\text{DM}}} \mathcal{F}(E) \times A_{\text{eff}}(E)\, dE, \qquad (9.18)$$

where A_{eff} is the effective area from Fig. 3 of Ref. [65]. Since the fluxes and effective areas of ν_μ and $\bar{\nu}_\mu$ are different, Eq. (9.18) has to be evaluated separately for ν_μ and $\bar{\nu}_\mu$. The median energy $E_{\text{med}}(\mathcal{F})$ is then defined through

$$\int_{E_{\text{threshold}}}^{E_{\text{med}}} \mathcal{F}(E) \cdot A_{\text{eff}}(E)\, \mathrm{d}E = \int_{E_{\text{med}}}^{m_{\text{DM}}} \mathcal{F}(E) \cdot A_{\text{eff}}(E)\, \mathrm{d}E. \qquad (9.19)$$

If the capture and annihilation processes in the Sun are in equilibrium, the neutrino flux, the capture/annihilation rate, as well as WIMP-nucleon scattering cross section all scale linearly with respect to each other. Thus the theoretical limit on the WIMP-nucleon scattering cross-section for a given flux prediction $\mathcal{F}_{\text{theory}}$ can be derived as

$$\sigma_{\text{theory}} = \sigma_{\text{benchmark}} \cdot \frac{n_s(\mathcal{F}_{\text{benchmark}})}{n_s(\mathcal{F}_{\text{theory}})} \cdot \frac{\Psi(E_{\text{med}}(\mathcal{F}_{\text{theory}}))}{\Psi(E_{\text{med}}(\mathcal{F}_{\text{benchmark}}))} \qquad (9.20)$$

where the first term in the RHS accounts for the variation in the level of signal events while the second term accounts for the variation in background due to the shift in median angular resolution. An analogous scaling relation can also be used to obtain theoretical limits on the annihilation rate Γ_{ann}.

The bounds on Γ_{ann} for the IceCube benchmark channels can be derived from the limits on σ by mean of the tools provided by WimpSim and DarkSUSY.

The IceCube limit on the neutrino flux $\mathcal{F}_{\text{limit}}$ requires knowledge of the neutrino spectrum per annihilation as it would be observed at Earth. The first step is to calculate the branching ratio to all relevant final states. Results are discussed in Sect. 9.3.6.

In order to convert these branching ratios into the required neutrino spectrum, we use the PPPC 4 DM ID code. This is combined with the results for the branching ratios to determine the final spectrum of muon neutrinos and antineutrinos per DM annihilation event. The Zh, γh and γZ final states are not available in the code, and so we use the average of the two pair-production spectra for each of these final states. We assume that the differences in the kinematical distributions, due to the different masses of Z, h and γ, have a minor impact on the shape of the final neutrino flux.

For the theoretical flux prediction thus obtained, the number of expected signal events as well as the median energy can be obtained from the expressions in Eqs. (9.18) and (9.19) for each of the three IceCube samples described in Ref. [65]. The integrals are evaluated separately for ν_μ and $\bar\nu_\mu$. These quantities can also be evaluated for the IceCube benchmark channel flux predictions ($\mathcal{F}_{\text{benchmark}}$) obtained from WimpSim 3.03 and nusigma 1.17. Subsequently, theoretical limits on σ and Γ_{ann} can be obtained using the scaling relation (9.20) and the analogous one for Γ_{ann}.

For a given $\mathcal{F}_{\text{theory}}$, σ_{theory} can be calculated w.r.t any of the three benchmark IceCube channels. The different calculations are consistent to within $\sim 30\%$ and are thus averaged.

9.3.6 Results for the Branching Ratios

In order to extract bounds from IceCube observations, the branching ratios for the annihilations of DM pairs into SM final states must be computed. In this section we present the final results for the BR's, based on the cross sections that are computed in detail in Appendix C.

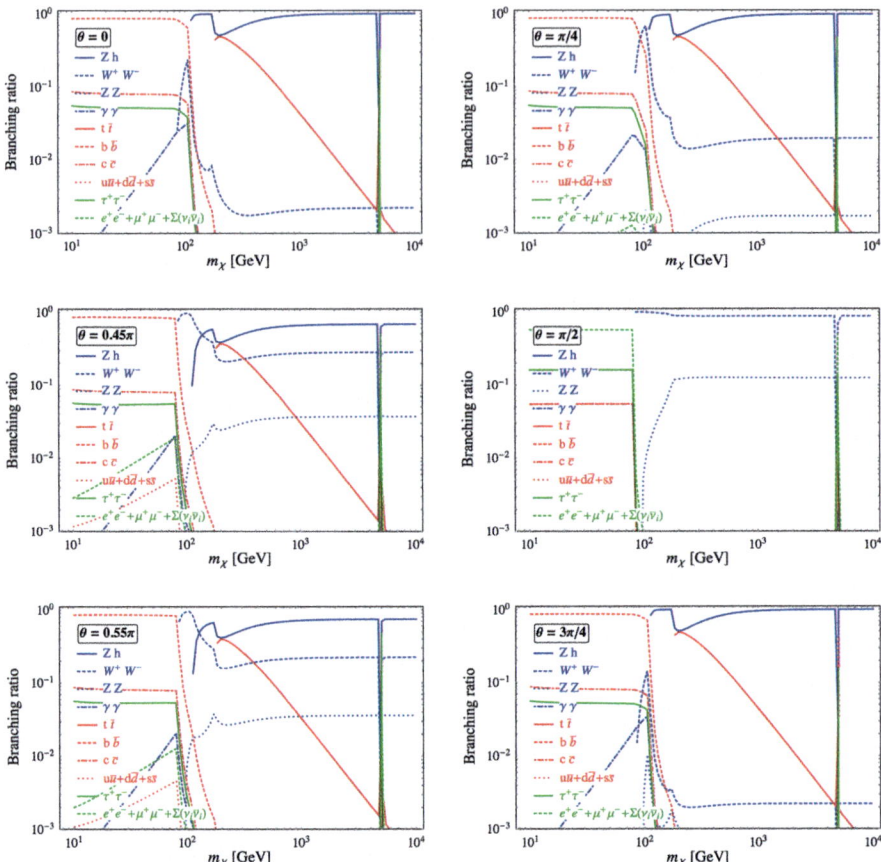

Fig. 9.6 Branching ratios for the annihilation of $\chi\chi$ into pairs of SM particles, for $m_{Z'} = 10$ TeV and a kinetic energy of the DM particles equal to the thermal one in the Sun core. There is no dependence on $g_{Z'}$ and g_χ. The six plots, from *left* to *right* and *top* to *bottom*, correspond to $\theta = 0$, $\pi/4$, 0.45π, $\pi/2$, 0.55π, $3\pi/4$. Only the channels with a BR greater than 10^{-3} are shown

Figure 9.6 shows the BR's into SM two-body final states as a function of m_χ, for $m_{Z'} = 10$ TeV and for six different values of θ (defined in Eq. (9.3)). The BR's for the final states shown on the plots do not depend at all on $g_{Z'}$ and g_χ, because they cancel in the ratios of cross sections, as can be seen from the formulæ in Appendix C. Leaving aside for a moment the pure $U(1)_{B-L}$, the main annihilation channels are the heavy fermion pairs ($t\bar{t}$, $b\bar{b}$ and $\tau^+\tau^-$) and Zh. These are indeed the only tree level channels for which $a \not\approx 0$ in the low velocity expansion $\sigma v_{\rm DM} \sim a + b\, v_{\rm DM}^2$.

The main annihilation channel below the kinematic threshold $m_\chi = 108$ GeV for Zh production is $b\bar{b}$, while the BR's into $c\bar{c}$ and $\tau^+\tau^-$ are less than 10% each. Even if its branching ratio is not the dominant one, the $\tau^+\tau^-$ channel dominates the IC bound below the Zh threshold because it yields more energetic neutrinos than the $b\bar{b}$ one.

In the region $m_\chi \sim 80\,\text{GeV} - 108\,\text{GeV}$ annihilation into W^+W^- may overcome the one into $b\bar{b}$, depending on the value of θ. Notice that, as will be explained at the end of this section, the one loop contribution to the W^+W^- cross section dominates over the tree level one, which is suppressed by the small $Z - Z'$ mixing angle and by the fact that, in the low velocity expansion, it has $a = 0$.

When the Zh channel opens up it overcomes all others and remains the only relevant channel unless $m_\chi \gtrsim m_{\text{top}}$, where the cross section into $t\bar{t}$ is comparable to the one to Zh. At higher DM masses, the cross section to Zh in the low v_{DM} limit is proportional to m_χ^2, while $\sigma(\chi\chi \to t\bar{t})$ is basically constant in m_χ. The former proportionality comes from the final state with a longitudinally polarised Z boson and a Higgs boson, and is ultimately due to the derivative coupling of would-be Goldstone bosons. This explains why Zh is the only relevant channel at large m_χ.

Around the resonance $\sigma(\chi\chi \to Zh)$ goes to 0 because the coefficient a in the low velocity expansion $\sigma v_{\text{DM}} \sim a + b\, v_{\text{DM}}^2$ vanishes. The reason is explained in Appendix C. Therefore, in a small window around the resonance, other channels dominate. The position of this window is basically the only way in which the branching ratios depend on $m_{Z'}$, as can be seen from Figs. 9.7 and 9.8. The effect of resonance excitation indeed cancel in the ratio when computing the branching ratios.

The previous picture applies for all values of θ, except $\theta = \pi/2$ which corresponds to the pure $U(1)_{B-L}$ case. In that case, there is no mixing between Z and Z', and the channel Zh disappears at tree level. Below the W^+W^- threshold $m_{\text{DM}} = m_W$, annihilation predominantly happens at tree level into fermionic channels. For $m_\chi > m_W$ the dominant channel is instead W^+W^-, with a $\mathcal{O}(10\%)$ contribution from ZZ. Annihilation into these channels is due to a diagram with a triangular fermionic loop, as discussed in Appendix C. The fermion channels, in the zero velocity limit, have a cross section proportional to $m_f^2 c_f^{A\,2}$, where m_f is the fermion mass and c_f^A is the coupling of Z' to the axial vector fermion bilinear. When $U(1)'$ is reduced to $U(1)_{B-L}$ the Z' couples to the vector current only. Thus in this limit the $\sigma(\chi\chi \to f\bar{f})$ has $a = 0$. The coefficient b is not proportional to m_f^2 (as it is a because of the helicity suppression), thus in the fourth plot of Fig. 9.6 the fermions contribute equally to the annihilation cross section (apart from a factor $B^2 \times (\#\,\text{colours}) = 1/3$ which penalises quarks with respect to leptons), unlike what happens for $\theta \neq \pi/2$.

Let us conclude this section by explaining why the tree level contribution for $\chi\chi \to W^+W^-$ has $a = 0$, which, together with the additional suppression by $\sin\psi$, selects Zh as the main channel at low velocities. The initial state $\chi\chi$, in the limit $v_{\text{DM}} \to 0$, has total angular momentum $J = 0$ and CP eigenvalue -1. The final state Zh is not a CP eigenstate, unlike W^+W^-. Now, a pair of vector bosons can have a CP eigenvalue -1 and a total angular momentum $J = 0$ only if they are both transversally polarized [6]. In this case, the tree level interaction $Z'WW$ (9.10) turns out to give a vanishing cross section. We notice that this argument does not apply to the W^+W^- and ZZ amplitudes when the triangular fermion loop is included (see Fig. C.1). In those cases, the effective $Z'W^+W^-$, $Z'ZZ$ vertices contain the terms

$$f_5^{Z'WW}\,\epsilon^{\mu\nu\rho\sigma}(k_1 - k_2)_\sigma Z'_\mu W_\nu^+ W_\rho^-, \qquad f_5^{Z'ZZ}\,\epsilon^{\mu\nu\rho\sigma}(k_1 - k_2)_\sigma Z'_\mu Z_\nu Z_\rho, \qquad (9.21)$$

where k_1, k_2 are the four-momenta of the outgoing bosons (see also [67, 68]). These terms lead to $a \neq 0$ in the cross section, making the W^+W^- and ZZ channels relevant despite of the loop suppression.

9.4 Summary of Results

We show in Fig. 9.7 bounds on $\sigma_{\chi p}^{SD}$ for the case $\theta = 0$, comparing direct detection results with those coming from IceCube LHC's mono-jet searches and γ-ray searches. Analogous results for different values of θ are presented in Fig. 9.8, in the plane Λ vs. m_χ. As representative values of θ we choose $\theta = 0$, $\pi/4$, $\pi/2$ and $3\pi/4$. We did not consider $\theta = \pi$ because it is exactly equivalent to $\theta = 0$. Since $\theta = \pi/2$ is quite a peculiar point, we added two values of θ in its vicinity, namely 0.45 and 0.55π.

As is clear from Eq. (9.14) and Table 9.2, for a generic angle θ the DM-nucleon scattering is mediated by a linear combination of $\mathcal{O}_4$, $\mathcal{O}_8$ and $\mathcal{O}_9$, with coefficients similar in magnitude. The contributions from $\mathcal{O}_8$ and $\mathcal{O}_9$ can be safely ignored for IceCube: given the composition of the solar environment, their nuclear form factors are between 100 and 1000 times smaller than the one for $\mathcal{O}_4$ [69]. This is not the case for DD experiments, where the three operators give a similar contribution, and the scattering cross section is not exactly $\sigma_{\chi p}^{SD}$, which is defined as being given by the operator $\mathcal{O}_4$ alone.

To obtain a bound in the usual $\sigma_{\chi p}^{SD}$ vs. m_χ plane, one needs to first obtain a bound on the total scattering cross section, which is easily done using the results of [43]. Then, this bound is translated into a bound on $\Lambda = m_{Z'}/(g_{Z'}\sqrt{g_\chi})$, using the

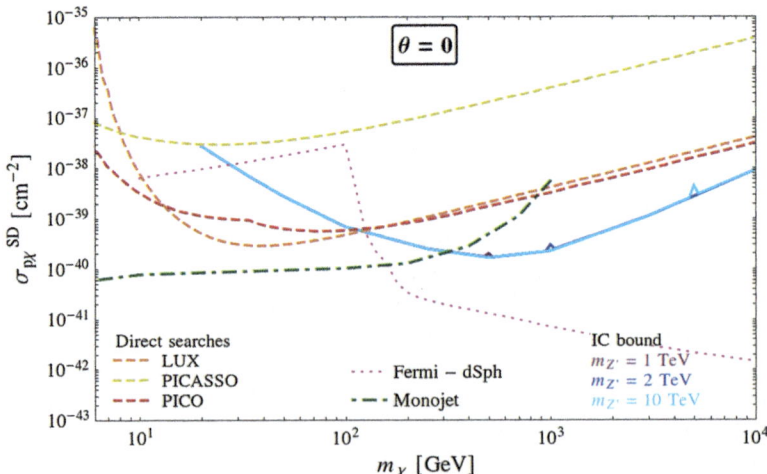

Fig. 9.7 Bound on $\sigma_{\chi p}^{SD}$ from direct detection, LHC's monojet analysis, IceCube and γ-ray searches, for $\theta = 0$

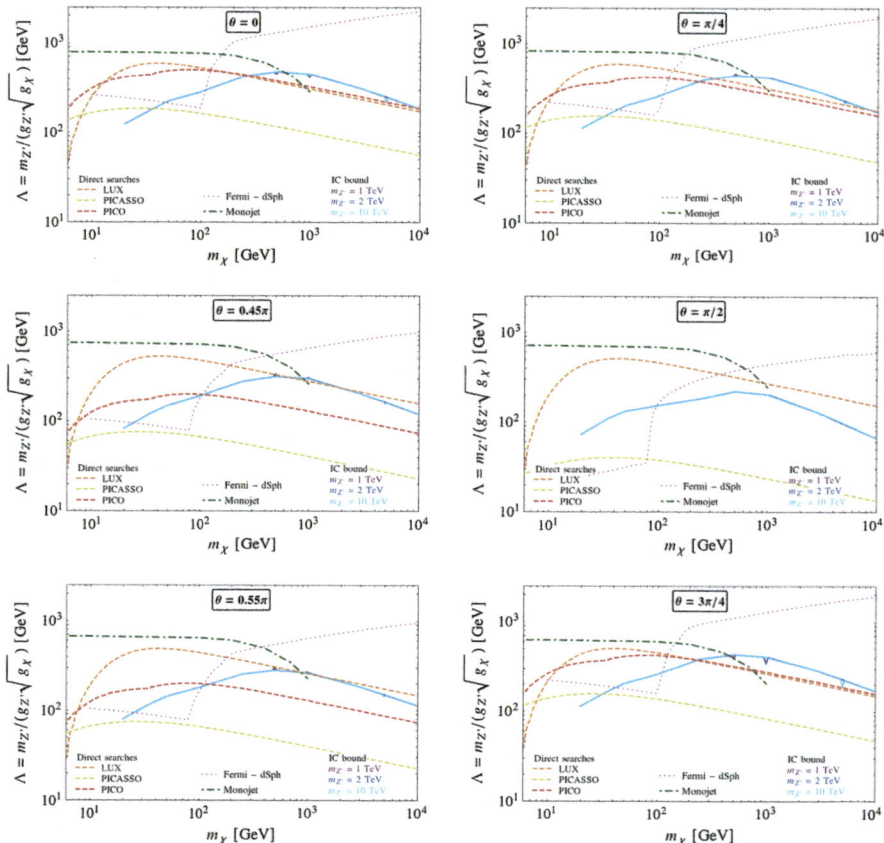

Fig. 9.8 Bound on $\Lambda = m_{Z'}/(g_{Z'}\sqrt{g_\chi})$ from direct detection, LHC's monojet analysis, and Ice-Cube, for different values of θ

expression for the cross section obtained from Eq. (9.12). Finally, the bound on Λ is translated into a bound on $\sigma_{\chi p}^{SD}$ using the expression

$$\sigma_{\chi p}^{SD} = \frac{3}{\pi}\left(-\frac{c_4}{\Lambda^2}\right)^2 \mu_p^2 \qquad (9.22)$$

which gives the scattering cross section when only the $\mathcal{O}_4$ operator is involved, where μ_p is the reduced mass of the proton–DM system, and the dimensionless coefficient is given by $c_4 = \frac{1}{4}\cos\theta(\Delta_u - \Delta_d - \Delta_s) \simeq \frac{1}{4}\cos\theta \cdot 1.35$, with Δ_q parametrizing the quark spin content of each nucleon, and is assumed to be equal for protons and neutrons [43].

For $\theta \neq 0$, we do not show bounds on $\sigma_{\chi p}^{SD}$ but only on Λ. The reason is that, since $\sigma_{\chi p}^{SD}$ is defined as the contribution to the scattering cross section due to the operator $\mathcal{O}_4$ only, given the limit on Λ we have $\sigma_{\chi p}^{SD} \propto \cos^2\theta/\Lambda_{\lim}^4$. When θ gets close to $\pi/2$, the computed value of $\sigma_{\chi p}^{SD}$ goes to 0 independently of Λ_{lim}, resulting in a spuriously strong bound.

The situation is slightly different when $\theta = \pi/2$. In this case, the coupling of the Z' to the vectorial current of the quark fields is identically 0, and therefore the coefficient of the $\mathcal{O}_4$ operator in the NR expansion vanishes. While for IceCube the contribution of $\mathcal{O}_9$ is subdominant with respect to that of $\mathcal{O}_8$ and can be ignored, both of them have to be taken into account to obtain DD bounds. A recast of the bound on Λ in terms of the usual $\sigma_{\chi p}^{SD}$ would make no sense in this scenario.

The PICO experiment gives interesting direct detection bounds. For $\theta = 0$ (i.e. for the usual spin-dependent operator $\mathcal{O}_4$), bounds on $\sigma_{\chi p}^{SD}$ can be read directly from Ref. [3], and translated into a bound on Λ. Bounds on Λ for other values of θ can not be obtained following the procedure we adopted for the other experiments, because PICO is not yet included between the Test Statistic functions given in [43]. Therefore, we obtained a conservative bound on Λ by rescaling the PICO limit on $\sigma_{\chi p}^{SD}$ as $\cos^2 \theta$ (since this enters the coefficient of the $\mathcal{O}_4$ operator) and then applying Eq. (9.22).

Let us now comment briefly on the results shown in Figs. 9.7 and 9.8. Bounds coming from LHC searches are typically the strongest ones in the low mass region, up to $m_\chi \sim 400 - 700$ GeV for large θ. IceCube searches have their maximal sensitivity in the region between a few hundred GeV and a few TeV. When θ is small, in this mass region they give a constraint on Λ which is stronger than the direct detection one. In particular, for $\theta = 0$ the constraint from IceCube is the dominant one from $m_\chi \sim 100$ GeV up to 10 TeV and beyond. The Fermi-LAT bound from dwarf spheroidal galaxies appears to be the dominant one for m_χ mass above 200–300 GeV. This is mainly due to the fact that, in our model, the dominant annihilation channel is Zh, which produces a large photon flux thanks to EW corrections. We expect that, in a leptophilic model in which DM annihilates copiously into τ and μ pairs, energetic neutrinos produced in their decay would make IceCube bounds the dominant ones.

We notice that the bounds shown in Fig. 9.8 fall in the region where the DM is underabundantly produced via the freeze-out mechanism (compare with Fig. 9.1). The difference between the two values of Λ goes up to one order of magnitude for large m_χ.

An important remark about IceCube results is that they are almost independent of $m_{Z'}$ and $\Gamma_{Z'}$ (together with $g_{Z'}$, g_χ, as stressed in Sect. 9.3.6). There are two reasons for this: first, as explained previously, when the equilibrium between annihilation and capture in the Sun is reached, neutrino fluxes at Earth only depend on branching ratios, which are not modified dramatically by the resonance except for a very narrow region around it. Electroweak corrections further dilute the difference, and as a result IceCube bounds for different values of $m_{Z'}$ are almost superimposed (except for a small bump around the resonant point, whose width is $\sim \Gamma_{Z'}$).

Previous studies of the constraints coming from the IceCube experiment were done in Refs. [70, 71]. In these works, the authors examined simplified models without considering the mixing of Z and Z' imposed by the Higgs boson charge under $U(1)'$, and therefore they do not include annihilation channels which turn out to be the dominant ones. Moreover, they weight the nominal IceCube benchmarks (which assume 100% annihilation into one channel) by the annihilation cross sections into different channels computed in their model, and they do not take into account EW corrections. In our work, we compute the BR's into various SM channels in a complete and consistent model, and we compute the neutrino fluxes including the EW corrections with the help of PPPC4DM ID. We infer the exclusion bounds with a recast of IceCube limits, as explained in Sect. 9.3.5, using the full shape of the neutrino fluxes to obtain the new bound.

9.5 Conclusions

While the spin-independent WIMP scenario has been probed experimentally to very high precision and direct detection experiments basically disfavor this possibility, the bounds on SD DM are much milder. WIMP-strength interactions between spin-dependent DM and baryonic matter are still perfectly allowed. Moreover, often the strongest bounds on SD DM do not come from direct detection experiments, but rather from LHC searches (direct or indirect) and Fermi-LAT. The IceCube experiment also produces interesting bounds, which are typically stronger than the direct detection ones above a few hundred GeV.

In this chapter we analyzed and compared these constraints, coming from different experiments. In order to be concrete, we concentrated on a particular set of spin-dependent DM models, in which the DM-baryon interaction is mediated by a heavy gauge Z'. If we restrict ourself to the models without spectators at the EW scale, the parameter space of gauge Z' models can be conveniently parametrized by three quantities: the mass of the heavy Z', the gauge coupling and the mixing angle θ between the hypercharge and $U(1)_{B-L}$ generators.

Although this study is not completely generic, as it does not cover all possible consistent models of SD DM, there are good reasons to believe that these models capture important phenomenological features that are generic to WIMP-like SD DM. We identified the region of parameter space favored by the observed thermal relic abundance and we have reanalyzed the existing LHC constraints, both direct (from the monojet searches) and indirect, on the Z' mass and coupling.

More importantly, we fully analyzed the low-temperature annihilation branching ratios of the Z'-mediated DM, which is crucial to derive IceCube constraints. In order to properly understand the expected neutrino fluxes from DM annihilation in the Sun's core, one has to know the annihilation channels of the DM in the Sun. Simply assuming that the dominant expected source of neutrinos is the WW or $b\bar{b}$ channel (as is done in the IceCube papers) is clearly insufficient. We found that, depending on the DM mass, one can divide the parameter space into three different regions. Below the Zh mass threshold (very light DM) the annihilations are indeed dominated by the $b\bar{b}$ channel, yielding very soft neutrino fluxes, although even in this case the IceCube constraints are dominated by small branching ratio to $\tau^+\tau^-$. Above this threshold Zh is a dominant annihilation channel and the dominant source of neutrinos almost in the entire parameter space. The third region is just above the

$t\bar{t}$ mass threshold. If the DM mass sits in this "island", one usually gets comparable annihilations into Zh and $t\bar{t}$, and both should be taken into account for the neutrino flux calculations. The W^+W^- channel can also become important, or even the leading channel, if the $U(1)'$ extension is very close to being $U(1)_{B-L}$. In this work we have properly recast the existing IceCube bounds including electroweak corrections, in order to derive reliable exclusion bounds on the secondary neutrinos coming from all annihilation channels.

We find that currently the strongest bounds on the SD Z'-mediated DM are imposed by LHC searches (for $m_{DM} \lesssim 400$ GeV) and by IceCube for heavier DM candidates, which are favored by thermal relic considerations. We also find that the best direct detection bounds come mostly from LUX (PICO becomes dominant only for very light ~ 20 GeV DM particles). These bounds are subdominant with respect to LHC ones in the case of light DM, but can be comparable to IC bounds for heavy DM if $U(1)'$ is close to being a $U(1)_{B-L}$ extension. We have also computed the values that yield the observed DM abundance through the freeze out mechanism, and we found that experimental exclusion limits fall in the slightly underabundant region.

Finally we notice that it would be interesting to see similar works for other SD DM candidates. It would be also useful to have bounds on the DM, annihilating into Zh reported directly by the IC and Fermi collaborations. We stress that this channel immediately arises when considering a consistent anomaly-free model for a $U(1)'$ extension which is not a pure $U(1)_{B-L}$. This important feature is not captured by the use of simplified DM models if these issues are not considered.

References

1. T. Jacques, A. Katz, E. Morgante, D. Racco, M. Rameez, A. Riotto, Complementarity of DM Searches in a Consistent Simplified Model: the Case of Z', JHEP **071**, 1610 (2016), arXiv:1605.06513
2. A. Manalaysay, L.U.X. The, dark matter search, Talk at IDM2016 (Sheffield, UK, 2016)
3. PICO Collaboration, C. Amole et al., Dark Matter Search Results from the PICO-60 CF$_3$ I Bubble Chamber, Submitted to: Phys. Rev. D (2015), arXiv:1510.07754
4. IceCube Collaboration, M.G. Aartsen et al., Search for dark matter annihilations in the Sun with the 79-string IceCube detector, Phys. Rev. Lett. **110**, no. 13 131302 (2013), arXiv:1212.4097
5. IceCube Collaboration, M.G. Aartsen et al., Improved limits on dark matter annihilation in the Sun with the 79-string IceCube detector and implications for supersymmetry, JCAP **022**(04), 1604 (2016), arXiv:1601.00653
6. M. Drees, M.M. Nojiri, The Neutralino relic density in minimal N=1 supergravity, Phys. Rev. **D47**, 376–408 (1993), arXiv:hep-ph/9207234
7. M. Cirelli, N. Fornengo, A. Strumia, Minimal dark matter, Nucl. Phys. **B753**, 178–194 (2006), arXiv:hep-ph/0512090
8. M. Cirelli, A. Strumia, Minimal Dark Matter: Model and results, New J. Phys. **11**, 105005 (2009), arXiv:0903.3381
9. N. Arkani-Hamed, S. Dimopoulos, Supersymmetric unification without low energy supersymmetry and signatures for fine-tuning at the LHC, JHEP **06**, 073 (2005), arXiv:hep-th/0405159
10. G.F. Giudice, A. Romanino, Split supersymmetry, Nucl. Phys. **B699**, 65–89 (2004), arXiv:hep-ph/0406088. [Erratum: Nucl. Phys. B **706**, 487 (2005)]

11. A. Arvanitaki, N. Craig, S. Dimopoulos, G. Villadoro, Mini-Split, JHEP **02** (2013), p. 126, arXiv:1210.0555
12. N. Arkani-Hamed, A. Gupta, D.E. Kaplan, N. Weiner, T. Zorawski, Simply Unnatural Supersymmetry, arXiv:1212.6971
13. Z. Bern, P. Gondolo, M. Perelstein, Neutralino annihilation into two photons, Phys. Lett. **B411**, 86–96 (1997), arXiv:hep-ph/9706538
14. L. Bergstrom, P. Ullio, Full one loop calculation of neutralino annihilation into two photons, Nucl. Phys. **B504**, 27–44 (1997), arXiv:hep-ph/9706232
15. O. Lebedev, Y. Mambrini, Axial dark matter: The case for an invisible Z', Phys. Lett. **B734**, 350–353 (2014), arXiv:1403.4837
16. F. Kahlhoefer, K. Schmidt-Hoberg, T. Schwetz, S. Vogl, Implications of unitarity and gauge invariance for simplified dark matter models, JHEP **02**, 016 (2016), arXiv:1510.02110. [JHEP02,016(2016)]
17. N.F. Bell, Y.Cai, R.K. Leane, Mono-W Dark Matter Signals at the LHC: Simplified Model Analysis, JCAP **1601**, no. 01 051 (2016), arXiv:1512.00476
18. O. Buchmueller, M.J. Dolan, S.A. Malik, C. McCabe, Characterising dark matter searches at colliders and direct detection experiments: Vector mediators, JHEP **037**, 1501 (2015), arXiv:1407.8257
19. M. Blennow, J. Herrero-Garcia, T. Schwetz, S. Vogl, Halo-independent tests of dark matter direct detection signals: local DM density, LHC, and thermal freeze-out, JCAP **1508**, no. 08 039 (2015), arXiv:1505.05710
20. A. Alves, S. Profumo, F.S. Queiroz, The dark Z' portal: direct, indirect and collider searches, JHEP **1404**, 063 (2014), arXiv:1312.5281
21. A. Alves, A. Berlin, S. Profumo, F.S. Queiroz, Dark Matter Complementarity and the Z' Portal, Phys. Rev. **D92**, no.8 083004 (2015), arXiv:1501.03490
22. A. Alves, A. Berlin, S. Profumo, F.S. Queiroz, Dirac-fermionic dark matter in U (1)$_X$ models, JHEP **10**, 076 (2015), arXiv:1506.06767
23. H. An, X. Ji, L.-T. Wang, Light Dark Matter and Z' Dark Force at Colliders, JHEP **07**, 182 (2012), arXiv:1202.2894
24. H. An, R. Huo, L.-T. Wang, Searching for Low Mass Dark Portal at the LHC, Phys. Dark Univ. **2**, 50–57 (2013), arXiv:1212.2221
25. M.T. Frandsen, F.Kahlhoefer, A.Preston, S.Sarkar, K. Schmidt-Hoberg, LHC and Tevatron Bounds on the Dark Matter Direct Detection Cross-Section for Vector Mediators, JHEP **07**, 123 (2012), arXiv:1204.3839
26. G. Arcadi, Y. Mambrini, M.H.G. Tytgat, B. Zaldivar, Invisible Z' and dark matter: LHC vs LUX constraints, JHEP **03**, 134 (2014), arXiv:1401.0221
27. I.M. Shoemaker, L. Vecchi, Unitarity and Monojet Bounds on Models for DAMA, CoGeNT, and CRESST-II, Phys.Rev. **D86**, 015023 (2012), arXiv:1112.5457
28. M.T. Frandsen, F.Kahlhoefer, S.Sarkar, K. Schmidt-Hoberg, Direct detection of dark matter in models with a light Z', JHEP **09**, 128 (2011), arXiv:1107.2118
29. P. Gondolo, P. Ko, Y. Omura, Light dark matter in leptophobic Z' models, Phys. Rev. **D85**, 035022 (2012), arXiv:1106.0885
30. M. Fairbairn, J. Heal, Complementarity of dark matter searches at resonance, Phys. Rev. **D90**, no. 11 115019 (2014), arXiv:1406.3288
31. P. Harris, V.V. Khoze, M. Spannowsky, C. Williams, Constraining Dark Sectors at Colliders: Beyond the Effective Theory Approach, Phys. Rev. D 91, 055009 (2015), arXiv:1411.0535
32. M. Chala, F. Kahlhoefer, M. McCullough, G. Nardini, K. Schmidt-Hoberg, Constraining Dark Sectors with Monojets and Dijets, JHEP **07**,089 (2015), arXiv:1503.05916
33. T. Jacques, K. Nordström, Mapping monojet constraints onto Simplified Dark Matter Models, JHEP **06**, 142 (2015), arXiv:1502.05721
34. A.J. Brennan, M.F. McDonald, J. Gramling, T.D. Jacques, Collide and Conquer: Constraints on Simplified Dark Matter Models using Mono-X Collider Searches, JHEP **1605** (112), (2016), arXiv:1603.01366

35. H. Dreiner, D. Schmeier, J. Tattersall, Contact Interactions Probe Effective Dark Matter Models at the LHC, Europhys. Lett. **102**, 51001 (2013), arXiv:1303.3348
36. K. Ghorbani, H. Ghorbani, Two-portal Dark Matter, Phys. Rev. **D91**, no. 12 123541 (2015), arXiv:1504.03610
37. S. Weinberg, The quantum theory of fields. Vol. 2: Modern applications (Cambridge University Press, 2013)
38. F. D'Eramo, B.J. Kavanagh, P. Panci, You can hide but you have to run: direct detection with vector mediators, JHEP **1608**, 111 (2016), arXiv:1605.04917
39. P. Gondolo, G. Gelmini, Cosmic abundances of stable particles: Improved analysis. Nucl. Phys. B **360**, 145–179 (1991)
40. K. Griest, D. Seckel, Three exceptions in the calculation of relic abundances. Phys. Rev. D **43**, 3191–3203 (1991)
41. Planck Collaboration, P.A.R. Ade et al., Planck 2015 results. XIII. Cosmological parameters, Astron. Astrophys. A 13, 594 (2016), arXiv:1502.01589
42. A.L. Fitzpatrick, W. Haxton, E. Katz, N. Lubbers, Y. Xu, The Effective Field Theory of Dark Matter Direct Detection, JCAP **1302**,004 (2013), arXiv:1203.3542
43. M. Cirelli, E. Del Nobile, P. Panci, Tools for model-independent bounds in direct dark matter searches, JCAP **1310**,019 (2013), arXiv:1307.5955
44. U. Haisch, F. Kahlhoefer, On the importance of loop-induced spin-independent interactions for dark matter direct detection, JCAP **1304**,050 (2013), arXiv:1302.4454
45. CMS Collaboration, V. Khachatryan et al., Search for physics beyond the standard model in dilepton mass spectra in proton-proton collisions at $\sqrt{s} = 8$ TeV, JHEP **04**, 025 (2015), arXiv:1412.6302
46. CMS Collaboration, S. Chatrchyan et al., Measurement of inclusive W and Z boson production cross sections in pp collisions at $\sqrt{s} = 8$ TeV, Phys. Rev. Lett. **112**, 191802 (2014), arXiv:1402.0923
47. J. Alwall, R. Frederix, S. Frixione, V. Hirschi, F. Maltoni, O. Mattelaer, H.S. Shao, T. Stelzer, P. Torrielli, M. Zaro, The automated computation of tree-level and next-to-leading order differential cross sections, and their matching to parton shower simulations, JHEP **07**, 079 (2014), arXiv:1405.0301
48. CMS Collaboration Collaboration, Search for a Narrow Resonance Produced in 13 TeV pp Collisions Decaying to Electron Pair or Muon Pair Final States, Technical Report CMS-PAS-EXO-15-005, CERN, Geneva, 2015
49. CMS Collaboration, V. Khachatryan et al., Search for dark matter, extra dimensions, and unparticles in monojet events in proton-proton collisions at $\sqrt{s} = 8$ TeV, Eur. Phys. J. **C75**, no. 5 235 (2015), arXiv:1408.3583
50. G. Busoni, A. De Simone, E. Morgante, A. Riotto, On the Validity of the Effective Field Theory for Dark Matter Searches at the LHC, Phys.Lett. **B728**, 412–421 (2014), arXiv:1307.2253
51. G. Busoni, A. De Simone, J. Gramling, E. Morgante, A. Riotto, On the Validity of the Effective Field Theory for Dark Matter Searches at the LHC, Part II: Complete Analysis for the s-channel, JCAP 1406, 060 (2014), arXiv:1402.1275
52. D. Racco, A. Wulzer, F. Zwirner, Robust collider limits on heavy-mediator Dark Matter, JHEP **05**, 009 (2015), arXiv:1502.04701
53. O. Buchmueller, M.J. Dolan, C. McCabe, Beyond effective field theory for dark matter searches at the LHC, JHEP 1401, 025 (2014), arXiv:1308.6799
54. DES, Fermi-LAT Collaboration, A. Drlica-Wagner et al., Search for Gamma-Ray Emission from DES Dwarf Spheroidal Galaxy Candidates with Fermi-LAT Data, Astrophys. J. **809**, no. 1 L4 (2015), arXiv:1503.02632
55. Fermi-LAT Collaboration, M. Ackermann et al., Searching for Dark Matter Annihilation from Milky Way Dwarf Spheroidal Galaxies with Six Years of Fermi Large Area Telescope Data, Phys. Rev. Lett. **115**, no. 23 231301 (2015), arXiv:1503.02641
56. J.F. Navarro, C.S. Frenk, S.D.M. White, A Universal density profile from hierarchical clustering, Astrophys. J. **490**, 493–508 (1997), arXiv:astro-ph/9611107

57. HESS Collaboration, H. Abdallah et al., Search for dark matter annihilations towards the inner Galactic halo from 10 years of observations with H.E.S.S, Phys. Rev. Lett. **117**(11), 111301 (2016), arXiv:1607.08142

58. Fermi-LAT Collaboration, M. Ackermann et al., Constraints on the Galactic Halo Dark Matter from Fermi-LAT Diffuse Measurements, Astrophys. J. **761**, 91 (2012), arXiv:1205.6474

59. T. Daylan, D.P. Finkbeiner, D. Hooper, T. Linden, S.K.N. Portillo, N.L. Rodd, T.R. Slatyer, The characterization of the gamma-ray signal from the central Milky Way: A case for annihilating dark matter, Phys. Dark Univ. **12**, 1–23 (2016), arXiv:1402.6703

60. T. Cohen, M. Lisanti, A. Pierce, T.R. Slatyer, Wino Dark Matter Under Siege, JCAP **1310**, 061 (2013), arXiv:1307.4082

61. J. Fan, M. Reece, In Wino Veritas? Indirect Searches Shed Light on Neutralino Dark Matter, JHEP **10**, 124 (2013), arXiv:1307.4400

62. M. Cirelli, G. Corcella, A. Hektor, G. Hutsi, M. Kadastik, P. Panci, M. Raidal, F. Sala, A. Strumia, PPPC 4 DM ID: A Poor Particle Physicist Cookbook for Dark Matter Indirect Detection, JCAP **1130**, 051 (2011), arXiv:1012.4515. [Erratum: JCAP1210, E01(2012)]

63. IceCube Collaboration, A. Achterberg et al., First Year Performance of The IceCube Neutrino Telescope, Astropart. Phys. **26**, 155–173 (2006), arXiv:astro-ph/0604450

64. J. Braun, J. Dumm, F. De Palma, C. Finley, A. Karle, T. Montaruli, Methods for point source analysis in high energy neutrino telescopes, Astropart. Phys. **29**, 299–305 (2008), arXiv:0801.1604

65. IceCube Collaboration, M. Rameez et al., Search for dark matter annihilations in the Sun using the completed IceCube neutrino telescope (2015), http://pos.sissa.it/archive/conferences/236/1209/ICRC2015_1209.pdf

66. G. Punzi, Comments on likelihood fits with variable resolution, eConf **C030908**, WELT002 235 (2004), arXiv:physics/0401045

67. K. Hagiwara, R.D. Peccei, D. Zeppenfeld, K. Hikasa, Probing the Weak Boson Sector in e+ e- –> W+ W-. Nucl. Phys. B **282**, 253–307 (1987)

68. G.J. Gounaris, J. Layssac, F.M. Renard, New and standard physics contributions to anomalous Z and gamma selfcouplings, Phys. Rev. **D62**, 073013 (2000), arXiv:hep-ph/0003143

69. R. Catena, B. Schwabe, Form factors for dark matter capture by the Sun in effective theories, JCAP **1504**, no. 04 042 (2015), arXiv:1501.03729

70. J. Blumenthal, P. Gretskov, M. Krämer, C. Wiebusch, Effective field theory interpretation of searches for dark matter annihilation in the Sun with the IceCube Neutrino Observatory, Phys. Rev. **D91**, no. 3 035002 (2015), arXiv:1411.5917

71. J. Heisig, M. Krämer, M. Pellen, C. Wiebusch, Constraints on Majorana Dark Matter from the LHC and IceCube, Phys. Rev. **D93**, no. 5 055029 (2016), arXiv:1509.07867

Chapter 10
Conclusions

The quest for Dark Matter is living extremely exciting times. As we have seen, the problem is one of the most long-standing ones in physics, the first hints dating back to the beginning of the XX century. After having realized, back in the '70s, that the solution to the puzzle probably lies in a new species of stable particles, a huge experimental effort has been developed in the last decades to unveil the nature of this new particle.

We have started this thesis by reviewing, in Sect. 1.1, the principal observational evidences for the existence of DM. We then turned our attention, in Sects. 1.2 and 1.3, to the many ways in which the present DM abundance could have been attained in the early universe and to the plethora of models that provide viable DM candidates. It would probably be impossible to write down a complete list, and we limited ourselves to the most studied cases, which, even if it is not guaranteed that any of them represent the correct solution, represent a good set of benchmarks spanning a wide variety of scenarios.

In the following chapters we focussed our attention to WIMP searches. The WIMP paradigm has been the leading one for many years, mostly due to two reasons: firstly, the fact that WIMP-like DM candidates arise naturally in Supersymmetric models, which have lead the field of beyond the Standard Model physics for decades providing an elegant solution to the SM Higgs hierarchy problem. Secondly, WIMPs can be strongly constrained by present experiments, as we discussed in this thesis.

In Chap. 2 the status of direct WIMP searches was summarized. In this kind of experiment, a large bag of target material is observed in underground laboratories, looking for the scattering of WIMPs of the galactic halo with target nuclei (or, in some cases, electrons). We briefly discussed how experiments are protected against cosmic ray contamination and the radioactive background, and most importantly how next generation experiments will face the so-called "ultimate" background of solar and cosmological neutrinos. Direct searches are now entering a transition phase. New ton scale experiments will soon start to produce results, reaching the sensitivity to exclude most of the parameter space of MSSM and extra dimensions DM and, at the same time, becoming sensitive to neutrino-nucleus interactions. A null result from these probes (which is somehow expected, given the lack of new physics signals

© Springer International Publishing AG 2017
E. Morgante, *Aspects of WIMP Dark Matter Searches at Colliders and Other Probes*, Springer Theses, https://doi.org/10.1007/978-3-319-67606-7_10

so far at the LHC) will be the starting of a new generation of experiments, which will both lower the energy threshold compared to present ones in order to probe the sub-GeV range, and introduce new experimental techniques in order to overcome the threat posed by the neutrino background, possibly by exploiting the directional information.

Indirect WIMP searches were discussed in Chap. 3. A flux of stable SM particles and their anti-particles (positrons, antiprotons, anti-deuterium, photons and neutrinos) can be produced in DM annihilations/decays in the galactic halo, and a search of an excess over the astrophysical background can lead to constraints on the properties of the DM particles. The three key ingredients in the determination of these fluxes are the DM annihilation cross section (which depends exclusively on the particle physics nature of DM), its distribution in the halo and the properties of cosmic rays propagation in the galaxy (the latter two informations, being astrophysical in nature, are the ones that dominate the uncertainties connected to these searches). The present status of indirect searches can be described in one word as confused. An excess in the positron fraction was detected by the PAMELA and AMS-02 collaborations, but after a great initial excitement a DM explanation is now disfavoured. A similar result was claimed by the AMS-02 collaboration in the antiproton channel, but careful reanalysis of the astrophysical background points towards the absence of any significant excess. Gamma rays searches are in a somehow similar situation. An excess was detected in FERMI data, compatible with an interpretation in terms of annihilations of a 30–50 GeV thermal relic DM particle, annihilating dominantly into a $b\bar{b}$ pair. While its statistical significance is, after a long debate, seemingly established, astrophysical explanations of the effect are still possible. In parallel, there exist claims of the detection of gamma ray lines from DM annihilations, which are still under an intense discussion in the literature.

In Chap. 4 we focussed on the antiprotons channel, questioning the capability of AMS-02 to distinguish between a DM interpretation and an astrophysical one for a hypothetical excess in the $\bar{p}/p$ flux. We showed that a signal of this kind can not be distinguished from an astrophysical one in which secondary antiparticles are produces by spallation in the shock region of Supernova remnants and then accelerated.

DM searches at the LHC were introduced in Chap. 5. We started by reviewing the basic tools and the relevant signatures of WIMP particles at the collider setup. Then, the prototypical DM search at the LHC, namely the mono-jet one, was described in some details. Finally, we discussed the issue of what theoretical model should be assumed when constraining the properties of DM at colliders. Three scenarios can be distinguished: in the first, the DM particle is considered as a part of a broader BSM theory, such as the MSSM, which was built as a solution to the Higgs hierarchy problem of the SM model and contains a good DM candidate quite as an accident. A number of different models exists, and even more are the possible searches. From the point of view of DM, this variety is partly a disadvantage, since it makes it harder to outline model independent conclusions on the DM properties from the results of these searches.

The second possibility is that of describing the DM interaction with quarks and gluons by means of a set of effective operators. This can be seen as the most economical and model independent approach, since, once the set of all operators of a given dimension has been written down, every DM model reduces to a combination of these operators, provided that the scale of the interaction (typically the mass of a mediator) is much larger than the DM mass and of the energy of the incoming partons. This condition enforces strong limitations on the validity of the EFT approach at the LHC, as it is described in details in Chap. 6. Intuitively the reason is clear: the EFT is expected to fail if its associated energy scale is not larger than the energy at which the interaction takes place, which is of the order of a few TeV at the LHC. Limits of hundreds of GeV on the effective scale fall in the region where the EFT is not valid, and should be considered with a grain of salt. In order to make sense of these limits, a recasting procedure has to be adopted, as described in Chap. 6.

There is then a third way, which is that of simplified models, described in Chap. 7. From a bottom-up viewpoint, the idea is to expand the effective operators including mediator particles in the description, thus avoiding the energy limitations of the EFT approach and adding a richer phenomenology, new search channels, etc. In a top-down framework instead, simplified models can be seen as a way to simplify the phenomenology of complex new physics models in such a way to restrict to the phenomena related to DM. It is important that simplified models are not oversimplified, in the sense that in many cases new particles have to be added and constraints have to be enforced on the couplings in order to avoid violations of perturbative unitarity and, consequently, unphysical predictions. Because of the enlarged parameter space of simplified models with respect to the EFT approach, very important questions arise on which is the best way of presenting results and on which are the most promising regions for a possible detection of DM, since these are the ones that should have the priority in the future LHC searches. A possible guidance in assigning priorities can be given by the relic abundance information, as discussed in Chap. 8.

The question of which of these frameworks is the most suitable one for DM searches is not well posed, since its answer depends on the experiment and on which kind of DM particle one has in mind. Each of the three should be pursued, and it is important that experimental collaborations present their results in such a way that they can be easily recast for any other DM model.

In Chap. 9 we considered an example simplified model and the bounds which can be set on it by a number of different searches. The first model was a Z' extension of the SM, in which the gauge group is enlarged by a new U(1) and the couplings are chosen in such a way to respect gauge invariance and cancel gauge anomalies. The second one was inspired by the resonance at 750 GeV seen last year in the di-photon channel by the ATLAS collaboration. We assumed that the signal was due to a pseudoscalar particle coupled, among the others, to DM, and considered what other experiments could help in constraining the coupling of the putative new scalar to DM. In both cases, we showed that stringent constraints can be derived on the spin dependent DM-nucleon cross section by the LHC, direct detection experiments and by indirect detection ones. For the latter, the knowledge of the couplings of the mediator to all SM particles is needed, and a comparison with LHC and direct detection can only be performed on a model dependent basis.

The quest for WIMP dark matter is now living an exciting phase. A huge experimental effort is being devoted to measure WIMPs' properties through their scattering with SM particles, their production at the LHC and their annihilations or decays in the galactic halo or in celestial bodies. Many models have already been excluded by data, and the next years will see a huge improvement both in terms of new analysis and of technological developments. While many theorists are starting considering alternatives to the standard WIMP scenario more and more seriously, it is time to analyse in depth the theoretical tools used in WIMP searches, in order to fully exploit the discovery potential that present day experiments provide: this thesis was intended as a contribution in this direction.

Appendix A
Mono-jet Cross Sections in the EFT Limit

A.1 Generalities

In this Appendix we show the details of the calculations of the tree-level cross sections for the hard scattering process $f(p_1) + \bar{f}(p_2) \rightarrow \chi(p_3) + \chi(p_4) + g(k)$, where f is either a quark (operators D1–D10 and DT1) or a gluon (D11–D14), and the final gluon is emitted from the initial state. For the t-channel operator DT1, we also computed the analytic cross section for the process with (anti-)quark emission, $q(p_1) + g(p_2) \rightarrow \chi(p_3) + \chi(p_4) + q(k)$.

The differential cross section is generically given by

$$d\hat{\sigma} = \frac{\sum \overline{|\mathcal{M}|^2}}{4(p_1 \cdot p_2)} d\Phi_3 \,, \tag{A.1}$$

where the three-body phase space is

$$d\Phi_3 = (2\pi)^4 \delta^{(4)}(p_1 + p_2 - p_3 - p_4 - k) \frac{d\mathbf{p}_3}{(2\pi)^3 2p_3^0} \frac{d\mathbf{p}_4}{(2\pi)^3 2p_4^0} \frac{d\mathbf{k}}{(2\pi)^3 2k^0} \,. \tag{A.2}$$

A.2 Matrix Elements

The matrix elements at the parton level are given by the sum of the Feynman diagrams in Fig. 6.1 for the operators D1–D14 and in Fig. 6.2 for the operator DT1.

In the limit of massless light quarks, they have definite helicity and it makes no difference for the cross sections whether there is q or $\gamma^5 q$ in the operator. Therefore the following identifications between pairs of operators hold:

$$D1' \leftrightarrow D3', \quad D2' \leftrightarrow D4', \quad D5 \leftrightarrow D7, \quad D6 \leftrightarrow D8, \quad D9 \leftrightarrow D10 \,, \tag{A.3}$$

© Springer International Publishing AG 2017
E. Morgante, *Aspects of WIMP Dark Matter Searches at Colliders and Other Probes*, Springer Theses, https://doi.org/10.1007/978-3-319-67606-7

while the "primed" and "unprimed" operators are related as in Eq. (6.28). For definiteness, we choose to work with $D1'$, $D4'$, $D5$, $D8$, $D9$ and $D11$–$D14$.

The amplitudes are given by

$$\mathcal{M}_{D1'} = -ig_s \frac{1}{\Lambda^2} \epsilon_\mu^{*a}(k) \left[\frac{\bar{v}(p_2)(\not{p}_1 - \not{k})\gamma^\mu T^a u(p_1)}{(p_1 - k)^2} - \frac{\bar{v}(p_2)\gamma^\mu T^a(\not{p}_2 - \not{k})u(p_1)}{(p_2 - k)^2} \right] \bar{u}(p_3)v(p_4),$$
(A.4)

$$\mathcal{M}_{D4'} = -ig_s \frac{1}{\Lambda^2} \epsilon_\mu^{*a}(k) \left[\frac{\bar{v}(p_2)\gamma^5(\not{p}_1 - \not{k})\gamma^\mu T^a u(p_1)}{(p_1 - k)^2} - \frac{\bar{v}(p_2)\gamma^\mu T^a(\not{p}_2 - \not{k})\gamma^5 u(p_1)}{(p_2 - k)^2} \right]$$
$$\times \bar{u}(p_3)\gamma^5 v(p_4),$$
(A.5)

$$\mathcal{M}_{D5} = -ig_s \frac{g_{\nu\rho}}{\Lambda^2} \epsilon_\mu^{*a}(k) \left[\frac{\bar{v}(p_2)\gamma^\nu(\not{p}_1 - \not{k})\gamma^\mu T^a u(p_1)}{(p_1 - k)^2} - \frac{\bar{v}(p_2)\gamma^\mu T^a(\not{p}_2 - \not{k})\gamma^\nu u(p_1)}{(p_2 - k)^2} \right]$$
$$\times \bar{u}(p_3)\gamma^\rho v(p_4),$$
(A.6)

$$\mathcal{M}_{D8} = -ig_s \frac{g_{\nu\rho}}{\Lambda^2} \epsilon_\mu^{*a}(k) \left[\frac{\bar{v}(p_2)\gamma^\nu\gamma^5(\not{p}_1 - \not{k})\gamma^\mu T^a u(p_1)}{(p_1 - k)^2} - \frac{\bar{v}(p_2)\gamma^\mu T^a(\not{p}_2 - \not{k})\gamma^\nu\gamma^5 u(p_1)}{(p_2 - k)^2} \right]$$
$$\times \bar{u}(p_3)\gamma^\rho\gamma^5 v(p_4),$$
(A.7)

$$\mathcal{M}_{D9} = -i\frac{g_s}{16} \frac{g_{\mu\rho}g_{\nu\sigma}}{\Lambda^2} \epsilon_\alpha^{*a}(k) \left[\frac{\bar{v}(p_2)\sigma^{\mu\nu}(\not{p}_1 - \not{k})\gamma^\alpha T^a u(p_1)}{(p_1 - k)^2} - \frac{\bar{v}(p_2)\gamma^\alpha T^a(\not{p}_2 - \not{k})\sigma^{\mu\nu} u(p_1)}{(p_2 - k)^2} \right]$$
$$\times \bar{u}(p_3)\sigma^{\rho\sigma} v(p_4),$$
(A.8)

$$\mathcal{M}_{D11} = \frac{g_s^3}{4\pi} \frac{1}{\Lambda^3} f_{abc} \epsilon_\mu(p_1)\epsilon_\nu(p_2)\epsilon_\rho^*(k) \bar{u}(p_3)v(p_4)$$
$$\left[\frac{(g^{\mu\sigma}(2p_1 - k)^\rho + g^{\rho\sigma}(2k - p_1)^\mu - g^{\mu\rho}(k + p_1)^\sigma)((p_1 - k)^\nu p_{2\sigma} - (p_1 - k) \cdot p_2 g_\sigma^\nu)}{(p_1 - k)^2} \right.$$
$$- \frac{(g^{\nu\sigma}(2p_2 - k)^\rho + g^{\rho\sigma}(2k - p_2)^\nu - g^{\nu\rho}(k + p_2)^\sigma)((p_2 - k)^\mu p_{1\sigma} - (p_2 - k) \cdot p_1 g_\sigma^\mu)}{(p_2 - k)^2}$$
$$- \frac{(g^{\mu\nu}(p_1 - p_2)^\sigma + g^{\nu\sigma}(p_1 + 2p_2)^\mu - g^{\mu\sigma}(2p_1 + p_2)^\nu)((p_1 + p_2)^\rho k_\sigma - k \cdot (p_1 + p_2)g_\sigma^\rho)}{(p_1 + p_2)^2}$$
$$\left. + g^{\mu\nu}(p_1 - p_2)^\rho + g^{\nu\rho}(p_2 + k)^\mu - g^{\mu\rho}(k + p_1)^\nu \right],$$
(A.9)

$$\mathcal{M}_{D12} = i\frac{g_s^3}{4\pi} \frac{1}{\Lambda^3} f_{abc} \epsilon_\mu(p_1)\epsilon_\nu(p_2)\epsilon_\rho^*(k) \bar{u}(p_3)\gamma^5 v(p_4)$$
$$\left[\frac{(g^{\mu\sigma}(2p_1 - k)^\rho + g^{\rho\sigma}(2k - p_1)^\mu - g^{\mu\rho}(k + p_1)^\sigma)((p_1 - k)^\nu p_{2\sigma} - (p_1 - k) \cdot p_2 g_\sigma^\nu)}{(p_1 - k)^2} \right.$$
$$- \frac{(g^{\nu\sigma}(2p_2 - k)^\rho + g^{\rho\sigma}(2k - p_2)^\nu - g^{\nu\rho}(k + p_2)^\sigma)((p_2 - k)^\mu p_{1\sigma} - (p_2 - k) \cdot p_1 g_\sigma^\mu)}{(p_2 - k)^2}$$
$$- \frac{(g^{\mu\nu}(p_1 - p_2)^\sigma + g^{\nu\sigma}(p_1 + 2p_2)^\mu - g^{\mu\sigma}(2p_1 + p_2)^\nu)((p_1 + p_2)^\rho k_\sigma - k \cdot (p_1 + p_2)g_\sigma^\rho)}{(p_1 + p_2)^2}$$
$$\left. + g^{\mu\nu}(p_1 - p_2)^\rho + g^{\nu\rho}(p_2 + k)^\mu - g^{\mu\rho}(k + p_1)^\nu \right],$$
(A.10)

$$\mathcal{M}_{D13} = -\frac{g_s^3}{4\pi}\frac{1}{\Lambda^3}f_{abc}\epsilon_\mu(p_1)\epsilon_\nu(p_2)\epsilon_\rho^*(k)\bar{u}(p_3)v(p_4)$$

$$\left[\frac{(g_\sigma^\mu(2p_1-k)^\rho + g_\sigma^\rho(2k-p_1)^\mu - g^{\mu\rho}(k+p_1)_\sigma)(\epsilon^{\sigma\nu\eta\chi}p_{2\eta}(p_1-k)_\chi))}{(p_1-k)^2}\right.$$

$$+\frac{(g_\sigma^\nu(2p_2-k)^\rho + g_\sigma^\rho(2k-p_2)^\nu - g^{\nu\rho}(k+p_2)_\sigma)(\epsilon^{\sigma\mu\eta\chi}p_{1\eta}(p_2-k)_\chi)}{(p_2-k)^2}$$

$$+\frac{(g^{\mu\nu}(p_1-p_2)_\sigma + g_\sigma^\nu(p_1+2p_2)^\mu - g_\sigma^\mu(2p_1+p_2)^\nu)(\epsilon^{\rho\eta\sigma\chi}k_\eta(p_1+p_2)_\chi)}{(p_1+p_2)^2}$$

$$\left. -\epsilon^{\mu\nu\rho\sigma}(p_1+p_2-k)_\sigma\right], \tag{A.11}$$

$$\mathcal{M}_{D14} = -i\frac{g_s^3}{4\pi}\frac{1}{\Lambda^3}f_{abc}\epsilon_\mu(p_1)\epsilon_\nu(p_2)\epsilon_\rho^*(k)\bar{u}(p_3)\gamma^5 v(p_4)$$

$$\left[\frac{(g_\sigma^\mu(2p_1-k)^\rho + g_\sigma^\rho(2k-p_1)^\mu - g^{\mu\rho}(k+p_1)_\sigma)(\epsilon^{\sigma\nu\eta\chi}p_{2\eta}(p_1-k)_\chi))}{(p_1-k)^2}\right.$$

$$+\frac{(g_\sigma^\nu(2p_2-k)^\rho + g_\sigma^\rho(2k-p_2)^\nu - g^{\nu\rho}(k+p_2)_\sigma)(\epsilon^{\sigma\mu\eta\chi}p_{1\eta}(p_2-k)_\chi)}{(p_2-k)^2}$$

$$+\frac{(g^{\mu\nu}(p_1-p_2)_\sigma + g_\sigma^\nu(p_1+2p_2)^\mu - g_\sigma^\mu(2p_1+p_2)^\nu)(\epsilon^{\rho\eta\sigma\chi}k_\eta(p_1+p_2)_\chi)}{(p_1+p_2)^2}$$

$$\left. -\epsilon^{\mu\nu\rho\sigma}(p_1+p_2-k)_\sigma\right]. \tag{A.12}$$

where p_1, p_2 are the initial momenta, k the momenta of the gluon, and p_3, p_4 the momenta of the DM particle/antiparticle, g_s is the SU(3) gauge coupling and T^a are the SU(3) generators in the fundamental representation, *i.e.* the standard QCD Gell-Mann matrices. The matrix elements with the operator DT1 are instead

$$\mathcal{M}_{DT1}^g = -i\frac{g^2 g_s}{M^2}\epsilon_\mu^* T_{ij}^a \times$$

$$\times\left\{\frac{\bar{u}(p_3)P_L(\not{p}_1-\not{p}_2)\gamma^\mu u(p_1)\bar{v}(p_2)P_R v(p_4)}{(p_1-k)^2} - \frac{\bar{u}(p_3)P_L u(p_1)\bar{v}(p_2)P_R\gamma^\mu(\not{p}_2-\not{k})v(p_4)}{(p_2-k)^2}\right\}$$

$$\mathcal{M}_{DT1}^q = -i\frac{g^2 g_s}{M^2}\epsilon_\mu T_{ij}^a \times$$

$$\times\left\{\frac{\bar{u}(k)P_R v(p_3)\bar{u}(p_4)P_L(\not{p}_1+\not{p}_2)\gamma^\mu u(p_1)}{(p_1+p_2)^2} - \frac{\bar{u}(k)\gamma^\mu(\not{p}_2-\not{k})P_R v(p_3)\bar{u}(p_4)P_L u(p_1)}{(p_2-k)^2}\right\}$$

$$\mathcal{M}_{DT1}^{\bar{q}} = -i\frac{g^2 g_s}{M^2}\epsilon_\mu T_{ij}^a \times$$

$$\times\left\{\frac{\bar{v}(k)P_L u(p_3)\bar{v}(p_4)P_R(\not{p}_1+\not{p}_2)\gamma^\mu v(p_1)}{(p_1+p_2)^2} - \frac{\bar{v}(k)\gamma^\mu(\not{p}_2-\not{k})P_L u(p_3)\bar{v}(p_4)P_R v(p_1)}{(p_2-k)^2}\right\}$$

$$\tag{A.13}$$

for the gluon, quark and anti-quark emission processes respectively. Here we denote the gluon polarization vector by ϵ_μ and the left and right projectors $(1 - \gamma_5)/2$ and $(1 + \gamma_5)/2$ with P_L and P_R respectively. The anti-quark matrix element is simply obtained from the quark one by exchanging quarks with anti-quarks and left with right projectors. The parton level cross sections for the two processes are thus the same, so here we only show the explicit derivation of the quark one.

The corresponding squared amplitudes, averaged over initial states (colour and spin) and summed over the final states are

$$\sum \overline{|\mathcal{M}_{D1'}|^2} = \frac{16}{9}\frac{g_s^2}{\Lambda^4}\frac{[(p_3 \cdot p_4) - m_{\rm DM}^2]\left[(k \cdot (p_1 + p_2))^2 - 2(p_1 \cdot p_2)(k \cdot p_1 + k \cdot p_2 - p_1 \cdot p_2)\right]}{(k \cdot p_1)(k \cdot p_2)},$$
(A.14)

$$\sum \overline{|\mathcal{M}_{D4'}|^2} = \frac{16}{9}\frac{g_s^2}{\Lambda^4}\frac{[(p_3 \cdot p_4) + m_{\rm DM}^2]\left[(k \cdot (p_1 + p_2))^2 - 2(p_1 \cdot p_2)(k \cdot p_1 + k \cdot p_2 - p_1 \cdot p_2)\right]}{(k \cdot p_1)(k \cdot p_2)},$$
(A.15)

$$\begin{aligned}
\sum \overline{|\mathcal{M}_{D5}|^2} = -\frac{32}{9}\frac{g_s^2}{\Lambda^4}\Bigg[& \frac{(k \cdot p_1)\left[(k \cdot p_1) + (k \cdot p_2) - 3(p_1 \cdot p_2) - m_{\rm DM}^2\right]}{(k \cdot p_2)} \\
& + \frac{(k \cdot p_2)\left[(k \cdot p_1) + (k \cdot p_2) - 3(p_1 \cdot p_2) - m_{\rm DM}^2\right]}{(k \cdot p_1)} - 4(p_1 \cdot p_2) \\
& -2\frac{(p_1 \cdot p_2)}{(k \cdot p_1)(k \cdot p_2)}\left[(k \cdot p_3)\left((p_1 \cdot p_3) + (p_2 \cdot p_3)\right) + (p_1 \cdot p_2)\left(m_{\rm DM}^2 + (p_1 \cdot p_2)\right)\right. \\
& \left. -2(p_1 \cdot p_3)(p_2 \cdot p_3)\right] \\
& +2\frac{(k \cdot p_3)(p_1 \cdot p_3) - (p_2 \cdot p_3)(p_1 \cdot p_3) + (p_2 \cdot p_3)^2 + 2(p_1 \cdot p_2)^2 + m_{\rm DM}^2(p_1 \cdot p_2)}{(k \cdot p_2)} \\
& +2\frac{(k \cdot p_3)(p_2 \cdot p_3) - (p_1 \cdot p_3)(p_2 \cdot p_3) + (p_1 \cdot p_3)^2 + 2(p_1 \cdot p_2)^2 + m_{\rm DM}^2(p_1 \cdot p_2)}{(k \cdot p_1)}\Bigg],
\end{aligned}$$
(A.16)

$$\begin{aligned}
\sum \overline{|\mathcal{M}_{D8}|^2} = \frac{32}{9}\frac{g_s^2}{\Lambda^4}\Bigg[& \frac{(k \cdot p_1)\left[(k \cdot p_1) + (k \cdot p_2) - 3(p_1 \cdot p_2) + m_{\rm DM}^2 + 2(p_3 \cdot p_4)\right]}{(k \cdot p_2)} \\
& + \frac{(k \cdot p_2)\left[(k \cdot p_1) + (k \cdot p_2) - 3(p_1 \cdot p_2) + m_{\rm DM}^2 + 2(p_3 \cdot p_4)\right]}{(k \cdot p_1)} - 4(p_1 \cdot p_2) \\
& +2\frac{(p_1 \cdot p_2)}{(k \cdot p_1)(k \cdot p_2)}\left[(p_1 \cdot p_2)\left(2(p_3 \cdot p_4) + m_{\rm DM}^2\right) + (k \cdot p_3)\left((p_1 \cdot p_3) + (p_2 \cdot p_3)\right)\right. \\
& \left. +2(p_1 \cdot p_3)(p_2 \cdot p_3) - (p_1 \cdot p_2)^2\right] \\
& +2\frac{(p_1 \cdot p_3)\left[-(k \cdot p_3) + (p_2 \cdot p_3)\right] - (p_2 \cdot p_3)^2 + (p_1 \cdot p_2)\left[2(p_1 \cdot p_2) - m_{\rm DM}^2 - 2(p_3 \cdot p_4)\right]}{(k \cdot p_2)} \\
& +2\frac{(p_2 \cdot p_3)\left[-(k \cdot p_3) + (p_1 \cdot p_3)\right] - (p_1 \cdot p_3)^2 + (p_1 \cdot p_2)\left[2(p_1 \cdot p_2) - m_{\rm DM}^2 - 2(p_3 \cdot p_4)\right]}{(k \cdot p_1)}\Bigg],
\end{aligned}$$
(A.17)

$$\sum \overline{|\mathcal{M}_{D9}|^2} = \frac{128}{9}\frac{g_s^2}{\Lambda^4}\left[-2[m_{\text{DM}}^2 - (k \cdot p_3)] + \frac{(k \cdot p_1)\left[-(k \cdot p_3) + (p_1 \cdot p_3) - (p_2 \cdot p_3) + m_{\text{DM}}^2\right]}{(k \cdot p_2)} \right.$$

$$-2\frac{(p_1 \cdot p_2)\left[-2(k \cdot p_3) + (p_1 \cdot p_3) + (p_2 \cdot p_3) + m_{\text{DM}}^2\right]}{(k \cdot p_2)}$$

$$-4\frac{[(k \cdot p_3) - (p_2 \cdot p_3)]\,[(p_1 \cdot p_3) - (p_2 \cdot p_3)]}{(k \cdot p_2)}$$

$$+\frac{(k \cdot p_2)\left[-(k \cdot p_3) + (p_2 \cdot p_3) - (p_1 \cdot p_3) + m_{\text{DM}}^2\right]}{(k \cdot p_1)}$$

$$-2\frac{(p_1 \cdot p_2)\left[-2(k \cdot p_3) + (p_1 \cdot p_3) + (p_2 \cdot p_3) + m_{\text{DM}}^2\right]}{(k \cdot p_1)}$$

$$-4\frac{[(k \cdot p_3) - (p_1 \cdot p_3)]\,[(p_2 \cdot p_3) - (p_1 \cdot p_3)]}{(k \cdot p_1)}$$

$$-2\frac{(p_1 \cdot p_2)\,[(k \cdot p_3) - (p_1 \cdot p_3) - (p_2 \cdot p_3)]\,[2(k \cdot p_3) + (p_1 \cdot p_2)]}{(k \cdot p_1)(k \cdot p_2)}$$

$$\left. +2\frac{(p_1 \cdot p_2)\left[-4(p_1 \cdot p_3)(p_2 \cdot p_3) + m_{\text{DM}}^2(p_1 \cdot p_2)\right]}{(k \cdot p_1)(k \cdot p_2)} \right], \tag{A.18}$$

$$\sum \overline{|\mathcal{M}_{D11}|^2} = \frac{3}{32\pi^2}\frac{g_s^6}{\Lambda^6}\left[(p_3 \cdot p_4) - m_{\text{DM}}^2\right]\left\{ \frac{(k \cdot p_1)^3}{(k \cdot p_2)(p_1 \cdot p_2)} + \frac{(k \cdot p_2)^3}{(k \cdot p_1)(p_1 \cdot p_2)} + \frac{(p_1 \cdot p_2)^3}{(k \cdot p_1)(k \cdot p_2)} \right.$$

$$+3\frac{(k \cdot p_1)(k \cdot p_2)}{(p_1 \cdot p_2)} + \frac{(k \cdot p_1)(p_1 \cdot p_2) - (k \cdot p_1)^2}{(k \cdot p_2)} + \frac{(k \cdot p_2)(p_1 \cdot p_2) - (k \cdot p_2)^2}{(k \cdot p_1)}$$

$$-\frac{(k_- \cdot p_1)(k \cdot p_2)^3}{(k \cdot k_-)(k \cdot p_1)(p_1 \cdot p_2)} - \frac{(k_- \cdot p_2)(k \cdot p_1)^3}{(k \cdot k_-)(k \cdot p_2)(p_1 \cdot p_2)}$$

$$+\frac{(k_- \cdot p_1)}{(k \cdot k_-)(p_1 \cdot p_2)}[(k \cdot p_1)^2 + (k \cdot p_1)(k \cdot p_2) - (k \cdot p_2)^2]$$

$$+\frac{(k_- \cdot p_2)}{(k \cdot k_-)(p_1 \cdot p_2)}[(k \cdot p_2)^2 + (k \cdot p_1)(k \cdot p_2) - (k \cdot p_1)^2]$$

$$+2\frac{(k_- \cdot p_1)}{(k \cdot k_-)(k \cdot p_1)}[(k \cdot p_2)^2 - (p_1 \cdot p_2)(k \cdot p_2)]$$

$$+2\frac{(k_- \cdot p_2)}{(k \cdot k_-)(k \cdot p_2)}[(k \cdot p_1)^2 - (p_1 \cdot p_2)(k \cdot p_1)]$$

$$+2\frac{(k_- \cdot p_1)}{(k \cdot k_-)}[(p_1 \cdot p_2) + (k \cdot p_1) - 2(k \cdot p_2)]$$

$$+2\frac{(k_- \cdot p_2)}{(k \cdot k_-)}[(p_1 \cdot p_2) + (k \cdot p_2) - 2(k \cdot p_1)]$$

$$\left. +(k \cdot p_1) + (k \cdot p_2) + 6(p_1 \cdot p_2) \right\}, \tag{A.19}$$

$$\sum \overline{|\mathcal{M}_{D12}|^2} = \frac{3}{32\pi^2}\frac{g_s^6}{\Lambda^6}\left[(p_3 \cdot p_4) + m_{\text{DM}}^2\right]\left\{ \frac{(k \cdot p_1)^3}{(k \cdot p_2)(p_1 \cdot p_2)} + \frac{(k \cdot p_2)^3}{(k \cdot p_1)(p_1 \cdot p_2)} + \frac{(p_1 \cdot p_2)^3}{(k \cdot p_1)(k \cdot p_2)} \right.$$

$$+3\frac{(k \cdot p_1)(k \cdot p_2)}{(p_1 \cdot p_2)} + \frac{(k \cdot p_1)(p_1 \cdot p_2) - (k \cdot p_1)^2}{(k \cdot p_2)} + \frac{(k \cdot p_2)(p_1 \cdot p_2) - (k \cdot p_2)^2}{(k \cdot p_1)}$$

$$-\frac{(k_- \cdot p_1)(k \cdot p_2)^3}{(k \cdot k_-)(k \cdot p_1)(p_1 \cdot p_2)} - \frac{(k_- \cdot p_2)(k \cdot p_1)^3}{(k \cdot k_-)(k \cdot p_2)(p_1 \cdot p_2)}$$

$$+\frac{(k_- \cdot p_1)}{(k \cdot k_-)(p_1 \cdot p_2)}[(k \cdot p_1)^2 + (k \cdot p_1)(k \cdot p_2) - (k \cdot p_2)^2]$$

$$+ \frac{(k_- \cdot p_2)}{(k \cdot k_-)(p_1 \cdot p_2)}[(k \cdot p_2)^2 + (k \cdot p_1)(k \cdot p_2) - (k \cdot p_1)^2]$$

$$+2 \frac{(k_- \cdot p_1)}{(k \cdot k_-)(k \cdot p_1)}[(k \cdot p_2)^2 - (p_1 \cdot p_2)(k \cdot p_2)]$$

$$+2 \frac{(k_- \cdot p_2)}{(k \cdot k_-)(k \cdot p_2)}[(k \cdot p_1)^2 - (p_1 \cdot p_2)(k \cdot p_1)]$$

$$+2 \frac{(k_- \cdot p_1)}{(k \cdot k_-)}[(p_1 \cdot p_2) + (k \cdot p_1) - 2(k \cdot p_2)]$$

$$+2 \frac{(k_- \cdot p_2)}{(k \cdot k_-)}[(p_1 \cdot p_2) + (k \cdot p_2) - 2(k \cdot p_1)]$$

$$+(k \cdot p_1) + (k \cdot p_2) + 6(p_1 \cdot p_2)\} , \tag{A.20}$$

$$\sum \overline{|\mathcal{M}_{D13}|^2} = \frac{3}{32\pi^2} \frac{g_s^6}{\Lambda^6} \left[(p_3 \cdot p_4) - m_{\mathrm{DM}}^2\right] \left\{ \frac{(k \cdot p_1)^3}{(k \cdot p_2)(p_1 \cdot p_2)} + \frac{(k \cdot p_2)^3}{(k \cdot p_1)(p_1 \cdot p_2)} + \frac{(p_1 \cdot p_2)^3}{(k \cdot p_1)(k \cdot p_2)} \right.$$

$$+3 \frac{(k \cdot p_1)(k \cdot p_2)}{(p_1 \cdot p_2)} + \frac{(k \cdot p_1)(p_1 \cdot p_2) - (k \cdot p_1)^2}{(k \cdot p_2)} + \frac{(k \cdot p_2)(p_1 \cdot p_2) - (k \cdot p_2)^2}{(k \cdot p_1)}$$

$$- \frac{(k_- \cdot p_1)(k \cdot p_2)^3}{(k \cdot k_-)(k \cdot p_1)(p_1 \cdot p_2)} - \frac{(k_- \cdot p_2)(k \cdot p_1)^3}{(k \cdot k_-)(k \cdot p_2)(p_1 \cdot p_2)}$$

$$+ \frac{(k_- \cdot p_1)}{(k \cdot k_-)(p_1 \cdot p_2)}[(k \cdot p_1)^2 - 3(k \cdot p_1)(k \cdot p_2) + 3(k \cdot p_2)^2]$$

$$+ \frac{(k_- \cdot p_2)}{(k \cdot k_-)(p_1 \cdot p_2)}[(k \cdot p_2)^2 - 3(k \cdot p_1)(k \cdot p_2) + 3(k \cdot p_1)^2]$$

$$+2 \frac{(k_- \cdot p_1)}{(k \cdot k_-)(k \cdot p_1)}[(k \cdot p_2)^2 - (p_1 \cdot p_2)(k \cdot p_2)]$$

$$+2 \frac{(k_- \cdot p_2)}{(k \cdot k_-)(k \cdot p_2)}[(k \cdot p_1)^2 - (p_1 \cdot p_2)(k \cdot p_1)]$$

$$+2 \frac{(k_- \cdot p_1)}{(k \cdot k_-)}[(p_1 \cdot p_2) + (k \cdot p_1) - 2(k \cdot p_2)]$$

$$+2 \frac{(k_- \cdot p_2)}{(k \cdot k_-)}[(p_1 \cdot p_2) + (k \cdot p_2) - 2(k \cdot p_1)]$$

$$\left. -3(k \cdot p_1) - 3(k \cdot p_2) + 2(p_1 \cdot p_2)\right\} , \tag{A.21}$$

$$\sum \overline{|\mathcal{M}_{D14}|^2} = \frac{3}{32\pi^2} \frac{g_s^6}{\Lambda^6} \left[(p_3 \cdot p_4) + m_{\mathrm{DM}}^2\right] \left\{ \frac{(k \cdot p_1)^3}{(k \cdot p_2)(p_1 \cdot p_2)} + \frac{(k \cdot p_2)^3}{(k \cdot p_1)(p_1 \cdot p_2)} + \frac{(p_1 \cdot p_2)^3}{(k \cdot p_1)(k \cdot p_2)} \right.$$

$$+3 \frac{(k \cdot p_1)(k \cdot p_2)}{(p_1 \cdot p_2)} + \frac{(k \cdot p_1)(p_1 \cdot p_2) - (k \cdot p_1)^2}{(k \cdot p_2)} + \frac{(k \cdot p_2)(p_1 \cdot p_2) - (k \cdot p_2)^2}{(k \cdot p_1)}$$

$$- \frac{(k_- \cdot p_1)(k \cdot p_2)^3}{(k \cdot k_-)(k \cdot p_1)(p_1 \cdot p_2)} - \frac{(k_- \cdot p_2)(k \cdot p_1)^3}{(k \cdot k_-)(k \cdot p_2)(p_1 \cdot p_2)}$$

$$+ \frac{(k_- \cdot p_1)}{(k \cdot k_-)(p_1 \cdot p_2)}[(k \cdot p_1)^2 - 3(k \cdot p_1)(k \cdot p_2) + 3(k \cdot p_2)^2]$$

$$+ \frac{(k_- \cdot p_2)}{(k \cdot k_-)(p_1 \cdot p_2)}[(k \cdot p_2)^2 - 3(k \cdot p_1)(k \cdot p_2) + 3(k \cdot p_1)^2]$$

$$+2 \frac{(k_- \cdot p_1)}{(k \cdot k_-)(k \cdot p_1)}[(k \cdot p_2)^2 - (p_1 \cdot p_2)(k \cdot p_2)]$$

$$+2 \frac{(k_- \cdot p_2)}{(k \cdot k_-)(k \cdot p_2)}[(k \cdot p_1)^2 - (p_1 \cdot p_2)(k \cdot p_1)]$$

$$+2\frac{(k_- \cdot p_1)}{(k \cdot k_-)}[(p_1 \cdot p_2) + (k \cdot p_1) - 2(k \cdot p_2)]$$

$$+2\frac{(k_- \cdot p_2)}{(k \cdot k_-)}[(p_1 \cdot p_2) + (k \cdot p_2) - 2(k \cdot p_1)]$$

$$-3(k \cdot p_1) - 3(k \cdot p_2) + 2(p_1 \cdot p_2)\} . \tag{A.22}$$

$$\sum \overline{|\mathcal{M}_{DT1}^g|^2} = \frac{1}{9}\frac{g_s^2}{\Lambda^4}\frac{1}{(k \cdot p_1)(k \cdot p_2)} \times$$

$$\left\{ p_1 \cdot p_3 \Big[(k \cdot p_4)(k \cdot p_1) - (k \cdot p_4)(p_1 \cdot p_2) - (k \cdot p_2)(p_1 \cdot p_4)\Big] + \right.$$

$$+ p_2 \cdot p_4 \Big[(k \cdot p_3)(k \cdot p_2) - (k \cdot p_3)(p_1 \cdot p_2) - (k \cdot p_1)(p_2 \cdot p_3)\Big] +$$

$$\left. + (p_1 \cdot p_3)(p_2 \cdot p_4)\Big[2p_1 \cdot p_2 - k \cdot p_1 - k \cdot p_2\Big]\right\} \tag{A.23}$$

$$\sum \overline{|\mathcal{M}_{DT1}^q|^2} = \frac{1}{6}\frac{g_s^2}{\Lambda^4}\frac{1}{(k \cdot p_1)(k \cdot p_2)} \times$$

$$\left\{ p_1 \cdot p_4 \Big[(k \cdot p_2)(p_1 \cdot p_3) - (k \cdot p_1)(p_2 \cdot p_3) + \right.$$

$$+(k \cdot p_2)(k \cdot p_3) + (k \cdot p_1)(k \cdot p_3) - (p_1 \cdot p_2)(k \cdot p_3)\Big] +$$

$$+ p_2 \cdot p_2 \Big[(p_1 \cdot p_4)(p_2 \cdot p_3) - (k \cdot p_4)(k \cdot p_3)\Big] +$$

$$\left. + (k \cdot p_3)\Big[(k \cdot p_1)(p_1 \cdot p_4) + (k \cdot p_1)(p_2 \cdot p_4) + (k \cdot p_2)(p_2 \cdot p_4)\Big]\right\} \tag{A.24}$$

where the polarization 4-vector in Eqs. (A.19)–(A.22) is defined as $k_- \equiv P(k^\nu)/\sqrt{k^\mu \cdot P(k_\mu)}$, where P is the parity operation.

A.3   Cross Sections

Now, the next step is to compute the cross sections in the lab frame. To this end we proceed by first evaluating the matrix elements and the phase space density in the centre-of-mass frame and then boosting the result to the lab frame. In the centre-of-mass (c.o.m) frame, let us parametrize the four-momenta involved in the process as

$$p_1 = x\frac{\sqrt{s}}{2}(1,0,0,1), \quad p_2 = x\frac{\sqrt{s}}{2}(1,0,0,-1), \quad k = x\frac{\sqrt{s}}{2}(z_0, z_0\hat{k}), \tag{A.25}$$

$$p_3 = x\frac{\sqrt{s}}{2}(1 - y_0, \sqrt{(1 - y_0)^2 - a^2}\hat{p}_3), \quad p_4 = x\frac{\sqrt{s}}{2}(1 + y_0 - z_0, \sqrt{(1 + y_0 - z_0)^2 - a^2}\hat{p}_4),$$

where the two colliding partons carry equal momentum fractions $x_1 = x_2 \equiv x$ of the incoming protons, $a \equiv 2m_{\mathrm{DM}}/(x\sqrt{s}) < 1$, $\hat{k} = (0, \sin\theta_0, \cos\theta_0)$, and θ_0 is the polar angle of $\hat{k}$ with respect to the beam line, in the c.o.m. frame. With the subscript $_0$ we will refer to quantities evaluated in the c.o.m. frame. The polarization 4-vector k_- in the c.o.m. frame simply reads $k_- = (1/\sqrt{2})(1, 0, -\sin\theta_0, -\cos\theta_0)$.

The conservation of three-momentum sets the angle θ_{03j} between $\hat{p}_3$ and $\hat{k}$ as: $\cos\theta_{03j} = (\mathbf{p}_4^2 - \mathbf{k}^2 - \mathbf{p}_3^2)/2|\mathbf{k}||\mathbf{p}_3|$. Integration over the azimuthal angle ϕ_0 of the outgoing jet simply gives a factor of 2π, while the matrix element does depend in the t-channel case on the azimuthal angle of the three-momentum $\mathbf{p}_3$ with respect to $\mathbf{k}$, ϕ_{03j}, and so it can not be integrated over at this stage, contrary to the s-channel case. Taking all of this into account, the phase space density simplifies to

$$d\Phi_3 = \frac{1}{(4\pi)^3} dE_3 \, dk \, d\cos\theta_0 = \frac{1}{(4\pi)^3} \frac{x^2 s}{4} dz_0 \, dy_0 \, d\cos\theta_0 . \tag{A.26}$$

$$d\Phi_3 = \frac{1}{8(2\pi)^4} dE_3 \, d|\mathbf{k}| \, d\cos\theta_0 \, d\phi_{03j} = \frac{x^2 s}{32(2\pi)^4} dy_0 \, dz_0 \, d\cos\theta_0 \, d\phi_{03j}. \tag{A.27}$$

in the s and t-channel respectively. The kinematical domains of y_0, z_0 and ϕ_{03j} are

$$\frac{z_0}{2}\left(1 - \sqrt{\frac{1 - z_0 - a^2}{1 - z_0}}\right) \leq y_0 \leq \frac{z_0}{2}\left(1 + \sqrt{\frac{1 - z_0 - a^2}{1 - z_0}}\right) \tag{A.28}$$

$$0 \leq z_0 \leq 1 - a^2 \tag{A.29}$$

$$0 \leq \phi_{03j} \leq 2\pi \tag{A.30}$$

The variables y_0 and ϕ_{03j} refer to the momentum $\mathbf{p}_3$ of an invisible DM particle; they are therefore not measurable, and we integrate over them. In the t-channel case, finding the total integrated cross section is useless, since these variables enter our definition of the momentum transfer Q_{tr}, and the condition $Q_{\mathrm{tr}} < \Lambda$ which we used to define the ratio R_Λ.

For the doubly-differential cross sections with respect to the energy and angle of the emitted gluon, in the c.o.m. frame, we obtain

$$\left.\frac{d^2\hat{\sigma}}{dz_0 d\cos\theta_0}\right|_{D1'} = \frac{\alpha_s}{36\pi^2} \frac{x^2 s}{\Lambda^4} \frac{\left[1 - z_0 - \frac{4m_{\mathrm{DM}}^2}{x^2 s}\right]^{3/2}}{\sqrt{1 - z_0}} \frac{[1 + (1 - z_0)^2]}{z_0 \sin^2\theta_0}, \tag{A.31}$$

$$\left.\frac{d^2\hat{\sigma}}{dz_0 d\cos\theta_0}\right|_{D4'} = \frac{\alpha_s}{36\pi^2} \frac{x^2 s}{\Lambda^4} \frac{\left[1 - z_0 - \frac{4m_{\mathrm{DM}}^2}{x^2 s}\right]^{1/2}}{\sqrt{1 - z_0}} \frac{[1 + (1 - z_0)^2]}{z_0 \sin^2\theta_0}, \tag{A.32}$$

$$\left.\frac{d^2\hat{\sigma}}{dz_0 d\cos\theta_0}\right|_{D5} = \frac{\alpha_s}{108\pi^2} \frac{x^2 s}{\Lambda^4} \frac{\sqrt{1 - z_0 - \frac{4m_{\mathrm{DM}}^2}{x^2 s}}}{\sqrt{1 - z_0}}$$

$$\times \frac{(1 - z_0 + \frac{2m_{DM}^2}{x^2 s})(8 - 8z_0 + (3 + \cos 2\theta_0)z_0^2)}{z_0 \sin^2 \theta_0} \,, \tag{A.33}$$

$$\left. \frac{d^2\hat{\sigma}}{dz_0 d\cos\theta_0} \right|_{D8} = \frac{\alpha_s}{108\pi^2} \frac{x^2 s}{\Lambda^4} \frac{[1 - z_0 - \frac{4m_{DM}^2}{x^2 s}]^{3/2}}{\sqrt{1 - z_0}} \frac{8 - 8z_0 + (3 + \cos 2\theta_0)z_0^2}{z_0 \sin^2 \theta_0} \,, \tag{A.34}$$

$$\left. \frac{d^2\hat{\sigma}}{dz_0 d\cos\theta_0} \right|_{D9} = \frac{\alpha_s}{27\pi^2} \frac{x^2 s}{\Lambda^4} \frac{\sqrt{1 - z_0 - \frac{4m_{DM}^2}{x^2 s}}}{[1 - z_0]^{3/2}}$$

$$\times \frac{(1 - z_0 + \frac{2m_{DM}^2}{x^2 s})(4 - 8z_0 + 6z_0^2 - (1 + \cos 2\theta_0)z_0^3)}{z_0 \sin^2 \theta_0} \,, \tag{A.35}$$

$$\left. \frac{d^2\hat{\sigma}}{dz_0 d\cos\theta_0} \right|_{D11} = \frac{3\alpha_s^3 x^4 s^2}{32768\pi^2 \Lambda^6} \frac{\left[1 - z_0 - \frac{4m_{DM}^2}{x^2 s}\right]^{3/2}}{z_0 \sqrt{1 - z_0} \sin^2 \theta_0} [128 - 128(1 + \cos 2\theta_0)z_0$$

$$+ (304 + 64 \cos 2\theta_0 + 16 \cos 4\theta_0)z_0^2 - 128(1 + \cos 2\theta_0)z_0^3$$

$$+ (79 + 44 \cos 2\theta_0 + 5 \cos 4\theta_0)z_0^4] \,, \tag{A.36}$$

$$\left. \frac{d^2\hat{\sigma}}{dz_0 d\cos\theta_0} \right|_{D12} = \frac{3\alpha_s^3 x^4 s^2}{32768\pi^2 \Lambda^6}$$

$$\times \frac{\sqrt{1 - z_0 - \frac{4m_{DM}^2}{x^2 s}}\sqrt{1 - z_0}}{z_0 \sin^2 \theta_0} [128 - 128(1 + \cos 2\theta_0)z_0$$

$$+ (304 + 64 \cos 2\theta_0 + 16 \cos 4\theta_0)z_0^2 - 128(1 + \cos 2\theta_0)z_0^3$$

$$+ (79 + 44 \cos 2\theta_0 + 5 \cos 4\theta_0)z_0^4] \,, \tag{A.37}$$

$$\left. \frac{d^2\hat{\sigma}}{dz_0 d\cos\theta_0} \right|_{D13} = \frac{3\alpha_s^3 x^4 s^2}{32768\pi^2 \Lambda^6} \frac{\left[1 - z_0 - \frac{4m_{DM}^2}{x^2 s}\right]^{3/2}}{z_0 \sqrt{1 - z_0} \sin^2 \theta_0} [128 - 128(1 + \cos 2\theta_0)z_0$$

$$+ (240 + 128 \cos 2\theta_0 + 16 \cos 4\theta_0)z_0^2$$

$$- 16(11 + 4 \cos 2\theta_0 + \cos 4\theta_0)z_0^3$$

$$+ (79 + 44 \cos 2\theta_0 + 5 \cos 4\theta_0)z_0^4] \,, \tag{A.38}$$

$$\left. \frac{d^2\hat{\sigma}}{dz_0 d\cos\theta_0} \right|_{D14} = \frac{3\alpha_s^3 x^4 s^2}{32768\pi^2 \Lambda^6} \frac{\sqrt{1 - z_0 - \frac{4m_{DM}^2}{x^2 s}}\sqrt{1 - z_0}}{z_0 \sin^2 \theta_0}$$

$$[128 - 128(1 + \cos 2\theta_0)z_0 (240 + 128 \cos 2\theta_0 + 16 \cos 4\theta_0)z_0^2$$

$$- 16(11 + 4 \cos 2\theta_0 + \cos 4\theta_0)z_0^3$$

$$+ (79 + 44 \cos 2\theta_0 + 5 \cos 4\theta_0)z_0^4] \,. \tag{A.39}$$

$$\left.\frac{\mathrm{d}^4\hat{\sigma}}{\mathrm{d}z_0\,\mathrm{d}\cos\theta_0\,\mathrm{d}y_0\,\mathrm{d}\phi_{03j}}\right|_{DT1}^g = \frac{1}{4608\pi^4}\frac{g_s^2}{\Lambda^4}\frac{1-z_0}{z_0^4}$$

$$\left\{ 4x(2-z_0)\csc\theta_0\cos\phi_{03j}(\cos\theta_0(z_0-2y_0)+z_0)\right.$$

$$\sqrt{s\left(sx^2y_0(z_0-1)(y_0-z_0)-m_{DM}^2z_0^2\right)}$$

$$-8m_{DM}^2z_0^2\cos^2\phi_{03j}+sx^2((z_0-2)z_0+2)\left(\sec^2(\theta_0/2)\,y_0^2\right.$$

$$\left.+\csc^2(\theta_0/2)\,(y_0-z_0)^2\right)$$

$$-2sx^2y_0^2((z_0-6)z_0+6)+4sx^2y_0(z_0-1)(y_0-z_0)\cos(2\phi_{03j})$$

$$\left.+2sx^2y_0((z_0-6)z_0+6)z_0-sx^2z_0^2((z_0-2)z_0+2)\right\}, \qquad (A.40)$$

$$\left.\frac{\mathrm{d}^4\hat{\sigma}}{\mathrm{d}z_0\,\mathrm{d}\cos\theta_0\,\mathrm{d}y_0\,\mathrm{d}\phi_{03j}}\right|_{DT1}^q = \frac{1}{98304\pi^4}\frac{g_s^2}{\Lambda^4}\frac{1-z_0}{z_0^3\cos^2\frac{\theta_0}{2}}$$

$$\left\{ 8x\sqrt{s}\left[z_0(z_0-y_0-1)-(z_0^2-(1+y_0)z_0+2y_0)\right.\right.$$

$$\left.\times\cos\theta_0\right]\cos\phi_{03j}\sin\theta_0\times\sqrt{sx^2y_0(z_0-y_0)(1-z_0)-m_{DM}^2z_0^2}$$

$$-2(1-\cos(2\theta_0))m_{DM}^2z_0^2$$

$$+4\left[sx^2y_0(z_0-y_0)(1-z_0)-m_{DM}^2z_0^2\right]\cos(2\phi_{03j})\sin^2\theta_0$$

$$+sx^2\left[11z_0^4-(6+22y_0)z_0^3+(11y_0^2+8y_0+3)z_0^2\right.$$

$$\left.-2y_0(1+y_0)z_0+2y_0^2\right]$$

$$+sx^2\left[z_0^4-2(1+y_0)z_0^3+(y_0^2+8y_0+1)z_0^2\right.$$

$$\left.-6y_0(1+y_0)z_0+6y_0^2\right]\cos(2\theta_0)$$

$$\left.-4sx^2z_0\left[z_0^3-2(1+y_0)z_0^2+(y_0^2+4y_0+1)z_0-2y_0(1+y_0)\right]\cos\theta_0\right\}.$$

$$(A.41)$$

Equation (A.31)–(A.34) agree with the findings in Refs. [1, 2], up to the factor of $1/9$, as we are considering coloured colliding particles.

To get the cross sections in the lab frame we perform a boost in the $\hat{z}$ axis, accounting for the generic parton momentum fractions x_1, x_2. The velocity of the c.o.m. of the colliding particles with respect to the lab frame is given by

$$\beta_{c.o.m.} = \frac{x_1-x_2}{x_1+x_2}, \qquad (A.42)$$

so that the relations between the quantities z_0, θ_0 and the analogous ones z, θ in the lab frame are

$$z_0 = \frac{(x_1 + x_2)^2 + (x_2^2 - x_1^2)\cos\theta}{4x_1 x_2} z$$

$$\sin^2\theta_0 = \frac{4x_1 x_2}{[(x_1 + x_2) + (x_2 - x_1)\cos\theta]^2}\sin^2\theta. \tag{A.43}$$

The Jacobian factor to transform $dz_0\, d\cos\theta_0 \to dz\, d\cos\theta$ is simply obtained using Eq. (A.43); the cross section in the lab frame is then

$$\frac{d^4\hat{\sigma}}{dz\, d\cos\theta\, dy_0\, d\phi_{03j}} = \frac{x_1 + x_2}{x_1 + x_2 + (x_1 - x_2)\cos\theta}$$

$$\times \left. \frac{d^4\hat{\sigma}}{dz_0\, d\cos\theta_0\, dy_0\, d\phi_{03j}}\right|_{\substack{z_0 \to z_0(z) \\ \theta_0 \to \theta_0(\theta)}}. \tag{A.44}$$

Expressing the energy of the emitted gluon or (anti-)quark in terms of the transverse momentum and rapidity, $k^0 = p_T \cosh\eta$, one finds

$$z = \frac{4 p_T \cosh\eta}{(x_1 + x_2)\sqrt{s}}, \qquad \cos\theta = \tanh\eta \tag{A.45}$$

which allows us to express the differential cross sections with respect to the transverse momentum and pseudo-rapidity of the emitted jet:

$$\frac{d^2\hat{\sigma}}{dp_T\, d\eta} = \frac{4}{(x_1 + x_2)\sqrt{s}\cosh\eta} \left. \frac{d^2\hat{\sigma}}{dz\, d\cos\theta\, dy_0}\right|_{\substack{z \to z(p_T, \eta) \\ \theta \to \theta(p_T, \eta)}}. \tag{A.46}$$

$$\frac{d^4\hat{\sigma}}{dp_T\, d\eta\, dy_0\, d\phi_{03j}} = \frac{4}{(x_1 + x_2)\sqrt{s}\cosh\eta} \left. \frac{d^4\hat{\sigma}}{dz\, d\cos\theta\, dy_0\, d\phi_{03j}}\right|_{\substack{z \to z(p_T, \eta) \\ \theta \to \theta(p_T, \eta)}}. \tag{A.47}$$

This way we get the translation of Eqs. (A.31)–(A.35) into the lab frame:

$$\left. \frac{d^2\hat{\sigma}}{dp_T d\eta}\right|_{D1'} = \frac{\alpha_s}{36\pi^2} \frac{x_1 x_2 s}{\Lambda^4} \frac{1}{p_T} \frac{\left[1 - f - \frac{4m_{DM}^2}{x_1 x_2 s}\right]^{3/2}\left[1 + (1 - f)^2\right]}{\sqrt{1 - f}}, \tag{A.48}$$

$$\left. \frac{d^2\hat{\sigma}}{dp_T d\eta}\right|_{D4'} = \frac{\alpha_s}{36\pi^2} \frac{x_1 x_2 s}{\Lambda^4} \frac{\sqrt{1 - f}}{p_T} \left[1 - f - \frac{4m_{DM}^2}{x_1 x_2 s}\right]^{1/2}\left[1 + (1 - f)^2\right], \tag{A.49}$$

$$\left.\frac{\mathrm{d}^2\hat{\sigma}}{\mathrm{d}p_{\mathrm{T}}\mathrm{d}\eta}\right|_{D5} = \frac{\alpha_s}{27\pi^2}\frac{x_1 x_2 s}{\Lambda^4}\frac{\sqrt{1 - f - \frac{4m_{\mathrm{DM}}^2}{x_1 x_2 s}}}{\sqrt{1 - f}}$$

$$\times\frac{\left[1 - f + \frac{2m_{\mathrm{DM}}^2}{x_1 x_2 s}\right]\left[1 + (1 - f)^2 - 2\frac{p_{\mathrm{T}}^2}{x_1 x_2 s}\right]}{p_{\mathrm{T}}}, \tag{A.50}$$

$$\left.\frac{\mathrm{d}^2\hat{\sigma}}{\mathrm{d}p_{\mathrm{T}}\mathrm{d}\eta}\right|_{D8} = \frac{\alpha_s}{27\pi^2}\frac{x_1 x_2 s}{\Lambda^4}\frac{[1 - f - \frac{4m_{\mathrm{DM}}^2}{x_1 x_2 s}]^{3/2}}{\sqrt{1 - f}}\frac{1 + (1 - f)^2 - 2\frac{p_{\mathrm{T}}^2}{x_1 x_2 s}}{p_{\mathrm{T}}}, \tag{A.51}$$

$$\left.\frac{\mathrm{d}^2\hat{\sigma}}{\mathrm{d}p_{\mathrm{T}}\mathrm{d}\eta}\right|_{D9} = \frac{2\alpha_s}{27\pi^2}\frac{x_1 x_2 s}{\Lambda^4}\frac{\sqrt{1 - f - \frac{4m_{\mathrm{DM}}^2}{s x_1 x_2}}}{[1 - f]^{3/2}}$$

$$\times\frac{(1 - f + \frac{2m_{\mathrm{DM}}^2}{x_1 x_2 s})\left[(1 - f)(1 + (1 - f)^2) + f\frac{4p_{\mathrm{T}}^2}{x_1 x_2 s}\right]}{p_{\mathrm{T}}}, \tag{A.52}$$

$$\left.\frac{\mathrm{d}^2\hat{\sigma}}{\mathrm{d}p_{\mathrm{T}}\mathrm{d}\eta}\right|_{D11} = \frac{3\alpha_s^3 x_1^2 x_2^2 s^2}{256\pi^2\Lambda^6}\frac{(1 - f - \frac{4m_{\mathrm{DM}}^2}{s x_1 x_2})^{3/2}}{p_{\mathrm{T}} f^2\sqrt{1 - f}}$$

$$\times\left[16\frac{p_{\mathrm{T}}^4}{x_1^2 x_2^2 s^2} + 8\frac{p_{\mathrm{T}}^2}{x_1 x_2 s}f + (1 - 8\frac{p_{\mathrm{T}}^2}{x_1 x_2 s} + 5\frac{p_{\mathrm{T}}^4}{x_1^2 x_2^2 s^2})f^2\right.$$

$$\left. + (-2 + 8\frac{p_{\mathrm{T}}^2}{x_1 x_2 s})f^3 + (3 - 4\frac{p_{\mathrm{T}}^2}{x_1 x_2 s})f^4 - 2f^5 + f^6\right], \tag{A.53}$$

$$\left.\frac{\mathrm{d}^2\hat{\sigma}}{\mathrm{d}p_{\mathrm{T}}\mathrm{d}\eta}\right|_{D12} = \frac{3\alpha_s^3 x_1^2 x_2^2 s^2}{256\pi^2\Lambda^6}\frac{\sqrt{1 - f - \frac{4m_{\mathrm{DM}}^2}{s x_1 x_2}}\sqrt{1 - f}}{p_{\mathrm{T}} f^2}\left[16\frac{p_{\mathrm{T}}^4}{x_1^2 x_2^2 s^2} + 8\frac{p_{\mathrm{T}}^2}{x_1 x_2 s}f\right.$$

$$+ (1 - 8\frac{p_{\mathrm{T}}^2}{x_1 x_2 s} + 5\frac{p_{\mathrm{T}}^4}{x_1^2 x_2^2 s^2})f^2 + (-2 + 8\frac{p_{\mathrm{T}}^2}{x_1 x_2 s})f^3$$

$$\left. + (3 - 4\frac{p_{\mathrm{T}}^2}{x_1 x_2 s})f^4 - 2f^5 + f^6\right], \tag{A.54}$$

$$\left.\frac{\mathrm{d}^2\hat{\sigma}}{\mathrm{d}p_{\mathrm{T}}\mathrm{d}\eta}\right|_{D13} = \frac{3\alpha_s^3 x_1^2 x_2^2 s^2}{256\pi^2\Lambda^6}\frac{(1 - f - \frac{4m_{\mathrm{DM}}^2}{s x_1 x_2})^{3/2}}{p_{\mathrm{T}} f^2\sqrt{1 - f}}\left[16\frac{p_{\mathrm{T}}^4}{x_1^2 x_2^2 s^2} + 8(\frac{p_{\mathrm{T}}^2}{x_1 x_2 s} - 2\frac{p_{\mathrm{T}}^4}{x_1^2 x_2^2 s^2})f\right.$$

$$+ (1 - 12\frac{p_{\mathrm{T}}^2}{x_1 x_2 s} + 5\frac{p_{\mathrm{T}}^4}{x_1^2 x_2^2 s^2})f^2 + (-2 + 8\frac{p_{\mathrm{T}}^2}{x_1 x_2 s})f^3$$

$$\left. + (3 - 4\frac{p_{\mathrm{T}}^2}{x_1 x_2 s})f^4 - 2f^5 + f^6\right], \tag{A.55}$$

$$\frac{\mathrm{d}^2\hat{\sigma}}{\mathrm{d}p_\mathrm{T}\mathrm{d}\eta}\bigg|_{D14} = \frac{3\alpha_s^3 x_1^2 x_2^2 s^2}{256\pi^2 \Lambda^6} \frac{\sqrt{1 - f - \frac{4m_{\mathrm{DM}}^2}{s x_1 x_2}}\sqrt{1 - f}}{p_\mathrm{T} f^2}$$

$$\times \left[16\frac{p_\mathrm{T}^4}{x_1^2 x_2^2 s^2} + 8(\frac{p_\mathrm{T}^2}{x_1 x_2 s} - 2\frac{p_\mathrm{T}^4}{x_1^2 x_2^2 s^2})f \right.$$

$$+ (1 - 12\frac{p_\mathrm{T}^2}{x_1 x_2 s} + 5\frac{p_\mathrm{T}^4}{x_1^2 x_2^2 s^2})f^2 + (-2 + 8\frac{p_\mathrm{T}^2}{x_1 x_2 s})f^3$$

$$\left. + (3 - 4\frac{p_\mathrm{T}^2}{x_1 x_2 s})f^4 - 2f^5 + f^6 \right], \tag{A.56}$$

where we have defined

$$f(p_\mathrm{T}, \eta, x_1, x_2) \equiv \frac{p_\mathrm{T}(x_1 e^{-\eta} + x_2 e^{\eta})}{x_1 x_2 \sqrt{s}}. \tag{A.57}$$

For the emission of a photon, rather than a gluon, from a quark with charge Q_q one simply replaces $(4/3)\,\alpha_s \rightarrow Q_q^2 \alpha$ in Eqs. (A.48)–(A.52). The cross section for the corresponding simplified models are simply obtained by replacing

$$\Lambda^4 = \frac{\left(Q_{\mathrm{tr}}^2 - M_{\mathrm{med}}^2\right)^2 + \Gamma^2 M_{\mathrm{med}}^2}{g_q^2 g_\chi^2}, \tag{A.58}$$

From these expressions one reproduces the results reported in Eqs. (6.17)–(6.25). We do not report the explicit expressions in the t-channel because they are too cumbersome and of limited interest.

References

1. Y.J. Chae, M. Perelstein, Dark matter search at a linear collider: effective operator approach. JHEP **1305**, 138 (2013), arXiv:1211.4008
2. H. Dreiner, M. Huck, M. Krämer, D. Schmeier, J. Tattersall, Illuminating dark matter at the ILC. Phys. Rev. **D87**(7) 075015 (2013), arXiv:1211.2254

.

Appendix B
Annihilation Cross Section with a Pure Vector/Axial Mediator (as in Chap. 8)

In this Appendix we collect the results of cross sections calculations for the process of DM annihilation into SM fermions

$$\chi\bar{\chi} \to f\bar{f} \tag{B.1}$$

We performed the calculation at zero temperature in the lab frame where χ is at rest, and the centre of mass energy $s = 2m_{\rm DM}^2\left(\frac{1}{\sqrt{1-v^2}}+1\right)$. This is equivalent to performing the calculation in the Moeller frame, and is the correct frame for the relic density calculations [1].

$$(\sigma v)_V = \frac{N_C(g_f^V)^2(g_{\rm DM}^V)^2}{2\pi} \frac{\sqrt{1-m_f^2/m_{\rm DM}^2}}{(M^2-4m_{\rm DM}^2)^2+\Gamma^2M^2}\Big[(m_f^2+2m_{\rm DM}^2)+$$
$$v^2\Big(\frac{11m_f^4+2m_f^2m_{\rm DM}^2-4m_{\rm DM}^4}{24m_{\rm DM}^2(1-\frac{m_f^2}{m_{\rm DM}^2})}+2\frac{m_{\rm DM}^2(m_f^2+2m_{\rm DM}^2)(M^2-4m_{\rm DM}^2)}{(M^2-4m_{\rm DM}^2)^2+\Gamma^2M^2}\Big)\Big], \tag{B.2}$$

$$(\sigma v)_A = \frac{N_C(g_f^A)^2(g_{\rm DM}^A)^2}{2\pi} \frac{\sqrt{1-m_f^2/m_{\rm DM}^2}}{(M^2-4m_{\rm DM}^2)^2+\Gamma^2M^2}\Big[m_f^2$$
$$+v^2\Big(\frac{23m_f^4-28m_f^2m_{\rm DM}^2+8m_{\rm DM}^4}{24m_{\rm DM}^2(1-\frac{m_f^2}{m_{\rm DM}^2})}+2\frac{m_f^2m_{\rm DM}^2(M^2-4m_{\rm DM}^2)}{(M^2-4m_{\rm DM}^2)^2+\Gamma^2M^2}\Big)\Big]. \tag{B.3}$$

© Springer International Publishing AG 2017
E. Morgante, *Aspects of WIMP Dark Matter Searches at Colliders and Other Probes*, Springer Theses, https://doi.org/10.1007/978-3-319-67606-7

In the limit $m_f \to 0$ these expressions become

$$(\sigma v)_V = \frac{N_C (g_f^V)^2 (g_{DM}^V)^2}{\pi} \frac{m_{DM}^2}{(M^2 - 4m_{DM}^2)^2 + \Gamma^2 M^2}$$
$$\times \left[1 + v^2 \left(-\frac{1}{12} + \frac{2m_{DM}^2 (M^2 - 4m_{DM}^2)}{(M^2 - 4m_{DM}^2)^2 + \Gamma^2 M^2} \right) \right], \quad \text{(B.4)}$$

$$(\sigma v)_A = \frac{N_C (g_f^A)^2 (g_{DM}^A)^2}{6\pi} \frac{m_{DM}^2}{(M^2 - 4m_{DM}^2)^2 + \Gamma^2 M^2} v^2. \quad \text{(B.5)}$$

In the limit $m_{DM} \ll M$ the effective operator approximation holds:

$$(\sigma v)_V = \frac{N_C m_{DM}^2}{2\pi \Lambda^4} \sqrt{1 - \frac{m_f^2}{m_{DM}^2}} \left[\left(\frac{m_f^2}{m_{DM}^2} + 2 \right) + v^2 \frac{11 m_f^4 / m_{DM}^4 + 2 m_f^2 / m_{DM}^2 - 4}{24(1 - m_f^2 / m_{DM}^2)} \right], \quad \text{(B.6)}$$

$$(\sigma v)_A = \frac{N_C}{2\pi \Lambda^4} \sqrt{1 - \frac{m_f^2}{m_{DM}^2}} \left[m_f^2 + v^2 \frac{23 m_f^4 / m_{DM}^2 - 28 m_f^2 + 8 m_{DM}^2}{24(1 - m_f^2 / m_{DM}^2)} \right]. \quad \text{(B.7)}$$

The widths for the vector mediator decay to a pair of fermions are given by

$$\Gamma_V = \frac{N_C (g_f^V)^2 (M^2 + 2m_f^2) \sqrt{1 - 4m_f^2 / M^2}}{12\pi M}, \quad \text{(B.8)}$$

$$\Gamma_A = \frac{N_C (g_f^A)^2 M (1 - 4m_f^2 / M^2)^{3/2}}{12\pi}. \quad \text{(B.9)}$$

Reference

1. P. Gondolo, G. Gelmini, Cosmic abundances of stable particles: improved analysis. Nucl. Phys. **B360**, 145–179 (1991)

Appendix C
Details of the Annihilation Rate Calculation in the Z' Model

In this appendix we present the calculation of the annihilation cross sections of the DM into the SM in detail. We also go in detail over the one-loop order annihilation into the WW. The results of this calculations are summarized on Fig. 9.6

The Feynman diagrams for the possible annihilation channels of $\chi\chi$ into the SM particles are shown on Fig. C.1. For each channel, we consider the leading order (tree level or one-loop), moreover we always restrict the calculation to the leading order in the mixing angle between Z and Z', ψ.

At the tree level the DM annihilates into $f\bar{f}$, and if $\theta \neq \pi/2$ also to W^+W^- and Zh. Equations (C.1)–(C.6) summarize the annihilation cross sections into all these channels at the tree level, distinguishing between the polarization of the vector bosons in the final states through a superscript (T) or (L) for transverse or longitudinal polarization, respectively.

$$
\sigma(\chi\chi \to f\bar{f}) = \frac{g_\chi^2 g_{Z'}^4 \cos^4\psi\, N_c^f}{3\pi s \left((s - m_{Z'}^2)^2 + \Gamma_{Z'}^2 m_{Z'}^2\right)}
$$
$$
\times \sqrt{\frac{s - 4m_f^2}{s - 4m_\chi^2}} \left((g_f^V)^2 (s - 4m_\chi^2)(s + 2m_f^2) \right.
$$
$$
\left. + (g_f^A)^2 \left(s(s - 4m_\chi^2) + 4m_f^2 \left(m_\chi^2 \left(7 - 6\frac{s}{m_{Z'}^2} + 3\frac{s^2}{m_{Z'}^2} \right) - s \right) \right) \right),
$$
$$
\tag{C.1}
$$

$$
\sigma(\chi\chi \to Z^{(L)}h) = g_\chi^2 g_{Z'}^4 \cos^2\theta \cos^4\psi \frac{\sqrt{(s - (m_h^2 + m_Z^2))^2 - 4m_h^2 m_Z^2}}{((s - m_{Z'}^2)^2) + \Gamma_{Z'}^2 m_{Z'}^2} \frac{1}{48\pi s^{5/2}} \cdot
$$
$$
\cdot \left(\left(m_h^4 (2m_\chi^2 + s) - 2m_h^2 (s + 2m_\chi^2)(s + m_Z^2) + 2m_\chi^2 (s^2 - 10sm_Z^2 + m_Z^4) \right. \right.
$$

© Springer International Publishing AG 2017

E. Morgante, *Aspects of WIMP Dark Matter Searches at Colliders and Other Probes*, Springer Theses, https://doi.org/10.1007/978-3-319-67606-7

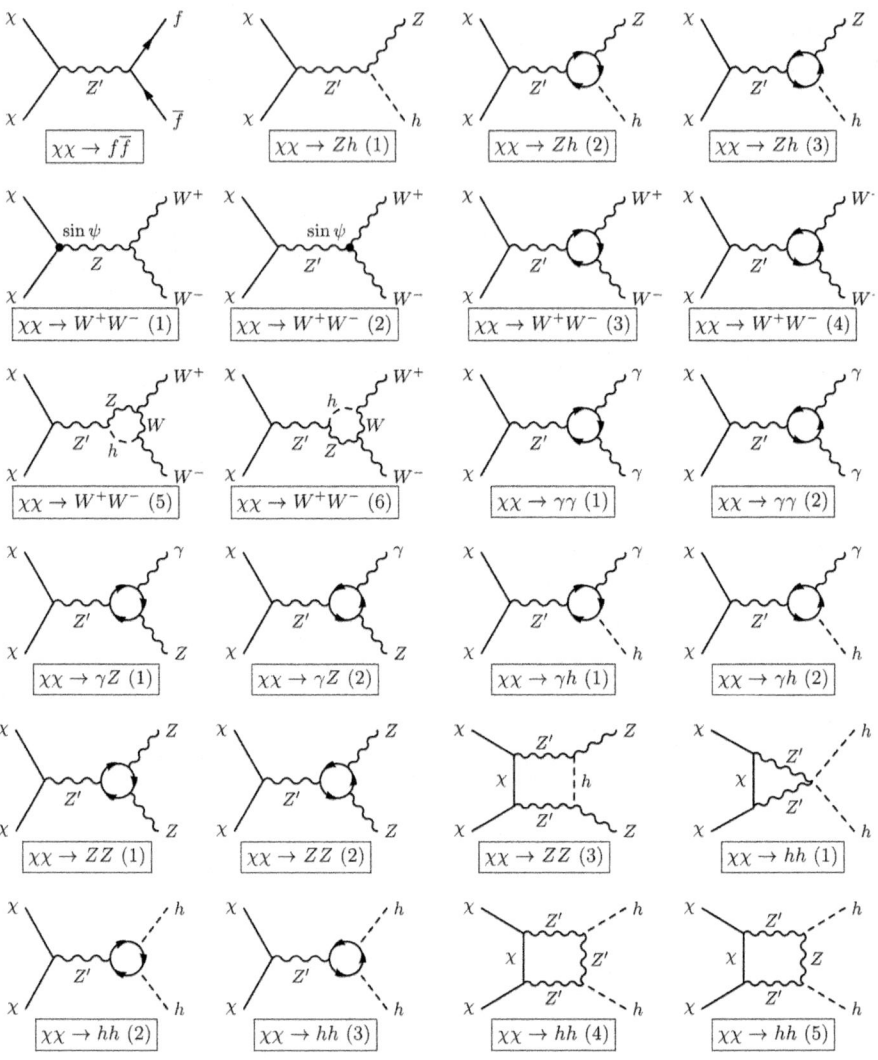

Figure C.1 Feynman diagrams for the annihilation of $\chi\chi$ into pairs of SM particles that have been considered in this work. In the fermion loops, the amplitude is summed over all the SM fermions

$$
\begin{aligned}
+ s(s + m_Z^2)^2\Big) &+ \frac{1}{m_{Z'}^2}\Big(- 6m_\chi^2 s(m_h^4 - 2m_h^2(m_Z^2 + s) + (m_Z^2 - s)^2)\Big) \\
&+ \frac{1}{m_{Z'}^4}\Big(3m_\chi^2 s^2(m_h^4 - 2m_h^2(s + m_Z^2) + (s - m_Z^2)^2)\Big)\Big),
\end{aligned} \tag{C.2}
$$

$$\sigma(\chi\chi \to Z^{(T)}h) = g_\chi^2 g_{Z'}^4 \cos^2\theta \cos^4\psi \frac{\sqrt{(s - (m_h^2 + m_Z^2))^2 - 4m_h^2 m_Z^2}}{((s - m_{Z'}^2)^2) + \Gamma_{Z'}^2 m_{Z'}^2} \frac{m_Z^2 (s - 4m_\chi^2)^{1/2}}{6\pi s^{3/2}},$$
$$(C.3)$$

$$\sigma(\chi\chi \to W^{+(L)}W^{-(L)}) = g_\chi^2 g_{Z'}^4 \alpha_W \cos^2\theta_W \cos^2\psi \sin^2\psi (s - 4m_W^2)^{3/2}(s - 4m_\chi^2)^{1/2}.$$
$$\cdot \frac{(m_{Z'}^2 - m_Z^2)^2 + (\Gamma_{Z'}m_{Z'} - \Gamma_Z m_Z)^2}{((s - m_Z^2)^2 - \Gamma_Z^2 m_Z^2)((s - m_{Z'}^2)^2 - \Gamma_{Z'}^2 m_{Z'}^2)} \frac{(2m_W^2 + s)^2}{12m_W^4 s}, \quad (C.4)$$

$$\sigma(\chi\chi \to W^{\pm(T)}W^{\mp(L)}) = g_\chi^2 g_{Z'}^4 \alpha_W \cos^2\theta_W \cos^2\psi \sin^2\psi (s - 4m_W^2)^{3/2}(s - 4m_\chi^2)^{1/2}.$$
$$\cdot \frac{(m_{Z'}^2 - m_Z^2)^2 + (\Gamma_{Z'}m_{Z'} - \Gamma_Z m_Z)^2}{((s - m_Z^2)^2 - \Gamma_Z^2 m_Z^2)((s - m_{Z'}^2)^2 - \Gamma_{Z'}^2 m_{Z'}^2)} \frac{4}{3 m_W^2}, \quad (C.5)$$

$$\sigma(\chi\chi \to W^{+(T)}W^{-(T)}) = g_\chi^2 g_{Z'}^4 \alpha_W \cos^2\theta_W \cos^2\psi \sin^2\psi (s - 4m_W^2)^{3/2}(s - 4m_\chi^2)^{1/2}.$$
$$\cdot \frac{(m_{Z'}^2 - m_Z^2)^2 + (\Gamma_{Z'}m_{Z'} - \Gamma_Z m_Z)^2}{((s - m_Z^2)^2 - \Gamma_Z^2 m_Z^2)((s - m_{Z'}^2)^2 - \Gamma_{Z'}^2 m_{Z'}^2)} \frac{2}{3s}. \quad (C.6)$$

A few clarifications are in order about the annihilation cross sections of the DM at tree level.

$f\overline{f}$: We denote the number of colours of the fermion f by N_c^f, and its vector and axial vector couplings to Z' by g_f^V, g_f^A respectively. The values of g_f^V, g_f^A are given in Table 9.2 In the zero velocity limit, corresponding to $s = 4m_\chi^2$, the cross section is proportional to m_f^2, because of the helicity suppression for pairs of annihilating fermions. In that limit, $\sigma \propto (g_f^A)^2 \propto \cos^2\theta$, i.e. the a coefficient in the low velocity expansion $\sigma v \simeq a + bv^2$ comes from the $U(1)_Y$ component of the $U(1)'$ extension.

Zh: The diagram on Fig. C.1 contains the tree level vertex $Z'Zh$ of Eq. (9.9). In the zero velocity limit, the only contribution comes from the production of a longitudinally polarized Z, since in Eq. (C.3) the factor $(s - 4m_\chi^2)^{1/2}$ vanishes.

WW: We denote $\alpha_W = g_W^2/(4\pi)$, where g_W is the weak coupling constant. The amplitude is the sum of the two diagrams on Fig. C.1: the annihilation occurs via the mixing of Z and Z' and the SM trilinear gauge vertex ZWW, see Eq. (9.10). For each of the final polarization states, the cross section is proportional to $(s - 4m_\chi^2)^{1/2}$.

Given that $\sigma(\chi\chi \to WW)v_{DM}$ (where v_{DM} is the DM velocity) is suppressed by $\sin^2\psi$ and by v_{DM}^2 at tree level it is worth checking whether contributions arising at one loop can become important. The contribution to the amplitude of diagrams 5 and 6 on Fig. C.1 is velocity suppressed because of the same argument reported at the end of Sec. 9.3.6. Therefore we only considered the contribution to the cross section coming from the sum of diagrams 3 and 4, also ignoring the interference terms with the other diagrams.

We also computed the cross sections for the annihilations into ZZ, $\gamma\gamma$, γh, γZ and hh at one loop. It is worth computing these corrections because of the velocity suppression of some tree level channels, and because in the pure $U(1)_{B-L}$ case the tree level annihilations into WW and Zh disappear. As for the Zh channel, we computed the one loop cross section only in the pure $U(1)_{B-L}$ case ($\theta = \pi/2$), in which the tree level amplitude vanishes, and the only remaining contribution comes from diagrams 2 and 3 in Fig. C.1. The results of the loop calculation are:

$$\sigma(\chi\chi \to W^{+\,(T)}W^{-\,(T)}) = \frac{g_{Z'}^4 g_\chi^2 N_c^i \left(s - 4m_W^2\right)^{3/2}\alpha_W^2}{768\pi^3 \left(\Gamma_{Z'}^2 m_{Z'}^2 + (m_{Z'}^2 - s)^2\right)s\sqrt{s - 4m_\chi^2}} \cdot$$

$$\cdot \left\{ \frac{48m_\chi^2\,(s - m_{Z'}^2)^2}{m_{Z'}^4\,(s - 4m_W^2)^2} \left| \sum_i N_c^i \Big\{ (g_{d^i}^L - g_{d^i}^R) B_0(s, m_{d^i}^2, m_{d^i}^2)m_{d^i}^2 - \right. \right.$$

$$- (g_{d^i}^L - g_{d^i}^R)(m_{d^i}^2 - m_{u^i}^2 - m_W^2)C_0(m_W^2, m_W^2, s, m_{d^i}^2, m_{u^i}^2, m_{d^i}^2)m_{d^i}^2 - \frac{1}{2}(g_{d^i}^L + g_{u^i}^L)(s - 4m_W^2) +$$

$$+ \left((g_{u^i}^R - g_{d^i}^L)m_{d^i}^2 + (g_{u^i}^R - g_{u^i}^L)m_{u^i}^2\right)B_0(m_W^2, m_{d^i}^2, m_{u^i}^2) + (g_{u^i}^L - g_{u^i}^R)m_{u^i}^2\,B_0(s, m_{u^i}^2, m_{u^i}^2) +$$

$$+ (g_{u^i}^L - g_{u^i}^R)m_{u^i}^2(m_{d^i}^2 - m_{u^i}^2 + m_W^2)C_0(m_W^2, m_W^2, s, m_{u^i}^2, m_{d^i}^2, m_{u^i}^2)\Big\} \Big|^2$$

$$+ \frac{32}{81}\frac{s - 4m_\chi^2}{m_W^4(s - 4m_W^2)^4}\left| \sum_i N_c^i \Big\{ \frac{1}{2}(s - 4m_W^2)((3m_{d^i}^2 - 3m_{u^i}^2 + 7m_W^2 - 5s/2)g_{d^i}^L + \right.$$

$$+ (3m_{d^i}^2 - 3m_{u^i}^2 - 7m_W^2 + 5s/2)g_{u^i}^L)m_W^2 + \frac{3}{2}\Big[3g_{d^i}^R(s - 4m_W^2)m_{d^i}^2 + g_{d^i}^L\big(-6m_{d^i}^4 + 2(6m_{u^i}^2 + 2m_W^2 + s)m_{d^i}^2 -$$

$$- 6m_{u^i}^4 + 6m_W^4 + s^2 - 3m_{u^i}^2 s - 7m_W^2 s\big)\Big]B_0(s, m_{d^i}^2, m_{d^i}^2)m_W^2 - \frac{3}{2}\Big[3g_{u^i}^R(s - 4m_W^2)m_{u^i}^2 +$$

$$+ g_{u^i}^L\big(-6m_{d^i}^4 - 3(s - 4m_W^2)m_{d^i}^2 - 6m_{u^i}^4 + 6m_W^4 + s^2 - 7m_W^2 s + 2m_{u^i}^2(s + 2m_W^2)\big)\Big]B_0(s, m_{u^i}^2, m_{u^i}^2)m_W^2 -$$

$$- \frac{9}{2}(m_W^2 - m_{d^i}^2 + m_{u^i}^2)\Big[-g_{d^i}^R(s - 4m_W^2)m_{d^i}^2 + g_{d^i}^L\Big(2m_{d^i}^4 - (s + 4m_{u^i}^2)m_{d^i}^2 +$$

$$+ 2(m_{u^i}^4 + (s - 2m_W^2)m_{u^i}^2 + m_W^4)\Big)\Big]C_0(m_W^2, m_W^2, s, m_{d^i}^2, m_{u^i}^2, m_{d^i}^2)m_W^2 +$$

$$+ 9\Big[-\frac{1}{2}(s - 4m_W^2)g_{u^i}^R m_{u^i}^2 + (m_{d^i}^4 + (s - 2m_{u^i}^2 - 2m_W^2)m_{d^i}^2 + m_{u^i}^4 + m_W^4 - m_{u^i}^2 s/2)g_{u^i}^L\Big].$$

$$\cdot (m_{d^i}^2 - m_{u^i}^2 + m_W^2)C_0(m_W^2, m_W^2, s, m_{u^i}^2, m_{d^i}^2, m_{u^i}^2)m_W^2 + \frac{3}{2}(m_{d^i}^2 - m_{u^i}^2)(g_{d^i}^L(m_{d^i}^2 - m_{u^i}^2 - m_W^2) -$$

$$- g_{u^i}^L(m_{d^i}^2 - m_{u^i}^2 + m_W^2))(s - 4m_W^2)B_0(0, m_{d^i}^2, m_{u^i}^2) + 3\Big[-\frac{3}{2}(s - 4m_W^2)(g_{d^i}^R m_{d^i}^2 - g_{u^i}^R m_{u^i}^2)m_W^2 +$$

$$+ g_{u^i}^L\Big(\frac{1}{2}(m_{d^i}^4 - 2(m_{u^i}^2 + m_W^2)m_{d^i}^2 + m_{u^i}^4 + m_W^4 + m_{u^i}^2 m_W^2)s - m_W^2(5m_{u^i}^4 + 2(m_W^2 - 5m_{u^i}^2)m_{d^i}^2 + 5m_{u^i}^4 +$$

$$+ 5m_W^4 - 4m_{u^i}^2 m_W^2)\Big) + g_{d^i}^L\Big(m_W^2(5m_{d^i}^4 - 2(5m_{u^i}^2 + 2m_W^2)m_{d^i}^2 + 5m_{u^i}^4 + 5m_W^4 + 2m_{u^i}^2 m_W^2) -$$

$$- \frac{1}{2}(m_{d^i}^4 + (m_W^2 - 2m_{u^i}^2)m_{d^i}^2 + (m_{u^i}^2 - m_W^2)^2)s)\Big]B_0(m_W^2, m_{d^i}^2, m_{u^i}^2)\Big\}\Big|^2 \right\} \tag{C.7}$$

We do not report the formula for $\sigma(\chi\chi \to W^{\pm\,(T)}W^{\mp\,(L)})$ because this channel turns out to have $a = 0$ (therefore it is irrelevant for the annihilation process in the Sun) and the corresponding formula is too cumbersome to be reported here.

$$\sigma(\chi\chi \to W^{+(L)}W^{-(L)}) = \frac{g_{Z'}^4 g_\chi^2 \alpha_W^2 \sqrt{s-4mx^2}}{62208\pi^3 m_W^8[(s-m_{Z'}^2)^2 + \Gamma_{Z'}^2 m_{Z'}^2]s(s-4m_W^2)^{5/2}} \cdot$$

$$\cdot \left| \sum_i N_c^i \left\{ (4m_W^2 - s)m_W^2 \left(g_{u^i}^L(-4m_W^4 + 12m_{u^i}^2 m_W^2 + 2sm_W^2 - s^2 - 6m_{d^i}^2(s - 2m_W^2)) \right. \right. \right.$$

$$\left. - g_{d^i}^L(-4m_W^4 + 12m_{d^i}^2 m_W^2 + 2sm_W^2 - s^2 - 6m_{u^i}^2(s - 2m_W^2)) \right)$$

$$+ 6\left(A_0(m_{d^i}^2)g_{d^i}^L - A_0(m_{u^i}^2)g_{u^i}^L \right)(s - 4m_W^2)^2 m_W^2$$

$$- 3B_0(0, m_{d^i}^2, m_{u^i}^2)(m_{d^i}^2 - m_{u^i}^2)(4m_W^2 - s)\left(g_{u^i}^L(4m_W^4 - 2sm_W^2 + m_{u^i}^2(8m_W^2 - s) + m_{d^i}^2(s - 8m_W^2)) + \right.$$

$$\left. g_{d^i}^L(4m_W^4 - 2sm_W^2 + m_{d^i}^2(8m_W^2 - s) + m_{u^i}^2(s - 8m_W^2)) \right)$$

$$+ 3B_0(m_W^2, m_{d^i}^2, m_{u^i}^2)\left(6(g_{d^i}^R m_{d^i}^2 - g_{u^i}^R m_{u^i}^2)(4m_W^2 - s)m_W^4 + g_{d^i}^L[8m_W^8 + 12sm_W^6 - 2s^2 m_W^4 + \right.$$

$$(m_{d^i}^2 - m_{u^i}^2)^2(32m_W^4 - 6sm_W^2 + s^2) + m_{u^i}^2(8m_W^4 + 6sm_W^2 + s^2)m_W^2 +$$

$$m_{d^i}^2(-16m_W^6 - 12sm_W^4 + s^2 m_W^2)] - g_{u^i}^L[8m_W^8 + 12sm_W^6 - 2s^2 m_W^4 +$$

$$(m_{d^i}^2 - m_{u^i}^2)^2(32m_W^4 - 6sm_W^2 + s^2) + m_{u^i}^2(-16m_W^4 - 12sm_W^4 + s^2)m_W^2 + m_{d^i}^2(m_W^2(8m_W^4 + 6sm_W^2 + s^2))] \right)$$

$$+ 3B_0(s, m_{u^i}^2, m_{u^i}^2)m_W^2 \left(6g_{u^i}^R m_{u^i}^2(4m_W^2 - s)m_W^2 + g_{u^i}^L[-24m_W^6 + 12sm_W^4 + 4s(s - 4m_{u^i}^2)m_W^2 + 6m_{d^i}^4 s \right.$$

$$\left. + s(6m_{u^i}^4 + sm_{u^i}^2 - s^2) + 3m_{d^i}^2(8m_W^4 - 2sm_W^2 + s(s - 4m_{u^i}^2))] \right)$$

$$- 3B_0(s, m_{d^i}^2, m_{d^i}^2)m_W^2 \left(6g_{d^i}^R m_{d^i}^2(4m_W^2 - s)m_W^2 + g_{d^i}^L[-24m_W^6 + 12sm_W^4 + 4s(s - 4m_{d^i}^2)m_W^2 + 6m_{u^i}^4 s \right.$$

$$\left. + s(6m_{d^i}^4 + sm_{d^i}^2 - s^2) + 3m_{u^i}^2(8m_W^4 - 2sm_W^2 + s(s - 4m_{d^i}^2))] \right)$$

$$+ 18C_0(m_W^2, m_W^2, s, m_{u^i}^2, m_{d^i}^2, m_{u^i}^2)m_W^2 \left(g_{u^i}^R m_{u^i}^2(m_{d^i}^2 - m_{u^i}^2 + m_W^2)(4m_W^2 - s)m_W^2 \right.$$

$$+ g_{u^i}^L[sm_{d^i}^6 + (4m_W^4 - 2sm_W^2 + s(s - 3m_{u^i}^2))m_{d^i}^4 + (-8m_W^6 + 5sm_W^4 + 3m_{u^i}^4 s$$

$$\left. - m_{u^i}^2(4m_W^4 + sm_W^2 + s^2))m_{d^i}^2 - (m_{u^i}^2 - m_W^2)(4m_W^6 - 2m_{u^i}^2 sm_W^2 + m_{u^i}^4 s)] \right)$$

$$+ 18C_0(m_W^2, m_W^2, s, m_{d^i}^2, m_{u^i}^2, m_{d^i}^2)m_W^2 \left(g_{d^i}^R m_{d^i}^2(m_{d^i}^2 - m_{u^i}^2 - m_W^2)(4m_W^2 - s)m_W^2 \right.$$

$$+ g_{d^i}^L[-4m_W^8 + m_{d^i}^6 s - m_{u^i}^6 s - 3m_{d^i}^4(m_{u^i}^2 + m_W^2)s + m_{u^i}^2(8m_W^6 - 5m_W^4 s)$$

$$\left. \left. \left. - m_{u^i}^4(4m_W^4 - 2sm_W^2 + s^2) + m_{d^i}^2(3sm_{u^i}^4 + (4m_W^4 + sm_W^2 + s^2)m_{u^i}^2 + 2m_W^4(2m_W^2 + s))] \right\} \right|^2 \right., \qquad (C.8)$$

$$\sigma(\chi\chi \to Z^{(T)}Z^{(T)}) = \frac{4g_\chi^2 g_{Z'}^4 m_\chi^2 m_Z^4 (m_{Z'}^2 - s)^2}{\pi^5 m_{Z'}^2 sv^4 \sqrt{s - 4m_\chi^2}\sqrt{s - 4m_Z^2}\left(\Gamma_{Z'}^2 m_{Z'}^2 + (m_{Z'}^2 - s)^2 \right)} \cdot$$

$$\cdot \left| \sum_f N_c^f \left\{ 4g_f^A m_Z^2(c_f^L)^2 - 4g_f^V m_Z^2(c_f^L)^2 - g_f^A s(c_f^L)^2 + g_f^V s(c_f^L)^2 + 4g_f^A(c_f^R)^2 m_Z^2 + 4(c_f^R)^2 g_f^V m_Z^2 \right. \right.$$

$$- g_f^A(c_f^R)^2 s - (c_f^R)^2 g_f^V s - \left[g_f^A\left((4m_f^2 - 4m_Z^2 + s)(c_f^L)^2 - 8c_f^R m_f^2 c_f^L + (c_f^R)^2(4m_f^2 - 4m_Z^2 + s) \right) \right.$$

$$\left. + ((c_f^L)^2 - (c_f^R)^2)g_f^V(4m_Z^2 - s) \right] B_0(m_Z^2, m_f^2, m_f^2) + \left[((c_f^L)^2 - (c_f^R)^2)g_f^V(4m_Z^2 - s) \right.$$

$$\left. + g_f^A\left((4m_f^2 - 4m_Z^2 + s)(c_f^L)^2 - 8c_f^R m_f^2 c_f^L + (c_f^R)^2(4m_f^2 - 4m_Z^2 + s) \right) \right] B_0(s, m_f^2, m_f^2)$$

$$\left. + 4g_f^A m_f^2\left((c_f^L)^2 m_Z^2 + (c_f^R)^2 m_Z^2 + c_f^L c_f^R(2m_Z^2 - s) \right) C_0(s, m_Z^2, m_Z^2, m_f^2, m_f^2, m_f^2) \right\} \right|^2, \qquad (C.9)$$

$$\sigma(\chi\chi \to Z^{(T)}Z^{(L)}) = \frac{4g_\chi^2 g_{Z'}^4 m_Z^2 \sqrt{s - 4m_\chi^2}}{3\pi^5 s v^4 \left(s - 4m_Z^2\right)^{3/2} \left(\Gamma_{Z'}^2 m_{Z'}^2 + \left(m_{Z'}^2 - s\right)^2\right)} \cdot$$

$$\cdot \left| \sum_f N_c^f \left\{ -4g_f^A (c_f^L)^2 m_Z^4 - 4g_f^A (c_f^R)^2 m_Z^4 + 4(c_f^L)^2 g_f^V m_Z^4 - 4(c_f^R)^2 g_f^V m_Z^4 \right. \right.$$

$$+ g_f^A (c_f^L)^2 s m_Z^2 + g_f^A (c_f^R)^2 s m_Z^2 - (c_f^L)^2 g_f^V s m_Z^2 + (c_f^R)^2 g_f^V s m_Z^2$$

$$+ \frac{1}{2}\left[g_f^A \left(-(20m_Z^4 - 6s m_Z^2 + s^2 + 4m_f^2(s - 4m_Z^2))(c_f^L)^2 + 8c_f^R m_f^2(s - 4m_Z^2)c_f^L \right. \right.$$

$$\left. - (c_f^R)^2 (20m_Z^4 - 6s m_Z^2 + s^2 + 4m_f^2(s - 4m_Z^2)) \right)$$

$$\left. + ((c_f^L)^2 - (c_f^R)^2) g_f^V (20m_Z^4 - 6s m_Z^2 + s^2) \right] B_0(m_Z^2, m_f^2, m_f^2)$$

$$- \frac{1}{2}\left[g_f^A \left(-(20m_Z^4 - 6s m_Z^2 + s^2 + 4m_f^2(s - 4m_Z^2))(c_f^L)^2 + 8c_f^R m_f^2(s - 4m_Z^2)c_f^L \right. \right.$$

$$\left. - (c_f^R)^2 (20m_Z^4 - 6s m_Z^2 + s^2 + 4m_f^2(s - 4m_Z^2)) \right)$$

$$\left. + ((c_f^L)^2 - (c_f^R)^2) g_f^V (20m_Z^4 - 6s m_Z^2 + s^2) \right] B_0(s, m_f^2, m_f^2)$$

$$+ 2\left[-\frac{1}{2}((c_f^L)^2 - (c_f^R)^2) g_f^V \left(-2m_Z^6 + 2s m_Z^4 + m_f^2(8m_Z^4 - 6s m_Z^2 + s^2) \right) \right.$$

$$- \frac{1}{2} g_f^A \left((2m_Z^6 - 2s m_Z^4 + 4m_f^2 s m_Z^2 - m_f^2 s^2)(c_f^L)^2 + 2c_f^R m_f^2(8m_Z^4 - 6s m_Z^2 + s^2)c_f^L \right.$$

$$\left. \left. \left. + (c_f^R)^2 \left(2m_Z^6 - 2s m_Z^4 + 4m_f^2 s m_Z^2 - m_f^2 s^2 \right) \right) \right] C_0(s, m_Z^2, m_Z^2, m_f^2, m_f^2, m_f^2) \right\} \right|^2,$$

$$\tag{C.10}$$

$$\sigma(\chi\chi \to Z^{(L)}Z^{(L)}) = 0, \tag{C.11}$$

$$\sigma(\chi\chi \to Z^{(T)}h)\big|_{\theta=\pi/2} = \frac{\alpha_W g_\chi^2 g_{Z'}^4}{24\pi^4 \left(m_{Z'}^4 + \Gamma_{Z'}^2 m_{Z'}^2 - 2s m_{Z'}^2 + s^2\right)} \cdot$$

$$\cdot \frac{\sqrt{s - 4m_\chi^2}\sqrt{(m_H^2 - m_Z^2)^2 + s^2 - 2s(m_H^2 + m_Z^2)}}{s^{3/2} v^2 \left((m_H - m_Z)^2 - s\right)^2 \left((m_H + m_Z)^2 - s\right)^2} \left| \sum_f g_f^V m_f^2 N_c^f (c_f^L + c_f^R) \cdot \right.$$

$$\cdot \left\{ C_0(m_H^2, m_Z^2, s, m_f^2, m_f^2, m_f^2) s^3 + 2(-m_H^2 - m_Z^2 + s) B_0(s, m_f^2, m_f^2) s \right.$$

$$- \left[2B_0(m_H^2, m_f^2, m_f^2) + (-4m_f^2 + 3m_H^2 + m_Z^2) C_0(m_H^2, m_Z^2, s, m_f^2, m_f^2, m_f^2) - 2\right] s^2$$

$$+ \left[-2B_0(m_Z^2, m_f^2, m_f^2) m_Z^2 + 2(m_H^2 + 2m_Z^2) B_0(m_H^2, m_f^2, m_f^2) \right.$$

$$\left. - \left((8m_f^2 - 3m_H^2 + m_Z^2) C_0(m_H^2, m_Z^2, s, m_f^2, m_f^2, m_f^2) + 4 \right)(m_H^2 + m_Z^2) \right] s$$

$$- (m_H^2 - m_Z^2)\Big[-2B_0(m_H^2, m_f^2, m_f^2)m_Z^2 + 2B_0(m_Z^2, m_f^2, m_f^2)m_Z^2$$

$$+ \Big((-4m_f^2 + m_H^2 - m_Z^2)C_0(m_H^2, m_Z^2, s, m_f^2, m_f^2, m_f^2) - 2\Big)(m_H^2 - m_Z^2)\Big]\Big\}\Big|^2, \qquad \text{(C.12)}$$

$$\sigma(\chi\chi \to Z^{(L)}h)\big|_{\theta=\pi/2} = \frac{\alpha_W g_\chi^2 g_{Z'}^4}{3\pi^4 \left(\Gamma_{Z'}^2 m_{Z'}^2 + m_{Z'}^4 - 2m_{Z'}^2 s + s^2\right)}\cdot$$

$$\cdot \frac{m_Z^2 \sqrt{s - 4m_\chi^2}}{\sqrt{s}v^2 \left(-2s\left(m_H^2 + m_Z^2\right) + \left(m_H^2 - m_Z^2\right)^2 + s^2\right)^{3/2}} \Big| \sum_f g_f^V m_f^2 N_c^f (c_f^L + c_f^R)\cdot$$

$$\cdot \Big\{ m_H^2\Big[(-m_H^2 + m_Z^2 + s)C_0(m_H^2, m_Z^2, s, m_f^2, m_f^2, m_f^2) - 2B_0(m_H^2, m_f^2, m_f^2)\Big]$$

$$+ (m_H^2 + m_Z^2 - s)B_0(m_Z^2, m_f^2, m_f^2) + (m_H^2 - m_Z^2 + s)B_0(s, m_f^2, m_f^2)\Big\}\Big|^2. \qquad \text{(C.13)}$$

$$\sigma(\chi\chi \to \gamma\gamma) = \frac{16\alpha_{em}^2 g_\chi^2 g_{Z'}^4}{\pi^3 \left(\Gamma_{Z'}^2 m_{Z'}^2 + (m_{Z'}^2 - s)^2\right)} \frac{m_\chi^2 \sqrt{s}(s - m_{Z'}^2)^2}{m_{Z'}^4 \sqrt{s - 4m_\chi^2}}\cdot$$

$$\cdot \Big| \sum_f g_f^A N_c^f Q_f^2\Big[2m_f^2 C_0(0, 0, s, m_f^2, m_f^2, m_f^2) + 1\Big]\Big|^2, \qquad \text{(C.14)}$$

$$\sigma(\chi\chi \to \gamma h) = \frac{\alpha_{em} g_\chi^2 g_{Z'}^4}{6\pi^4 \left(\Gamma_{Z'}^2 m_{Z'}^2 + m_{Z'}^4 - 2m_{Z'}^2 s + s^2\right)} \frac{\sqrt{s - 4m_\chi^2}}{s^{3/2}v^2 \left(s - m_H^2\right)}\cdot$$

$$\cdot \Big| \sum_f g_f^V m_f^2 N_c^f Q_f\Big\{ -2s B_0(m_H^2, m_f^2, m_f^2) + 2s B_0(s, m_f^2, m_f^2)$$

$$+ (s - m_H^2)\Big[(4m_f^2 - m_H^2 + s)C_0(m_H^2, 0, s, m_f^2, m_f^2, m_f^2) + 2\Big]\Big\}\Big|^2, \qquad \text{(C.15)}$$

$$\sigma(\chi\chi \to \gamma Z^{(T)}) = \frac{\alpha_{em}\alpha_W g_\chi^2 g_{Z'}^4}{3\pi^3 \left([(s - m_{Z'}^2)^2 + \Gamma_{Z'}^2 m_{Z'}^2] m_{Z'}^4 s^{5/2} (s - m_Z^2) \sqrt{s - 4m_\chi^2}\right)}\cdot$$

$$\cdot \Big\{ \frac{m_{Z'}^4}{2}(s - 4m_\chi^2)\Big| \sum_f N_c^f Q_f \Big\{ m_Z^2 s(g_f^A(c_f^L + c_f^R) + g_f^V(c_f^R - c_f^L))(B_0(m_Z^2, m_f^2, m_f^2) - B_0(s, m_f^2, m_f^2))$$

$$- (s - m_Z^2)\Big[2m_f^2 C_0(m_Z^2, 0, s, m_f^2, m_f^2, m_f^2)(g_f^A c_f^L m_Z^2 + g_f^A c_f^R m_Z^2 - c_f^L g_f^V s + c_f^R g_f^V s)$$

$$+ m_Z^2 (g_f^A (c_f^L + c_f^R) + g_f^V (c_f^R - c_f^L))\Big]\Big|^2 + 3m_\chi^2 (s - m_Z^2)^4 (s - m_{Z'}^2)^2 \cdot$$

$$\cdot \Big| \sum_f N_C^f Q_f \Big\{ 2g_f^A m_f^2 (c_f^L + c_f^R) C_0(m_Z^2, 0, s, m_f^2, m_f^2, m_f^2) + g_f^A (c_f^L + c_f^R) + g_f^V (c_f^R - c_f^L) \Big\} \Big|^2, \Big\}$$

$$\tag{C.16}$$

$$\sigma(\chi\chi \to \gamma Z^{(L)}) = \frac{\alpha_{em}\alpha_W g_\chi^2 g_{Z'}^4}{6\pi^3 \Big[[(s - m_{Z'}^2)^2 + \Gamma_{Z'}^2 m_{Z'}^2] m_Z^2 s^{3/2} \Big(s - m_Z^2\Big)} \frac{\sqrt{s - 4m_\chi^2}}{} \cdot$$

$$\cdot \Big| \sum_f N_C^f Q_f \Big\{ m_Z^2 s (g_f^A (c_f^L + c_f^R) + g_f^V (c_f^R - c_f^L)) (B_0(m_Z^2, m_f^2, m_f^2) - B_0(s, m_f^2, m_f^2))$$

$$- (s - m_Z^2) \Big[2m_f^2 C_0(m_Z^2, 0, s, m_f^2, m_f^2, m_f^2)(g_f^A c_f^L m_Z^2 + g_f^A c_f^R m_Z^2 - c_f^L g_f^V s + c_f^R g_f^V s)$$

$$+ m_Z^2 (g_f^A (c_f^L + c_f^R) + g_f^V (c_f^R - c_f^L)) \Big] \Big\} \Big|^2, \tag{C.17}$$

$$\sigma(\chi\chi \to hh) = \frac{g_\chi^4 g_{Z'}^8 m_\chi^2 \cos^4\theta}{2048\pi^5} \frac{\sqrt{s - 4m_h^2}}{s(s - 4m_\chi^2)^{3/2}} \cdot$$

$$\cdot \Big| B_0(m_\chi^2, m_\chi^2, m_{Z'}^2) - B_0(s, m_{Z'}^2, m_{Z'}^2) + (2m_\chi^2 + m_{Z'}^2 - s) C_0(m_\chi^2, m_\chi^2, s, m_{Z'}^2, m_\chi^2, m_{Z'}^2) \Big|^2. \tag{C.18}$$

In the cross sections involving a fermionic loop, we denoted by $\sum_f$ a sum over all the SM fermion species. In the WW cross section, instead, we denoted by $\sum_i$ a sum over the six fermion families (three families of quarks and three of leptons), with m_{u^i}, m_{d^i} the upper and lower component of the doublet, respectively, and with $g_{u^i}^L$, $g_{u^i}^R$, $g_{d^i}^L$, $g_{d^i}^R$ the combinations

$$g_{u^i}^L = g_{u^i}^V - g_{u^i}^A, \qquad g_{u^i}^R = g_{u^i}^V + g_{u^i}^A, \qquad g_{d^i}^L = g_{d^i}^V - g_{d^i}^A, \qquad g_{d^i}^R = g_{d^i}^V + g_{d^i}^A. \tag{C.19}$$

The functions A_0, B_0 and C_0 are the standard Passarino-Veltman one loop one-, two- and three-points scalar integrals [1]:

$$A_0\left(m_0^2\right) = \frac{\mu^{4-D}}{i\pi^{D/2}\gamma_\Gamma} \int \frac{d^D k}{k^2 - m_0^2}, \tag{C.20}$$

$$B_0\left(p^2, m_1^2, m_2^2\right) = \frac{\mu^{4-D}}{i\pi^{D/2}\gamma_\Gamma} \int \frac{d^D k}{\left(k^2 - m_1^2\right)\left((k+p)^2 - m_2^2\right)}, \tag{C.21}$$

Table C.1 Coupling of the Z boson to SM fermions

SM fermion f	c_f^L	c_f^R
Leptons	$-\frac{1}{2} + \sin^2 \theta_W$	$\sin^2 \theta_W$
Neutrinos	$\frac{1}{2}$	0
Up quarks	$\frac{1}{2} - \frac{2}{3} \sin^2 \theta_W$	$-\frac{2}{3} \sin^2 \theta_W$
Down quarks	$-\frac{1}{2} + \frac{1}{3} \sin^2 \theta_W$	$\frac{1}{3} \sin^2 \theta_W$

$$
C_0 \left(p_1^2, p_2^2, (p_1 + p_2)^2, m_1^2, m_2^2, m_3^2 \right)
$$
$$
= \frac{\mu^{4-D}}{i \pi^{D/2} \gamma_\Gamma} \int \frac{d^D k}{\left(k^2 - m_1^2 \right) \left((k + p_1)^2 - m_2^2 \right) \left((k + p_1 + p_2)^2 - m_3^2 \right)}, \quad \text{(C.22)}
$$

with

$$
\gamma_\Gamma = \frac{\Gamma^2(1 - \epsilon)\Gamma(1 + \epsilon)}{\Gamma(1 - 2\epsilon)}, \quad D = 4 - 2\epsilon, \quad \text{(C.23)}
$$

where γ_Γ approaches 1 in the limit $\epsilon \to 0$. In these equations, we denoted by $v \simeq 246\,\text{GeV}$ the vacuum expectation value of the Higgs field, and by c_f^L, c_f^R the SM coupling of the Z boson to left- and right-handed fermions respectively (see Table C.1).

Let us briefly comment on the one loop cross sections:

WW: The bosonic loop diagrams 5 and 6 in Fig. C.1 are velocity suppressed as expected.

ZZ: The box diagram (number 3 on Fig. C.1) is suppressed at low energies by the two heavy propagators in the loop, and gives only a minor effect. Therefore we ignored it in our calculations.

$\gamma\gamma$: The cross section for annihilation into $\gamma\gamma$ vanishes on resonance, due to the factor of $(s - m_{Z'}^2)^2$ in the numerator. This is a consequence of the Landau-Yang theorem [2, 3] that states that a spin-1 particle can not decay into two photons, and is a reassuring cross-check of our results.

Also notice that the $\gamma\gamma$ cross section is proportional to g_f^A, the axial coupling of the fermions to the Z', and vanishes in the limit of pure $B - L$. This is due to the Dirac structure of the fermion loop in the very same way in which the cross section for annihilation into γh is proportional to the vectorial coupling g_f^V, and can be seen as a realization of the Furry theorem [4], which states that any physical amplitude involving an odd number of photons vanishes (in our case one of the photons is replaced by the vectorial part of the Z').

hh: The two diagrams with a fermionic loop (diagrams 2 and 3 in Fig. C.1) sum to zero, while the two box diagrams (numbers 4 and 5) give a contribution at most comparable to that of the triangular diagram (number 1). Since the cross section for annihilation into hh including only the triangular diagram is subdominant by several orders of magnitude, we can safely ignore the contribution of the two box diagrams.

Results of our calculations show that in the low kinetic energy regime that is relevant for DM annihilation in the Sun loop channels are usually subdominant. Some of the cross sections receive a velocity suppression ($\sigma v \simeq b v^2$) and precisely vanish in the zero velocity limit. Those are $W^{(T)} W^{(T)}$, $Z^{(T)} Z^{(L)}$, $\gamma Z^{(L)}$, γh, hh, $Z^{(T)} h$ and, for $\theta = \pi/2$, $Z^{(L)} h$. We do not explicitly show the analytical expansion around $v = 0$ because the velocity appears as an argument of the Passarino-Veltman functions. The only process, which acquires a relevant contribution at the one-loop level is $W^{(L)} W^{(L)}$.

All the cross sections we computed, except for hh that has no s-channel exchange of a Z' boson, vanish around $m_{\text{DM}} = m_{Z'}$ in the $v_{\text{DM}} \rightarrow 0$ limit. Again, this is a cross-check of the correctness of our calculations. Indeed, close to the resonance the cross section for $\chi\chi \rightarrow XX$ is proportional to the product $\Gamma(Z' \rightarrow \chi\chi) \cdot \Gamma(Z' \rightarrow XX)$, but $\Gamma(Z' \rightarrow \chi\chi)$ vanishes if $m_{\text{DM}} = m_{Z'}/2$, which is implied by a resonant production of Z' with $v_{\text{DM}} = 0$.

References

1. G. Passarino, M.J.G. Veltman, One loop corrections for e+ e-annihilation into mu+ mu- in the Weinberg model. Nucl. Phys. **B160**, 151 (1979)
2. L.D. Landau, On the angular momentum of a system of two photons. Dokl. Akad. Nauk Ser. Fiz. **60**(2), 207–209 (1948)
3. C.-N. Yang, Selection rules for the dematerialization of a particle into two photons. Phys. Rev. **77**, 242–245 (1950)
4. S. Weinberg, The Quantum Theory of Fields. Vol. 1: Foundations (Cambridge University Press, 2005)

Appendix D
General Formalism for the Calculation of the Relic Density

Our technique to compute the abundance and notation follow Refs. [1] and [2]. First we find the freezeout temperature by solving

$$e^{x_F} = \frac{\sqrt{\frac{45}{8}} g_{\mathrm{DoF}} m_{\mathrm{DM}} M_{\mathrm{Pl}} c (c+2) \langle \sigma v \rangle}{2\pi^3 g_\star^{1/2} \sqrt{x_F}}, \qquad (\mathrm{D}.1)$$

where $x = m_{\mathrm{DM}}/T$ and subscript F denotes the value at freezeout, $g_{\mathrm{DoF}} = 2$ is the number of degrees of freedom of the DM particle, c is a matching constant usually taken to be $1/2$, $g_\star$ is the number of relativistic degrees of freedom, $M_{\mathrm{Pl}} = 1/\sqrt{G_N}$ is the Planck mass. Usually, it is safe to expand in powers of the velocity and use the approximation

$$\langle \sigma v \rangle = a + b \langle v^2 \rangle + \mathcal{O}(\langle v^4 \rangle) \simeq a + 6b/x_F. \qquad (\mathrm{D}.2)$$

However, when the mediator width is small, this approximation can down near the s-channel resonance in the annihilation rate at $M \simeq 2m_{\mathrm{DM}}$ [1, 3, 4] if the width is small. Around this point it becomes more accurate to use the full expression,

$$\langle \sigma v \rangle = \frac{x}{8 m_{\mathrm{DM}}^5 K_2^2[x]} \int_{4m_{\mathrm{DM}}^2}^\infty ds \, \sigma (s - 4m_{\mathrm{DM}}^2) \sqrt{s} K_1 [\sqrt{s}\, x / m_{\mathrm{DM}}]. \qquad (\mathrm{D}.3)$$

With this information, one can calculate the relic abundance,

$$\Omega_{\mathrm{DM}} h^2 = \Omega_\chi h^2 + \Omega_{\bar\chi} h^2 = \frac{2 \times 1.04 \times 10^9 \, \mathrm{GeV}^{-1} m_{\mathrm{DM}}}{M_{\mathrm{Pl}} \int_{T_0}^{T_F} g_\star^{1/2} \langle \sigma v \rangle \mathrm{d}T}, \qquad (\mathrm{D}.4)$$

where the factor of 2 is for Dirac DM. When the non-relativistic approximation to the annihilation rate holds, this simplifies to

$$\Omega_{\mathrm{DM}} h^2 = \frac{2 \times 1.04 \times 10^9 \, \mathrm{GeV}^{-1} x_F}{\overline{g_\star}^{1/2} M_{\mathrm{Pl}} (a + 3b/x_F)} \qquad (\mathrm{D}.5)$$

© Springer International Publishing AG 2017
E. Morgante, *Aspects of WIMP Dark Matter Searches at Colliders and Other Probes*, Springer Theses, https://doi.org/10.1007/978-3-319-67606-7

where $\overline{g_\star}^{1/2}$ is a typical value of $g_\star^{1/2}(T)$ in the range $T_0 \leq T \leq T_F$. We have tested the validity of this approximation and find that there is a negligible difference to the full relativistic calculation, since the widths we consider are relatively large. If the physical widths are used, then care should be taken that this approximation still holds when the width becomes small, especially when the annihilation rate has a larger p-wave component.

References

1. P. Gondolo, G. Gelmini, Cosmic abundances of stable particles: improved analysis. Nucl. Phys. **B360**, 145–179 (1991)
2. G. Bertone, D. Hooper, J. Silk, Particle dark matter: evidence, candidates and constraints. Phys. Rept. **405**, 279–390 (2005), arXiv:hep-ph/0404175
3. K. Griest, M. Kamionkowski, Unitarity limits on the mass and radius of dark matter particles. Phys. Rev. Lett. **64**, 615 (1990)
4. K. Griest, D. Seckel, Three exceptions in the calculation of relic abundances. Phys. Rev. **D43**, 3191–3203 (1991)

Printed by Printforce, the Netherlands